DYNAMICS OF SATELLITES

DYNAMIQUE DES SATELLITES

INTERNATIONAL UNION OF THEORETICAL
AND APPLIED MECHANICS

DYNAMICS OF SATELLITES

SYMPOSIUM PARIS, MAY 28–30, 1962

DYNAMIQUE DES SATELLITES

SYMPOSIUM PARIS, 28–30 MAI 1962

EDITED BY

MAURICE ROY

WITH 94 FIGURES

1963

SPRINGER-VERLAG BERLIN HEIDELBERG GMBH

SPRINGER-VERLAG BERLIN HEIDELBERG GMBH

Library of Congress Catalog Card Number 63—10993

ISBN 978-3-642-48132-1   ISBN 978-3-642-48130-7 (eBook)
DOI 10.1007/978-3-642-48130-7

Préface

Depuis le lancement de SPOUTNIK I par l'Union Soviétique le 4 Octobre 1957, des expériences humaines de Mécanique céleste de cette sorte ont été répétées à de nombreuses reprises en U.R.S.S. et aux U.S.A.

En 1961, sur ma proposition, l'Union Internationale de Mécanique théorique et appliquée retint l'idée de consacrer en 1962 un Symposium spécial à la confrontation des résultats des expériences soviétiques et américaines en vue d'en tirer le maximum d'enseignements sur la question fondamentale suivante concernant la « Dynamique des satellites artificiels » de la Terre : quelles sont la nature et les lois des forces réelles qui agissent sur ces mobiles au voisinage de notre planète, et qui déterminent par conséquent leur mouvement?

En d'autres termes, il s'agissait de faire le point de nos connaissances sur le problème du mouvement des Astres, magistralement résolu par NEWTON il y a plus de trois siècles pour des astres quasi-ponctuels et assez éloignés. Les moyens d'observation utilisés pour connaître avec la meilleure précision possible le mouvement des satellites artificiels lancés depuis 1957, et le fait de la proximité relative de ces satellites par rapport à la Terre sont par eux-mêmes de nature à révéler soit des altérations de la loi classique de l'attraction newtonienne, dont la signification serait à rechercher, soit l'intervention de forces perturbatrices, dont l'origine et l'expression seraient à préciser.

Ce sont l'examen et la discussion de ces problèmes qui constituèrent l'objet du Symposium en question, dont le Comité Scientifique fut composé du Dr. H. L. DRYDEN, du Dr. D. G. KING-HELE, du Pr. G. C. McVITTIE, du Pr. LÉONID I. SEDOV, et du signataire de ces lignes, à qui en fut confiée, avec l'honneur de la présidence, la responsabilité de l'organisation à Paris.

Par sa conception même, ce Symposium postulait l'accord des savants soviétiques et américains, dont la participation était essentielle. C'est un heureux privilège de I.U.T.A.M. que d'avoir pu obtenir sans difficulté cet accord, et je tiens ici à exprimer au Dr. H. DRYDEN et au Pr. L. SEDOV, l'un et l'autre si connus des savants de la Mécanique en tous les pays, pour la contribution si active et si efficace qu'ils ont apportée à notre Comité organisateur.

Comme tous les Symposiums I.U.T.A.M., celui-ci était réservé à un nombre restreint de participants, auteurs ou observateurs, tous invités par le Comité. En fait, s'agissant d'un problème fondamental mais étroitement limité, il fut décidé de borner en définitive à 38 le nombre des participants et à 25 le nombre des communications, dont 1 d'origine A. (All.), 3 d'origine U.K., 10 d'origine U.R.S.S. et 11 d'origine U.S.A.

Les séances de travail se sont tenues les 28, 29 et 30 Mai 1962, à Paris, dans des locaux réservés et spécialement aménagés de la Maison de la Chimie, offrant aux participants un cadre d'intimité de travail favorable à des échanges de vues animés.

Sans entrer dans le détail des études que le Lecteur trouvera plus loin, bornons-nous à mentionner ici quelques conclusions.

La loi de NEWTON de l'attraction universelle reste inébranlée, étant retenue comme l'expression la plus sûre de la force de gravitation pour prévoir la trajectoire des satellites artificiels, et pour rattacher à d'autres causes les altérations constatées de leurs trajectoires.

L'effet de relativité générale a été traité par un auteur et reconnu négligeable.

Le couplage entre les mouvements du centre d'inertie et autour de celui-ci a été discuté et son très minime effet réel a été apprécié, tandis qu'une intervention non négligeable d'un couple d'origine magnétique dans le mouvement autour du centre d'inertie a été reconnu possible, concurremment avec celle du très faible couple d'origine gravitationnelle.

Les forces extérieures de surface, ou « résistances » telles que la résistance de milieu ou celle de charge électrique, ont été discutées et leurs appréciations comparées. Ces forces sont extrêmement minimes, mais l'effet plus ou moins séculaire des perturbations qu'elles occasionnent est notable, surtout pour les satellites de basse altitude.

L'interprétation des résultats accumulés depuis quatre ans et demi, et déjà excessivement nombreux, entraîne une profonde révision des supputations antérieures sur la nature, la composition et les propriétés, notamment de densité et de température, de la très haute atmosphère et de l'espace environnant.

Le Symposium de Paris marquera sans doute une étape dans le progrès des connaissances scientifiques relatives à ces problèmes fondamentaux de la Dynamique des Satellites artificiels, problèmes dont l'importance est primordiale pour la future exploration humaine de l'Espace extra-atmosphérique.

Le présent volume, dont la Maison Springer a assuré l'édition avec le soin qui a fondé sa réputation et avec une diligence dont je tiens à la remercier ici, contient la totalité des communications présentées au Symposium. La délégation russe a bien voulu fournir des traductions, en anglais ou en français, de ses contributions et je l'en remercie hautement.

Pour sa part, l'éditeur aurait souhaité pouvoir publier toutes les discussions, si intéressantes, qui ont marqué les séances de travail. Mais, ceux qui connaissent par expérience les inévitables difficultés que l'on rencontre pour coordonner, avec le concours des intéressés et en vue de leur publication sans retard, les différentes parties de telles discussions comprendront certainement pourquoi, notamment pour hâter l'édition, je n'ai pu réaliser le projet envisagé. Cependant, à la demande des intéressés, quelques interventions ont pu être reproduites dans ce volume, à la suite des communications concernées.

La plupart des participants, dont la liste alphabétique est la suivante:

MM. B. ANDERSON, R. BAKER, Y. BATRAKOV, L. BROER, D. BROUWER, P. CONTENSOU, M. DAVIES, G. DOUBOCHINE, H. DRYDEN, B. GAR-FINKEL, L. GAUTHIER, S. HERRICK, W. IRVINE, L. JACCHIA, W. KAULA, D. KING-HELE, Y. KITAEV, D. KULIKOV, J. KOVALEVSKY, Y. KOZAI, K. MAGNUS, B. MAY, D. McDERMOTT, G. McVITTIE, R. MERSON, A. MOLTCHANOV, P. MUSEN, R. NAUMANN, W. PRIESTER, L. PUYEO, H. ROBE, M. ROY, W. SCHULZ, L. SEDOV, I. SHAPIRO, G. TEMPLE, S. WYATT,

ont prise une part trés active aux échanges de vues et discussions, et une trace invisible en subsiste dans les retouches apportées, en définitive, par plusieurs auteurs à leurs textes primitifs.

Ce sont à tous ces participants, auteurs de rapports ou intervenants des discussions, que j'exprime à nouveau la gratitude du Comité pour leur active coopération, pour la valeur scientifique de leur concours, enfin pour leur esprit de collaboration cordiale et directe. En particulier, je remercie à nouveau le Dr. KOVALEVSKY et le Dr. MUSEN qui ont bien voulu opérer aimablement, lorsqu'il apparaissait utile, la conversion de la langue russe en anglais ou en français, et réciproquement.

Le Symposium a été honoré de la présence du Pr. G. TEMPLE, Président de I.U.T.A.M., et du Pr. L. GAUTHIER, Vice-Doyen de la Faculté des Sciences de Paris. Les Dames ont pu tirer quelque plaisir de visites organisées à leur intention pour occuper chaque jour une partie de leur aprés-midi. Un dîner intime mais cordial, organisé le 29 Mai au Bois de Boulogne grâce à une aide très obligeante de la Direction des Relations Culturelles du Ministère français des Affaires Etrangères, m'a permis d'exprimer à nos hôtes l'espoir qu'ils garderaient bon et utile souvenir de leur séjour studieux à Paris.

Je voudrais encore, en terminant cette introduction, citer les organisation industrielles françaises qui, en outre d'une aide particulière accordée par le C.N.R.S., ont bien voulu m'apporter une assistance bénévole, généreuse, et d'une très haute utilité pour notre organisation matérielle: Société française Babcock et Wilcox, Chambre Syndicale des Producteurs d'Aciers Fins et Spéciaux, Compagnie de Péchiney, Cie Générale de Télégraphie Sans Fil, Société d'Electro-Métallurgie d'Ugine, et Cie Fse Thomson-Houston.

Paris, Mars 1963

Maurice Roy

Liste des participants — List of participants

Prof. BENGT JOEL ANDERSSON	Royal Institute of Technology, Stockholm, Sweden
Prof. ROBERT M. L. BAKER JR.	University of California—and Lockheed California Company, Los Angeles, Calif., U.S.A.
Dr. Y. V. BATRAKOV	Institute for Theoretical Astronomy, Leningrad, U.S.S.R.
Dr. V. V. BELETSKY	Institute of Mathematics-Academy of Sciences of U.S.S.R. Moscow
Prof. Dr. LAMBERTUS J. F. BROER	Technological University Department of Physics, Eindhoven (Pays-Bas), Netherlands
Prof. DIRK BROUWER	Yale University Observatory, New Haven, Conn., U.S.A.
Prof. P. CONTENSOU	O.N.E.R.A., Chatillon-Sous-Bagneux (Seine), France
Mr. MORTON J. DAVIES	University College of Wales, Aberystwyth, Cards., England
Prof. G. N. DOUBOCHINE	Sternberg Institute of Astronomy, Moscow, U.S.S.R.
Dr. H. L. DRYDEN	N.A.S.A., Washington (D.C), U.S.A.
Dr. BORIS GARFINKEL	Ballistic Research Laboratories, Aberdeen Proving Ground, Maryland, U.S.A.
Prof. LUC GAUTHIER	Faculté des Sciences de Paris, Pt du Département de Mécanique, Sceaux, France
Prof. SAMUEL HERRICK	University of California, Los Angeles, Calif., U.S.A.; consultant to Lockheed California Company
Dr. WILLIAM M. IRVINE	Sterrewacht te Leiden, Leiden, Nederland
Dr. LUIGI G. JACCHIA	Smithsonian Astrophysical Observatory, Cambridge, Mass., U.S.A.
Mr. WILLIAM M. KAULA	Goddard Space Flight Center, N.A.S.A. Greenbelt, Maryland, U.S.A.
Mr. DESMOND G. KING-HELE	Royal Aircraft Establishment, Farnborough, Hants, England
Dr. Y. K. KITAEV	Academy of Sciences, Moscow, U.S.S.R.
Dr. D. K. KULIKOV	Institute for Theoretical Astronomy, Leningrad, U.S.S.R.
Dr. JEAN KOVALEVSKY	Bureau des Longitudes, Paris, France
Dr. YOSHIHIDE KOZAI	Smithsonian Astrophysical Observatory, Cambridge, Mass., U.S.A.
Dr. V. N. LEBEDEV	Computation Centre of the U.S.S.R. Academy of Sciences, Moscow
Dr. M. L. LIDOV	Institute of Mathematics-Academy of Sciences of U.S.S.R. Moscow

Prof. KURT MAGNUS	Technische Hochschule Stuttgart, Deutschland
Mr. BRIAN RICHARD MAY	Radioresearch Station, Slough, Bucks, England
Mr. DENIS P. MCDERMOTT	University College, London, England
Prof. G. C. MCVITTIE	University of Illinois Observatory, Urbana, Ill., U.S.A.
Mr. ROBERT H. MERSON	Royal Aircraft Establishment Farnborough, Hants, England
Prof. N. N. MOISSEYEV	Computation Centre of the U.S.S.R. Academy of Sciences, Moscow
Dr. A. M. MOLTCHANOV	Institut Steklov de Mathématiques Moscou, U.S.S.R.
Dr. PETER MUSEN	Goddard Space Flight Center, Theoretical Division, N.A.S.A., Greenbelt, Maryland, U.S.A.
Mr. ROBERT J. NAUMANN	Marshall Space Flight Center, N.A.S.A. Huntsville, Alabama, U.S.A.
Prof. WOLFGANG PRIESTER	Sternwarte der Universität Bonn, Bonn, Deutschland
Dr. LUIS PUEYO	Instituto Nacional de Tecnica Aeronautica, Madrid, España
Monsieur HENRI ROBE	Institut d'Astrophysique, Cointe-Sclessin, Belgique
Prof. MAURICE ROY	Académie des Sciences, Paris, France
Dr. WERNER SCHULZ	Deutsche Forschungsanstalt für Luft- und Raumfahrt, Braunschweig, Deutschland
Prof. L. I. SEDOV	Academy of Sciences, Moscow, U.S.S.R.
Dr. IRWIN IRA SHAPIRO	MIT Lincoln Laboratory, Lexington, Mass., U.S.A.
Dr. A. S. SOCHILINA	Institute for Theoretical Astronomy, Leningrad, U.S.S.R.
Prof. GEORGE TEMPLE	Oxford University, Mathematical Institute, Oxford, England
Prof. STANLEY PORTER WYATT JR.	University of Illinois Observatory, Urbana, Ill., U.S.A.

Sommaire — Contents

5ème séance

mardi 29 Mai, après-midi

Président: Professeur G. TEMPLE

6ème séance

mercredi 30 Mai, matinée

Président: Professeur G. C. McVITTIE

7ème séance

mercredi 30 Mai, après-midi

Président: Professeur G. N. DOUBOCHINE

Clôture du colloque par M. MAURICE ROY

The elimination of spurious data in the process of preliminary and definitive orbit determination

By

Robert M. L. Baker, Jr.

University of California, Department of Astronomy and Lockheed California
.   Company, Los Angeles, Calif., U.S.A.

Abstract. Since the advent of high-precision electronic and electro-optical instruments, considerable research has been carried out in new methods of preliminary orbit determination. Early procedures for processing mini-track, pulsed-radar, and BAKER-NUNN camera data are briefly summarized. A more novel method, employing high-accuracy, range-rate data together with low-accuracy, angle data is also presented and its application to a recent Ranger lunar probe orbit, as observed by GOLDSTONE, is noted.

The process of eliminating spurious observational data on the basis of these preliminary orbits and through the use of a filtering technique is considered in detail. The potential advantages of the Gibbsian method for correlating observed data are also noted.

I. Historical background

An exciting requirement for orbit determination arose in 1801 when GAUSS was challenged by the problem of "rediscovering" the minor planet Ceres. The track of this first minor planet had been lost after a few initial observations. Up to this time, there had existed no great fear of "losing" a major planet, and the demands for rigorous methods of orbit determination from terrestrial observations were not really apparent.

As might be expected, even NEWTON had concerned himself with the problem of orbit determination and had devised graphical solutions; while EULER in 1744 published the first complete non-graphical solution. OLBERS' method for determining parabolic orbits (1797) has stood the test of time and still finds general application. In spite of these earlier approaches, including even the initial efforts of LAMBERT, LAGRANGE, and LAPLACE, we owe to GAUSS the most exhaustive and useful early research into the field of orbit determination. More recently GIBBS, LEUSCHNER, and HERRICK have provided us with significant advances in precision numerical techniques for definitive orbit determination and improvement.

Roy, Satellites   1

The classical problem facing astronomers of the past was rather simple to formulate: given a limited number of tediously reduced angular measurements of an object from one or more observation stations on the Earth made at specific times, what were the elements of the orbit upon which the object was travelling? Provided with these orbital elements, the astronomer could generate an ephemeris accurately by the method of general or special perturbations or approximately by utilizing two-body formulae.

However, complete position fixes of the object werenever available to the astronomer; i. e., no range or range-rate observations were available. These data had somehow to be derived or estimated in the orbit computation technique itself. Such a situation may well be the reverse of the one confronting modern astrodynamics, in which a vast quantity of relatively easily reduced range and range-rate data are often accurately known, whereas angular position data may be unknown or known only to an inferior degree of precision.

Classical orbit determination procedures can generally be categorized as Gaussian, Gibbsian, Lagrangian, or Laplacian. For the purposes of this paper, only the method due to LAPLACE is dealt with in detail, and the interested reader's attention is directed to Ref. [1, 2] and [3] for a more complete exposition of this and the other methods.

The application of classical orbit-determination schemes such as the Laplacian to the preliminary determination of orbits from range and/or range-rate data is not particularly obvious. Many special-purpose schemes have been devised by scientists faced with the immediate problem of processing DOPPLER data. Some of these procedures make use of the remarkable inflection point that is found in a plot of range rate versus time. Typical of this method are the versions discussed by V. A. KOTELNIKOV [4] and C. S. LORENS [5]. The time of the inflection point (i. e., when $\ddot{\varrho} = 0$) coincides with the moment when the distance between the satellite and the observation station is minimum. This time can be graphically estimated to perhaps 0.2 second and to a somewhat higher accuracy numerically; e. g., on p. 261 of reference 5 a quartic interpolation formula is utilized to find the time and distance of closest approach at the inflection point. Given the slope $\dddot{\varrho}$ of the $\dot{\varrho}$-versus-time plot at this time and the assumption of either a horizontally rectilinear or a circular orbit, the speed of the vehicle and its minimum distance from the DOPPLER station can be estimated.

W. A. KEMPER [6] has utilized a least-squares fit to obtain a position fix of an object by triangulation, using simultaneous (i. e., synoptic) radar data obtained from two different stations. A smoothing process based upon a number of these complete "fixes" is employed to obtain satellite position, velocity, and acceleration.

I. HARRIS and R. JASTROW [7] establish angular measures of a satellite passing by a mini-track station with antennae placed along E–W and N–S baselines. From the phase difference of incoming radio waves from a satellite measured between these antennae, the angle to the satellite, α, is obtained to an accuracy of H/r_c (where H is the height of the satellite, and r_c is the geocentric distance of the mini-track station). An iterative process is used to solve for α, H, and $\dot{s}$; substitution of these quantities into the *vis-viva* integral and into the equation of a conic yields the semi-major axis a, the eccentricity e, and the argument of perifocus ω.

Other methods, somewhat similar to the foregoing, are described in Ref. [6] and [8]. Useful as these methods may be for carrying out the tasks for which they were designed, they are only stop-gap, special-purpose procedures and should be replaced by more carefully devised, general-purpose schemes that take optimum advantage of the carefully contrived methods devised by LAPLACE, LAGRANGE, GAUSS, GIBBS, and others.

In more specific terms most present systems for orbit determination (see Ref. [9–17]) suffer from one or more of the following difficulties:

1. Accurate range or range-rate data are degraded or contaminated by inferior angular data (e. g., ± 40 feet in range error for a space vehicle at a height of one Earth's radius is roughly equivalent to plus or minus one microradian or 0.2 second arc error in angle; moreover, the ratio increases with height).

2. Synoptic data are required when observations are made from two or more stations in order to "simplify" or render possible the orbit determination.

3. Dynamical relationships are ignored entirely when the orbit is determined from geometrical fixes obtained, e. g., by the intersection of three surfaces of revolution in space.

4. If two or more uncooperative or passive satellites (or empty boosters or carriers, etc.) make passes simultaneously, then there exists the possibility of ambiguity in the correlation of the data collected.

5. Some systems, although proved to be satisfactory for the determination of low-altitude satellites, are not necessarily applicable to high-altitude satellites and space probes. As C. S. LORENS [5] states, "The use of (D)oppler tracking is limited principally to satellites or projectiles which come close to the Earth."

6. An uninterrupted sequence of data over a complete pass, i. e., data sufficient to define the inflection point, is required.

In addition to the foregoing deficiencies, many researchers fail to distinguish clearly between orbit determination and orbit prediction.

The initial conditions for orbit prediction are provided by orbit determination. Thus the question of the realism with which a certain procedure can describe the future orbit of a space vehicle, i. e., produce an ephemeris, is a question apart from that of the initial orbit determination. Certainly orbit determination and orbit prediction go hand in hand to produce an ephemeris; but the capability and accuracy of each should be assessed individually.

An analogous problem to that of the confusion between orbit determination and prediction is that in many orbit-determination schemes the differences between the preliminary orbit and subsequent, improved orbits are not clearly comprehended. All too frequently many details are incorrectly included in the generation of the preliminary orbit. It should be kept in mind that the preliminary orbit is indeed preliminary, and that it would often be inconsistent to take into account such details as the correction for the index of refraction and other deleterious ionospheric effects[1], the complex smoothing or normal-place procedures for processing data, or the inclusion of the secular perturbation in certain orbital elements in the preliminary orbit. The proper place for most of these refinements is in the differential correction procedure.

More recently Deutsch [18] has considered the use of the Doppler method for determining the orbits of artificial satellites. He calls upon the spectroscopic, binary-star technique and utilizes it in a new sense for the determination of a planetary satellite carrying a frequency-stabilized transmitter or transponder. Deutsch can obtain orbital elements (including the planetary mass) to ± 1 per cent. Paul Koskela [19] and Bruce Douglas [20] are independently extending this work into elliptical orbits with perturbations. Also studying this new area of electronic data are Harris and Cahill [21] who used a Gauss-Olbers procedure based on three or more complete fixes in two to four minutes of observation time.

Patton [22] also considers orbit determination from a single-pass Doppler observation, but instead of remaining with the preliminary orbit problem, he goes into a standard least-squares fit in a differential correction technique. He assumes that a preliminary orbit has somehow already been conceived. One of his comments is particularly noteworthy: "It has been found that parameters consisting of position and velocity components for a given time readily yield a convergent solution . . ." The proper choice of parameters assumes considerable importance in the differential correction procedure subsequent to the preliminary orbit generation.

---

[1] In this same regard, it is noted that ionospheric refraction usually causes greater inaccuracies in angular than in range or range-rate data — another reason for not utilizing these data together.

II. Classical methods

Laplacian method. Perhaps the most well-known classical method for preliminary orbit determination is the one due to LAPLACE. The method depends upon the acquisition of three sets of α and δ obtained, e. g., by the reduction of BAKER-NUNN camera films. From these, by numerical differentiation, one generates the first and second derivatives of the unit vector L reckoned at the second date. (See Vol. II of HERRICK's Astrodynamics [2], or see pages 131 through 135 of Ref. [1] for a derivation of this method.) The four fundamental equations of the Laplacian method are

$$\left(\frac{\mu\dot{\varrho}}{r^3} + \ddot{\varrho}\right) L + 2\dot{\varrho}\,\dot{L} + \varrho\,\ddot{L} = \ddot{R} + \frac{\mu R}{r^3} \tag{1}$$

(called the POINCARÉ form and having three components), and

$$r^2 = \varrho^2 - 2\varrho\,(L \cdot R) + R^2. \tag{2}$$

(All notation in these equations is for the most part standard and is defined in Ref. [1].)

These four equations include the four unknowns ϱ, $\dot{\varrho}$, $\ddot{\varrho}$, and r. The quantities $\dot{\varrho}$ and $\ddot{\varrho}$ can be eliminated from Eq. (1) to reduce these three equations in four unknowns to one equation in two unknowns:

$$D\varrho = A' - \frac{B'}{r^3}.$$

Eqs. (2) and (3) can then be solved simultaneously, e. g., by NEWTON's method, for ϱ and r at the second date. Three major difficulties are found in the application of this method to space vehicle observations:

1. For nearly great circle paths on the celestial sphere, $\dot{L}$ is perpendicular to L and $\ddot{L}$ so that D goes to zero, and A' and B' are poorly defined. As suggested by S. HERRICK, this difficulty can often be partially circumvented by solving Eq. (3) directly for r under the assumption that D equals zero.

2. Because the numerical differentiation is expanded about the second date, the Laplacian orbit represents observations well at this date, but not at the first and third dates. This difficulty can be circumvented through the use of a LEUSCHNER differential correction (see pages 146–150 of Ref. [1]).

3. If the angular data are noisy, as indicated by erratic and often large second differences, the $\dot{L}$ and especially $\ddot{L}$ are poorly determined. There is no way to avoid this difficulty, and because of it the Laplacian method is ordinarily abandoned in satellite work when α and δ are not known to a high degree of accuracy. Thus, except for high-precision

reduction of BAKER-NUNN camera films or astronomical photographic plates, the method is ordinarily not useful.[1]

Lagrange-Gauss-Gibbs first approximation (after S. Herrick). Prior to the employment of the Lagrangian, Gaussian, or Gibbsian preliminary orbit method, a first-approximation procedure must be employed to find the ϱ's at three dates. A unified first-approximation procedure has been developed by HERRICK that is termed the "LAGRANGE-GAUSS-GIBBS first approximation." The fundamental equations are

$$c_1 \varrho_1 \, \boldsymbol{L} - \varrho_2 \, \boldsymbol{L}_2 + c_3 \, \varrho_3 \, \boldsymbol{L}_3 = c_1 \, \boldsymbol{R}_1 - \boldsymbol{R}_2 + c_3 \, \boldsymbol{R}_3$$

$$c_i = A_i + \frac{B_i'}{r_2^3} + \cdots, \tag{5}$$

and Eq. (2). (See Ref. [2] for the derivation.)

The $A_i' s$ and $B_i' s$ are functions of the time intervals between observations. Equations (2), (4), and (5) may be solved for ϱ_1, ϱ_2, ϱ_3, and r_2. Many of the major difficulties encountered in the Laplacian method are absent in the LAGRANGE-GAUSS-GIBBS first approximation. For deep space probes, however, in which $\boldsymbol{L}_1$, $\boldsymbol{L}_2$, and $\boldsymbol{L}_3$ are close together and inaccurately known, the method is sometimes unusable. Furthermore, Eqs. (5) are truncated series expessions which may lead to errors if the observations are spaced far apart in time and if r_2 is not large.

III. Selected new methods

As an illustrative example of the special demands occassioned by satellite probe orbits, let us consider the definition of an orbit given relatively inaccurate and only slightly changing values of α and δ and a relatively accurate value of $\dot{\varrho}$. This is precisely the type of information available from many of the United States deep space tracking stations such as GOLDSTONE (operated by the Jet Propulsion Laboratory of NASA) and e. g., obtained from the Ranger lunar probe tracking.

Modified Lapacian method [23]. If one eliminates $\dot{\varrho}$ rather than ϱ and $\ddot{\varrho}$ from Eq. (1), by taking the dot product of Eq. (1) with $\boldsymbol{L}^2$,

---

[1] JEAN KOVALESKI commented that angular data from more than the minimum three dates have been used effectively by him to circumvent the use of intermediate elements. He represents all of the angular data (as obtained from photographic plates) by a polynomial i. e.,

$$\boldsymbol{L}_t = \boldsymbol{L}_2 + \dot{\boldsymbol{L}}_2 \, \tau_t + \ddot{\boldsymbol{L}}_2 \, \tau_t^2 /2! + \dddot{\boldsymbol{L}}_2 \, \tau_t^3 /3! + \cdots,$$

and obtains very accurate $\dot{\boldsymbol{L}}$ for the Laplacian method from which point he proceeds directly to the differential correction [30].

[2] Note that $\boldsymbol{L} \cdot \dot{\boldsymbol{L}}$ always remains equal to zero, $\dfrac{d(\boldsymbol{L} \cdot \dot{\boldsymbol{L}})}{d\tau} = \dot{\boldsymbol{L}} \cdot \dot{\boldsymbol{L}} + \boldsymbol{L} \cdot \ddot{\boldsymbol{L}} = 0$ or $\boldsymbol{L} \cdot \ddot{\boldsymbol{L}} \equiv -\dot{\boldsymbol{L}} \cdot \dot{\boldsymbol{L}}$.

one can develop a form of the Laplacian method in which range-rate data may easily be introduced. The fundamental equations are

$$\left(\dot{L} \cdot \dot{L} - \frac{\mu}{r^3}\right) \varrho + \frac{\mu}{r^3} L \cdot R - \ddot{\varrho} + L \cdot \ddot{R} = 0 \qquad (6)$$

and Eq. (2). All these quantities are defined at the second date, and $\ddot{\varrho}$ is obtained by numerical differentiation from a series of ϱ's. (See Ref. [19] for a derivation of the equations.) Eqs. (2) and (6) are solved simultaneously for ϱ and r by NEWTON's approximation. The method is particularly advantageous when there are rather large time intervals between obsevations, but one encounters difficulties due to an inaccurate determination of $\dot{L}$ and especially $\ddot{\varrho}$ by numerical differentiation. Numerical differentiation of ϱ, yielding $\ddot{\varrho}$, can be improved slightly by the following procedure (also developed in Ref. [24]):

$$\ddot{\varrho} = \frac{d}{d\tau}\dot{r}(\cdot L) + (\ddot{R} \cdot L + \dot{R} \cdot \dot{L}) \quad \text{at time,} \quad t_2, \qquad (6a)$$

where

$$\frac{d}{d\tau}(\dot{r} \cdot L) = [-\tau_{23}^2(\dot{r} \cdot L) + (\tau_{23}^2 - \tau_{12}^2)(\dot{r} \cdot L) + \tau_{12}^2(\dot{r}_3 \cdot L_3)]/\tau_{12}\tau_{23}\tau_{13},$$

with

$$\dot{r}_i \cdot L_i = \dot{\varrho}_i - \dot{R}_i \cdot L_i \quad \text{and} \quad \tau_{ij} \triangleq k(t_j - t_i), \quad i, j = 1, 2, 3;$$

but the influence of a poorly defined $\dot{L}$ (i. e., $\dot{L}_2$) is still present.

Modified Lagrange-Gauss-Gibbs method [24]. There are two characteristics of deep space probe observations that, as we have seen, create difficulties in most preliminary orbit methods:

1. $\dot{L}_2$ and $\ddot{L}_2$ are poorly defined.
2. The unit vectors L_1, L_2, and L_3 are all quite close together (nearly parallel).

In the modified LAGRANGE-GAUSS-GIBBS method these disadvantages are turned into advantages in that one factor tends to offset the other. The fundamental equations of of the method are

$$\varrho_2 = A + B\dot{\varrho}_1 + C\dot{\varrho}_2 + D\dot{\varrho}_3, \qquad (7)$$

$$r_2\dot{r}_2 = \varrho_2\dot{\varrho}_2 - R_2 \cdot (\dot{\varrho}_2 L_2 + \varrho_2\dot{L}_2) + \varrho_2 L_2 \cdot \dot{R}_2, \qquad (8)$$

and Eq. (2). (The derivation of these equations is presented in Ref.[24].) The method simply involves an original guess of ϱ_2, a solution of Eq. (8) [to gain values for the coefficients in Eq. (7)], and then a *direct* (not an iterative) solution of Eq. (8). As can be seen by inspection of Eq. (8) and from the coefficients of Eq. (7), the influence of $\dot{L}_2$ is greatly reduced. (See Appendix A for definition of the coefficients.) In Eq. (8), although

$\dot{\varrho}_2 \, \boldsymbol{L}_2$ and $\varrho_2 \, \dot{\boldsymbol{L}}_2$ are of about the same magnitude for the deep space probes (observed near the observer's zenith), $\boldsymbol{R}_2$ is nearly perpendicular to $\boldsymbol{L}_2$, so that the adverse influence of an inaccurate value of $\boldsymbol{L}_2$ is attenuated. Furthermore, only the product $(\boldsymbol{L}_1 - \boldsymbol{L}_3) \cdot \dot{\boldsymbol{L}}_2$ is found in the coefficients of Eq. (7), so that the influence of an inaccurate value of $\dot{\boldsymbol{L}}_2$ and the near equality of $\boldsymbol{L}_1$ and $\boldsymbol{L}_3$ cancel. Another advantage is that the method yields a solution for ϱ_2 directly and does not require the generation of a derivative for use in NEWTON's approximation. Disadvantages of the method include limitations on the time between observations. For space probes near the lunar distance, observations spaced further apart than about three days involve the accumulation of too much truncation error.

As in the case of the Laplacian method, a modified LEUSCHNER differential correction can be employed to improve agreement at the first and third dates [24].

General comments. Numerical comparisons of the foregoing methods for the Ranger III space probe may be found in Ref. [24]. It should be noted that the modified LAGRANGE-GAUSS-GIBBS method reduces to the modified Laplacian if the time intervals between observations approach zero.

Most preliminary orbit methods such as these discussed above and in Ref. [25] suffer from three classes of difficulties:

1. Observational error, both random and systematic.

2. Deviation from the usually assumed two-body, circular, or parabolic orbits, etc.

3. Ambiguities and indeterminacies inherent in the data or in the method.

The first difficulty can render a preliminary orbit method unusable if the errors are too large; very little can be done about this problem. With respect to 2, often an extension of the f and g series to include both in-plane perturbations such as drag and the aspherical Earth can be accomplished, as in Ref. [1, 26] and [27]. Furthermore, the f and g series can be extended, e. g.,

$$\boldsymbol{r}_i = f_i \, \boldsymbol{r}_0 + g_i \, \dot{\boldsymbol{r}}_0 + h_i \, \boldsymbol{r}_0 \times \dot{\boldsymbol{r}}_0,$$

to take into account out-of-plane perturbations. Problems of ambiguities and indeterminacies, 3, are usually more subtle, and their solution involves considerable insight into the method being used. The vanishing of the D determinant in the Laplacian method for great circle orbital paths on the celestial sphere is one good example, and the four possible solutions for range-only orbit determination (see Ref. [27]) represents another interesting case. If one can settle for the solution of fewer than the six orbital elements, many of these problems can be eliminated or reduced. Often a rearrangement of formulas will alleviate the difficulty as well.

Aside from the range-rate and angle preliminary orbit determination methods discussed in the foregoing, other methods such as range-only and range and elevation-angle only are receiving considerable attention. In this regard, the interested reader is directed to AVERILL's work [27] and MILSTEAD's work [28], respectively.

IV. Elimination of spurious data

Perhaps the most vexing problem of preliminary orbit analysis is the elimination of spurious data. Since the data used in the preliminary orbit methods discussed in the foregoing sections are often selected arbitrarily from a larger collection of data, there always exists a chance that one of the three angular fixes for, say, the Laplacian method, might be erroneous. The combination of this bad point with two good points might lead to a preliminary orbit that would at best require an exorbitant amount of differential correction for improvement. At worst it might lead to a complete divergence. Often the solution to the problem is the application of smoothing or the collection of neighboring data into "normal points" as has been the custom in conventional astronomical data reduction. On the other hand, when it is not clear that all the data refer to the same satellite or indeed refer to an orbit of an object primarily under the influence of gravity, the problem is more severe and an averaging technique is often not very valuable.

Analysis of a continuous sequence of data by filtering techniques. It has been suggested by THEODORE I. FINE [29] that certain techniques in modern filter theory could be made applicable if there existed a reasonably continuous set of sequential data. I quote Mr. FINE's suggestion:

"A sequential collection of raw data is first put through a low-pass filter which separates the high frequency and low frequency components of the data. The cutoff and rolloff frequencies of the filter can be varied easily to permit experimental determination of optimum values for given data. The filter is designed by first determining a frequency characteristic that is desired and then determining the Z transform of the corresponding transfer function. The coefficients of the Z transform then provides the desired weights. This method is quite general and provides a straightforward technique for the design of numerical filters with desired frequency characteristics. The type of filter response to be employed will depend strongly on the characteristics of the data sources with which this filter is to be used.

Since satellite orbital frequency is quite low only the low frequency output need be examined to determine whether the vehicle is moving under the influence of gravitational forces only. The low frequency data (are) used to compute a preliminary orbit by one of the standard methods ... This preliminary orbit is then used to represent the other data points and a set of residuals are formed. These residuals are then examined statistically to determine whether the data should be attributed to a vehicle that is probably moving under the influence of gravitational forces. This decision is based upon *a priori* descriptions of instrument accuracy and a set of confidence levels for various types of errors. That is, a decision of this type can have two types of errors. — 1. It may conclude that the vehicle was orbiting when no such vehicle was actually present, or 2. it may conclude that the vehicle was not in orbit when indeed it was. The decision is made differently depending upon the relative significance of these two errors.

If it is concluded that the vehicle is probably moving only under the influence of gravitational forces, standard orbital prediction methods can be applied to predict the future position of the vehicle.

If it is concluded that the object is moving predominantly under the influence of forces other than gravity, it is still desirable to predict the vehicle's future position. If some *a priori* knowledge of the forces acting on the vehicle is available this can be used in the prediction scheme.

In order to analyze the high frequency output it will probably be most useful to perform a FOURIER analysis of this data and utilize the resulting FOURIER coefficients. If some physical hypothesis is available to explain the high frequency phenomena this can be correlated with the FOURIER coefficients. The high frequency output can also be compared to the *a priori* estimate of the instrument noise to determine if the high frequency output is due to system noise.

If it is concluded that there is a physical basis for the high frequency output, the FOURIER series can then be used for prediction."

Analysis of a discrete collection of data points by a Gibbsian technique. Because of its independence of time of observation, a particularly promising method for testing and correlating discrete sets of data is the Gibbsian procedure. If, for example, complete observational fixes were available (from, say, radar skin tracking) all sets of three vectors that were contained in a single plane, i. e., that obey the equations

$$r_2 = c_1\, r_1 + c_3\, r_3,\qquad(9)$$

could be collected together. Among these collected sets one would solve for the semilatus rectum (see page 120 of Ref. [1] for the derivation of these relations); i. e., one would find

$$p = \frac{c_1 r_1 + c_3 r_3 - r_2}{(c_1 + c_3 - 1)},\qquad(10)$$

wherein the c's had been solved for each of the sets of r's obeying (9). Those sets of r's, being not only co-planer (to, say, three standard deviations, in their estimated error) but also having the same value of p (again to within, say, three standard deviations) could be counted on "probably" to refer to the same object. Those r's that could not be so classified would then be subject to rejection. The problem would, of course, require solution on a more sophisticated statistical baisis; but the broad concept is clear enough. The principal problem lies not in the solution of hundreds of equations like Eq. (9), for obviously the solution involves simple cross-products that could be rapidly accomplished on a high-speed digital computer i. e.,

$$c_1 = |\boldsymbol{r}_2 \times \boldsymbol{r}_3| / |\boldsymbol{r}_1 \times \boldsymbol{r}_3|$$

and   (11)

$$c_3 = |\boldsymbol{r}_2 \times \boldsymbol{r}_1| / |\boldsymbol{r}_1 \times \boldsymbol{r}_3|$$

or of the simple Eq. (10), but lies rather in the estimation of the vectors $\boldsymbol{r}_1$ together with the estimated standard deviation of $\boldsymbol{r}$ if these data are *not* given completely to us by radar. In consequence of this we find that we have now turned full circle. In order to select the proper set of incomplete data (e. g., α_i, δ_i, $\dot{\varrho}_i$, where $i = 1, 2, 3$) to use in our preliminary orbit to gain estimates of, say, $\boldsymbol{r}_1, \boldsymbol{r}_2, \boldsymbol{r}_3, \ldots$ we require these very end-product quantities. Clearly, a cleverer approach is required—brute force guessing seems to be an undesirable answer, and the problem remains quite challenging.

References

[1] BAKER, R. M. L., Jr., and M. W. MAKEMSON (1960): An Introduction to Astrodynamics, New York: Academic Press.

[2] HERRICK, S. (1963): Astrodynamics New York: Van Nostrand.

[3] HERGET, P. (1948): The Computation of Orbits (published privately by the author).

[4] KOTELNIKOV, V. A. (1958): "Determination of the Elements of the Orbit of a Satellite Using Doppler Effect," paper presented at the Geophysique Internationale held in Moscow, July 30 to August 9.

[5] LORENS, C. S. (1959): "The Doppler Method of Satellite Tracking," California Institute of Technology Jet Propulsion Laboratory Report 30-2, March 30.

[6] KEMPER, W. A. (1958): "Satellite Trajectory and Orbit Calculations from Mini-track Triangulation," U.S. Naval Proving Grounds, Dalgren Va., Report No. 1633, December 8.

[7] HARRIS, I., and R. JASTROW (1958): "A Short Program for Satellite Orbit Prediction," Report of NRL Progress, February.

[8] SKINNER, T. J. (1958): "Tracking of a Moving Transmitter by the Doppler Effect," Air Force Cambridge Research Center, Bedford, Mass., AFCRC-TR-58-364.

[9] McDONALD, W. S. (1958): "Satellite Tracking from Several Coordinated Doppler Receiving Stations," California Institute of Technolgy Jet Propulsion Laboratory External Publication No. 554, Sept. 5.

[10) GUIER, W. H., and G. C. WEIFFENBACH (1958): "Theoretical Analysis of Doppler Radio Signals from Earth Satellites," Johns Hopkins University Applied Research Laboratory, Bumblebee Report R-276.

[11] MENGEL, J. T., and P. HERGET (1958): "Tracking Satellites by Radio," Scientific American 198, 23—29.

[12] FISHMAN, M. (1958): "Satellite Tracking Techniques," Vistas in Astronautics 1, 67—70.

[13] CROOKS, J. W. (1959): "A Simplified High Precision 200-millionmile Tracking Guidance and Communication System," Fourth Symposium on Ballistic Missile and Space Technology, UCLA August 24—27.

[14] STERNKE, G. et al (1958): "Doppler Velocity and Position Study," Aeronutronic Report No. U-356.

[15] WERNER, R. V. (1958): "Advanced Space Instrumentation Techniques," paper presented at IRE National Symposium on Telemetering, Miami Beach, Fla., September 22—24.

[16] "Interpolation and Allied Tables," 1936, H. M. Stationary Office, London.

[17] IZSAK, I. G. (1960): "Orbit Determination from Simultaneous Doppler-Shift Measurements," Smithsonian Inst. Astrophys. Obs. Special Repor No. 38, January 15.

[18] DEUTSCH, A. J. (1961): "A Doppler Method for Determining the Orbit of an Artifical Satellite Around Another Planet," paper presented at the First International Symposium on Analytical Astrodynamics at UCLA, June.

[19] KOSKELA, P. (1961): Personal communication, Orbit Analysis Department, Aeronutronic Division, Ford Motor Company, Newport Beach, California, USA.

[20] DOUGLAS, B. C. (1962): Internal memorandum, Astrodynamics Research Center, Lockheed California Company, West Los Angeles, California, USA.

[21] CAHILL, W. F., and I. HARRIS (1961): "Determination of Satellite Orbits from Radar Data," Proc. IRE, 18, No. 9, Sept.

[22] PATTON, R. B. (1960): "Orbit Determination from Single Pass Doppler Observations," Space Trajectories, New York: Academic Press, Inc. pp. 251—268.

[23] Hughes Aircraft Company, "Proposal for Satellite Surveillance System," FDL 916-R, Circa (1960).

[24] BAKER, R. M. L., Jr. (1960): "Preliminary Orbit Methods for Radar Observations of Space Probes," submitted for possible publication in Astronom. J.

[25] BAKER, R. M. L., Jr. (1960): "Orbit Determination From Range and Range-Rate Data," paper presented at the ARS Semi-Annual Meeting, Los Angeles, California, May 9—12, Preprint No. 1220—60. (1960).

[26] BAKER, R. M. L., Jr. (1960): "Novel Orbit-Determination Techniques as Applied to Air Force Systems," paper presented to the Seventh Annual ARDC Science and Engineering Symposium, Boston, Mass. November 30.

[27] BAKER, R. M. L., Jr. (1962), and L. AVERILL: "Numerical Analyses of Range-Only Orbit Determination," submitted for publication as an Astrodynamics Report of the University of California, Los Angeles.

[28] MILSTEAD, A. H. (1962): "Range and Elevation-Angle Only Orbit Determination," report presented to the Department of Astronomy University of California, Los Angeles, for Astronomy 299, May.

[29] FINE, TH. I.: Internal memorandum, Astrodynamics Research Center, Lockheed California Company, West Los Angeles, California, USA.

[30] KOVALESKY, I. (1961), and F. BARLIER: "Determination of the Osculating Elements of an Artificial Satelite Orbity," Extrait des c. R. Acad. Sci., Paris 252, 1273—1275.

Appendix

Formula coefficients

The coefficients of the formulas found in the modified LAGRANGE-GAUSS-GIBBS method are presented below:

$$\varrho_2 = A + \dot{\varrho}_1 B + \dot{\varrho}_2 C + \dot{\varrho}_3 D$$

$$A \triangleq [\dot{f}_1 \dot{g}_3 \boldsymbol{L}_1 \cdot \boldsymbol{R}_2 - \dot{g}_1 \dot{f}_3 \boldsymbol{L}_3 \cdot \boldsymbol{R}_2 + \dot{g}_1 \dot{g}_3 \dot{\boldsymbol{R}}_2 \cdot (\boldsymbol{L}_1 - \boldsymbol{L}_3) - \dot{\boldsymbol{R}}_1 \cdot \boldsymbol{L}_1 \dot{g}_3 +$$
$$+ \dot{\boldsymbol{R}}_3 \cdot \boldsymbol{L}_3 \dot{g}_1]/E,$$

$$B \triangleq \dot{g}_3/E,$$

$$C \triangleq \dot{g}_1 \dot{g}_3 \boldsymbol{L}_2 \cdot (\boldsymbol{L}_3 - \boldsymbol{L}_1)/E \quad (C \text{ is very small usually}),$$

$$D \triangleq -\dot{g}_1/E$$

and

$$E \triangleq \dot{f}_1 \dot{g}_3 \boldsymbol{L}_1 \cdot \boldsymbol{L}_2 - \dot{g}_1 \dot{f}_3 \boldsymbol{L}_3 \cdot \boldsymbol{L}_2 + \dot{g}_1 \dot{g}_3 \boldsymbol{L}_2 \cdot (\boldsymbol{L}_1 - \boldsymbol{L}_3),$$

where

$$\dot{f}_i = -\frac{\mu \tau_i}{r_2^3} + \frac{3\mu \tau_i^2}{2 r_2^4} \dot{r}_2 - \frac{\mu}{6 r_2^6} \left(4\mu - 15 r_2 \dot{r}_2^2 + 3\mu \frac{r_2}{a} \right) \tau_i^3 - \frac{5 \mu^2 \dot{r}_2}{8 r^7} \cdot$$
$$\left(4 - 7 \frac{r_2 \dot{r}_2^2}{\mu} - \frac{3 r_2}{a} \right) \tau_i^4 + \cdots . \quad i = 1,3$$

and

$$\dot{g}_i = 1 - \frac{\mu \tau_i^2}{2 r_2^3} + \frac{\mu \tau_i^3}{r_2^4} \dot{r}_2 + \frac{\mu \left(10 - \dfrac{45 r_2 \dot{r}_2^2}{\mu} - \dfrac{9 r_2}{a} \right)}{24 r_2^6} \tau_i^4$$

$$i = 1, 3$$

with $\tau_i \triangleq k(t_i - t_0)$, $\quad i = 1, 2, 3$.

Sur le mouvement de translation-rotation des corps célestes artificiels

Par

G. N. Doubochine[1]

Université de Moscou, U.S.S.R.

Il est bien connu qu'on regarde séparément en mécanique céleste classique les mouvements de translation et de rotation des corps célestes.

On considère d'abord le problème du mouvement de translation du corps comme le mouvement d'un point matériel et on regarde ensuite le mouvement de rotation de ce corps autour de son centre d'inertie, en regardant les coordonnées de ce centre comme des fonctions connues.

On suppose de cette manière que le mouvement de rotation d'un corps (le mouvement autour de son centre des masses) dépend de son mouvement de translation, tandis qu'on néglige toujours l'effet de rotation sur le mouvement de ce centre.

Je crois qu'il est maintenant nécessaire d'étudier soigneusement ce problème. Pour cela il faut examiner d'abord le mouvement général du corps sans diviser ce mouvement en deux mouvements séparés de translation et de rotation.

Cette opinion est complètement partagée par des savants soviétiques spécialistes en mécanique céleste, dont quelques-uns participent avec succès à l'étude de ce problème.

D'autre part, la théorie du mouvement de translation-rotation des corps a pris ces dernières années une signification nouvelle en rapport avec la nécessité évidente de l'étude plus approfondie du mouvement de rotation des corps célestes artificiels (satellites artificiels, vaisseaux cosmiques, etc. . .).

VLADIMIR CONDOURAR est le premier qui s'est mis à étudier ces questions et c'est lui qui avait soutenu sa thèse sur ce thème il y a déjà vingt ans, et depuis a souvent publié des articles intéressants et sur la théorie générale du problème et sur ses applications à la théorie de libration de la Lune, et à la théorie du mouvement des corps célestes artificiels.

Je m'occupe de ces questions depuis 1955, surtout au point de vue de la théorie générale du mouvement de translation-rotation.

---

[1] Ce rapport est un résumé de quelques travaux de l'auteur sur la théorie du mouvement de translation-rotation des corps célestes, publiés dans les revues scientifiques soviétiques.

NINA MAGNARADZÉ (Université de Tbilissi), MICHEL YAROV-YAROVOY (Université de Moscou), et d'autres ont aussi ces dernières années consacré leur attention au problème du mouvement de translation-rotation des corps.

VLADIMIR BÉLETSKY (Institut des Mathématiques de l'Académie des Sciences de l'URSS.) s'occupe principalement de la théorie du mouvement de rotation des satellites artificiels de la Terre et vient de soutenir une thèse très intéressante sur cette question.

Dans mon rapport je résumerai en peu de mots quelques résultats de mes propres recherches sur la théorie du mouvement de translation-rotation des corps célestes considérés comme des corps rigides, dont les particules élémentaires s'attirent d'après la loi de NEWTON.

Tout d'abord j'ai étudié les équations différentielles du mouvement du système des corps identiques, écrits dans un certain système des coordonnées absolues.

Soit ξ_i, η_i, ζ_i les coordonnées absolues du centre de masse du corps $M_i (i = 0, 1, 2, \ldots n)$ et ψ_i, ϑ_i, φ_i-angles d'EULER ordinaires, déterminant l'orientation des axes d'inertie du corps M_i dans le même système des coordonnées.

Désignons maintenant par U_{ij} la fonction de force de deux corps M_i et M_j. Alors

$$U_{ij} = f \int\limits_{(M_i)} dm_i \int\limits_{(M_j)} \frac{dm_j}{\Delta_{ij}}$$

et

$$U = \sum_{ij} U_{ij}$$

est la fonction des forces de tout le système des corps.

La fonction U_{ij} est une certaine fonction, bien déterminée, de quantités

$$\xi_i - \xi_j, \ \eta_i - \eta_j, \ \zeta_i - \zeta_j, \ \psi_i, \ \vartheta_i, \ \varphi_i, \ \psi_j, \ \vartheta_j, \ \varphi_j$$

et par conséquent, la fonction U dépend de toutes coordonnées rectangulaires et de tous angles d'EULER.

Il n'est pas difficile de vérifier que U satisfait aux relations suivantes:

$$
\left.
\begin{aligned}
&\sum_i \frac{\partial U}{\partial \xi_i} = 0; \\
&0 = \sum_i \left(\eta_i \frac{\partial U}{\partial \zeta_i} - \zeta_i \frac{\partial U}{\partial \eta_i} + \frac{\sin\psi_i}{\sin\vartheta_i} \frac{\partial U}{\partial \varphi_i} - \frac{\sin\psi_i \cos\vartheta_i}{\sin\vartheta_i} \frac{\partial U}{\partial \psi_i} + \cos\psi_i \frac{\partial U}{\partial \vartheta_i} \right) \\
&\sum_i \frac{\partial U}{\partial \eta_i} = 0; \\
&0 = \sum_i \left(\zeta_i \frac{\partial U}{\partial \xi_i} - \xi_i \frac{\partial U_i}{\partial \zeta_i} - \frac{\cos\psi_i}{\sin\vartheta_i} \frac{\partial U}{\partial \varphi_i} + \frac{\cos\psi_i \cos\vartheta_i}{\sin\vartheta_i} \frac{\partial U}{\partial \psi_i} + \sin\psi_i \frac{\partial U}{\partial \vartheta_i} \right) \\
&\sum_i \frac{\partial U}{\partial \zeta_i} = 0; \quad 0 = \sum_i \left(\xi_i \frac{\partial U}{\partial \eta_i} - \eta_i \frac{\partial U}{\partial \xi_i} + \frac{\partial U}{\partial \psi_i} \right).
\end{aligned}
\right\} \quad (1')
$$

Les équations différentielles du problème ont la forme:

$$
\left.
\begin{aligned}
& m_i\,\ddot{\xi} = \frac{\partial U}{\partial \xi_i}; \\[4pt]
& A_i\,\dot{p} - (B_i - C_i)\,q_i\,r_r = \left(\frac{\partial U}{\partial \psi_i} - \cos\vartheta_i\,\frac{\partial U}{\partial \vartheta_i}\right)\frac{\sin\varphi_i}{\sin\vartheta_i} - \cos\varphi_i\,\frac{\partial U}{\partial \vartheta_i} \\[4pt]
& m_i\,\ddot{\eta}_i = \frac{\partial U}{\partial \eta_i}; \\[4pt]
& B_i\,\dot{q}_i - (C_i - A_i)\,r_i\,p_i = \left(\frac{\partial U}{\partial \psi_i} - \cos\vartheta_i\,\frac{\partial U}{\partial \vartheta_i}\right)\frac{\cos\varphi_i}{\sin\vartheta_i} - \sin\varphi_i\,\frac{\partial U}{\partial \vartheta_i} \\[4pt]
& m_i\,\ddot{\zeta}_i = \frac{\partial U}{\partial \zeta_i}; \quad C_i\,\dot{r}_i - (A_i - B_i)\,p_i\,q_i = \frac{\partial U}{\partial \varphi_i} \\[4pt]
& p_i = \sin\varphi_i \sin\vartheta_i\,\dot{\psi}_i + \cos\varphi_i\,\dot{\vartheta}_i, \quad q_i = \cos\varphi_i \sin\vartheta_i\,\dot{\psi}_i - \sin\varphi_i\,\dot{\vartheta}_i, \\[4pt]
& r_i = \cos\vartheta_i\,\dot{\psi}_i + \dot{\varphi}_i \qquad\qquad\qquad\quad (i = 0, 1, 2, \ldots, n)
\end{aligned}
\right\} \quad (1)
$$

où m_i, A_i, B_i, C_i désignent la masse et les moments d'inertie du corps M_i.

Les équations (1) forment un sytème d'équations simultanées, dont l'intégration complète avec les conditions initiales données déterminera le mouvement de translation et de rotation de chacun des corps M_i. Il est évident en outre que ces mouvements dépendent respectivement l'un de l'autre.

Cependant il est clair qu'on ne peut pas, en général, intégrer complètement les équations (1). On peut seulement affirmer que les équations (1) ont dix intégrales premières analogues aux intégrales classiques du problème du mouvement des points matériels s'attirant suivant la loi de Newton.

Neuf de ces intégrales — les intégrales du mouvement de centre des masses et les intégrales de moment cinétique — se déduisent des équations (1) à l'aide des relations (1').

La dixième intégrale est l'intégrale d'énergie et est une conséquence immédiate du principe de la conservation de l'énergie.

Ces dix intégrales ont la forme:

$$
\left.
\begin{aligned}
& \sum_i m_i\,\dot{\xi}_i = a_1, \quad \sum_i m_i\,\dot{\eta}_i = a_2, \quad \sum_i m_i\,\dot{\zeta}_i = a_3 \\[4pt]
& \sum_i m_i\,\xi_i = a_1\,t + b_1, \quad \sum_i m_i\,\eta_i = a_2\,t + b_2, \quad \sum_i m_i\,\zeta_i = a_3\,t + b_3, \\[4pt]
& \sum_i \left[m_i(\eta_i\,\dot{\zeta}_i - \zeta_i\,\dot{\eta}_i) + A_i\,p_i\,a_{11}^{(i)} + B_i\,q_i\,a_{12}^{(i)} + C_i\,r_i\,a_{13}^{(i)}\right] = c_1, \\[4pt]
& \sum_i \left[m_i(\zeta_i\,\dot{\xi}_i - \xi_i\,\dot{\zeta}_i) + A_i\,p_i\,a_{21}^{(i)} + B_i\,q_i\,a_{22}^{(i)} + C_i\,r_i\,a_{23}^{(i)}\right] = c_2, \\[4pt]
& \sum_i \left(m_i(\xi_i\,\dot{\eta}_i - \eta_i\,\dot{\xi}_i) + A_i\,p_i\,a_{31}^{(i)} + B_i\,q_i\,a_{32}^{(i)} + C_i\,r_i\,a_{33}^{(i)}\right] = c_3, \\[4pt]
& T = U + h; \quad T = \sum_i \left[m_i(\dot{\xi}_i^2 + \dot{\eta}_i^2 + \dot{\zeta}_i^2) + A_i\,p_i^2 + B_i\,q_i^2 + C_i\,r_i^2\right]
\end{aligned}
\right\} \quad (2)
$$

où $a_{s\sigma}^{(i)}$, dépendants des ψ_i, ϑ_i, φ_i, sont les cosinus directeurs des axes de l'inertie du corps M_i.

Si les corps M_i sont tout à fait arbitraires, les équations (1) n'ont pas d'autres intégrales que les dix indiquées. Mais dans des cas particuliers les équations (1) peuvent avoir encore quelques intégrales, distinctes des (2). Par exemple, si le corps M_0 a la symétrie axiale, alors $A_0 = B_0$ et U ne dépend pas de l'angle φ_0.

Alors, les équations (1) admettent une intégrale $r_0 = $ const et il en est ainsi pour chaque corps à la symétrie axiale.

La fonction des forces U_{ij} de deux corps peut être développée en série infinie de même que dans le cas du corps et du point matériel.

Cette série, obtenue à l'aide des polynomes de GEGENBAUER, a la forme suivante:

$$U_{ij} = f \sum_{\varkappa = 0}^{\infty} \frac{U_{i\xi j}^{(\varkappa)}}{R_{ij}^{2\varkappa + 1}},$$

où R_{ij} est la distance entre les centres des masses de deux corps et $U_{ij}^{(\varkappa)}$ — les polynomes de degré $\varkappa$ par rapport à $\xi_i - \xi_j, \eta_i - \eta_j, \zeta_i - \zeta_j$ dont les coefficients dépendent des angles d'EULER de deux corps.

Les termes premiers de ces séries ont la forme:

$$U_{ij} = f \frac{m_i m_j}{R_{ij}} + f m_j \frac{A_i + B_i + C_i - 3 I_i^{(ij)}}{2 R_{ij}^3} + f m_i \frac{A_j + B_j + C_j - 3 I_j^{(ij)}}{2 R_{ij}^3} + \cdots$$

ou $I_i^{(ij)}$ est le moment d'inertie du corps M_i par rapport à la droite joignant les centres des masses de deux corps.

Remarquons maintenant que si dans le développement (3) on néglige des termes du troisième ordre au moins par rapport aux distances inverses, alors la fonction U sera réduite à la fonction des forces du système des points matériels. Alors, les équations (1) seront désagrégées en deux systèmes indépendants, dont l'un définit les mouvements des centres de masses (comme des points matériels) et l'autre la rotation de chaque corps autour de son centre de masses (suivant les lois d'EULER-POINSOT).

Si nous conservons dans chaque fonction U_{ij} les termes du 3ème ordre (par rapport aux distances inverses), les équations (1) ne seront pas désagrégées en deux systèmes et devront être intégrées simultanément.

Si pourtant l'on néglige les termes d'ordres supérieurs non dans la fonction U, mais dans les parties droites des équations (1) (c'est-à-dire dans les dérivées de U), nous aurons un résultat tout autre.

En effet, en conservant les termes du 3ème ordre (en négligeant tous les autres), dans les équations $m_i \ddot{\xi}_i = \dfrac{\partial U}{\partial \xi_i}, \ldots$ nous obtiendrons de nouveau les équations du mouvement des $n + 1$ points matériels et le mouvement de translation des corps M_i ne dépendra pas de leurs mouvements de rotation.

Mais les équations définissant les angles d'Euler (écrites avec la même précision) contiendront des parties, dépendant des coordonnées des centres de masses, et c'est pourquoi les mouvements de rotation des corps dépendront de leurs mouvements de translation.

Maintenant nous considérerons le cas particulier du problème général où $n = 1$, c'est-à-dire quand nous n'avons que deux corps M_0 et M_1.

Le système (1) est dans ce cas d'ordre 24, que nous pouvons réduire au système d'ordre 18, en utilisant six intégrales du mouvement du centre de masses. Par exemple, nous pouvons déduire les équations déterminant le mouvement de translation du corps M_1 par rapport au corps M_0.

Si en outre M_0 est une sphère, possédant la distribution sphérique des densités, alors la fonction des forces respective ne dépendra pas des φ_0, ψ_0, ϑ_0 et nous aurons définitivement le système d'ordre 12, déterminant le mouvement de translation-rotation du corps M_1 par rapport au M_0. Le dernier problème, le plus simple dans le théorie du mouvement de translation-rotation, est encore si complexe qu'il ne peut pas être résolu en termes finis (même pour les hypothèses les plus simples sur la forme et la structure du corps M_1).

C'est pourquoi nous sommes obligés de nous borner ici à la recherche de quelques solutions particulières et de trouver d'autres solutions (plus générales) à l'aide de séries infinies.

Le problème du mouvement de translation-rotation de deux corps (le problème général de deux corps) peut avoir des solutions particulières simples, où le centre de masse d'un corps décrit une trajectoire plane (par rapport au centre de masse de l'autre corps) et chacun des deux corps conserve une orientation invariable par rapport à cette trajectoire.

C'est V. Condourar qui a trouvé le premier des solutions particulères semblables dans le problème du mouvement de deux ellipsoïdes homogènes de rotation.

J'ai analysé cette question au point de vue général, en posant le problème de définir les conditions auxquelles le centre de masse M_1 décrit une orbite plane par rapport au centre de masse M_0. En utilisant pour la résolution de ce problème les intégrales du moment cinétique, j'ai trouvé plusieurs cas où le centre de masse de M_1 décrit l'orbite circulaire par rapport à M_0.

Le cas le plus général de ce genre est celui où chacun des deux corps possède la symétrie géométrique-dynamique par rapport à un certain axe passant par le centre de masse du corps.

Pour un tel corps l'ellipsoïde d'inertie central est un ellipsoïde de révolution et c'est pourquoi la fonction des forces, en ce cas, ne dépend pas des angles φ_0, φ_1.

Alors, dans les solutions particulières avec les orbites circulaires chacun des deux corps conserve une orientation invariable par rapport à cette orbite, et notamment l'un des axes de l'ellipsoïde d'inertie de chaque corps coïncide avec la direction du rayon-vecteur du centre de masse du corps M_1, et l'un des deux autres est normal au plan de l'orbite.

Outre les orbites circulaires il existe encore les trajectoires rectilignes et dans ce cas chacun des deux corps conserve une orientation invariable par rapport à la droite que décrit le centre de masse de M_1 dans cette solution.

Il est possible que le mouvement circulaire existe encore dans des cas plus compliqués, mais particuliers.

AINSI BÉLETSKY a montré dans sa thèse que si le corps M_0 est le point matériel, le corps M_1 peut avoir l'ellipsoïde d'inertie à trois axes inégaux.

Les solutions particulières du problème des deux corps, dont nous parlons ici et que j'appelle «solutions régulières», sont très simples.

En effet dans ces solutions nous aurons

$$\varrho = a, \quad v = n\,t, \quad z = 0, \quad \dot{\psi} = \text{const}, \quad \dot{\varphi} = \text{const}, \quad \vartheta = \text{const},$$

où ϱ, v, z sont les coordonnées cylindriques du corps (par rapport à M_1) et ψ, φ, ϑ les angles d'EULER de ce corps.

En partant de ces solutions régulières, nous pouvons trouver, à l'aide des méthodes de A. LIAPOUNOFF, de multiples solutions voisines, exprimables en séries de puissances des variations des conditions initiales.

J'ai examiné, en même temps, le problème de stabilité du mouvement régulier (au sens de LIAPOUNOFF) pour le cas où M_0 est le point matériel (ou la sphère) et M_1 le corps ayant la symétrie dynamique par rapport à un certain axe.

Dans ce cas on est arrivé aux résultats suivants: La solution régulière, que j'appelle «flèche» et qui est définie par les conditions:

$$\varrho = a, \quad v = n\,t, \quad z = 0; \quad \psi = n\,t + (\pi, 0), \quad \vartheta = \vartheta^{(0)}, \quad \cos\vartheta^{(0)} = \frac{\varepsilon}{n},$$

avec

$$\varepsilon = (1 - \sigma)\,\omega, \quad \omega = \dot{\varphi}^{(0)} + \cos\vartheta^{(0)}\,\dot{\psi}^{(0)}, \quad \sigma = 1 - \frac{C}{A},$$

est stable pour $\sigma < 0$ et instable pour $\sigma > 0$.

La solution régulière que j'appelle «flotte» peut être stable ou instable. Ces solutions sont définies par les conditions

$$\varrho = a, \quad v = n\,t, \quad z = 0, \quad \vartheta = \vartheta^{(0)} = \frac{\pi}{2}$$

2*

et sont stables, par exemple, pour $\sigma = -1$ et instables pour $\sigma = +1$. Pour les autres ϱ les conditions de stabilité ont une forme plus compliquée.

Enfin, la solution régulière que j'appelle «rais» et qui est définie par les formules:

$$\varrho = a, \quad v = n\,t, \quad z = 0, \quad \psi = nt + \left(\frac{3\pi}{2}, \frac{\pi}{2}\right), \quad \vartheta = \vartheta^{(0)},$$

$$\cos \vartheta^{(0)} = \frac{\varepsilon}{n(3\sigma + 1)}$$

est toujours stable pour $\sigma > 0$, et pour $\sigma < 0$ peut être stable ou instable selon des conditions complémentaires.

J'ai étudié plus soigneusement un cas limite, ayant une signification importante dans la théorie du mouvement des satellites artificiels qui ont la forme d'un cylindre allongé.

C'est le cas où le corps M_1 est le segment matériel homogène de longueur $2l$.

J'indiquerai seulement l'un des résultats obtenus. La solution régulière «rais» qui est définie dans ce problème par les formules:

$$\varrho = a, \quad v = n\,t, \quad z = 0; \quad \psi = n\,t + \left(\frac{3\pi}{2}, \frac{\pi}{2}\right), \quad \vartheta = \vartheta^{(0)} = 90°$$

où

$$n^2 = \frac{\mu}{a\,(a^2 - l^2)} > \frac{\mu}{a^3}$$

peut être considérée, si l'on veut, comme un certain mouvement troublé avec l'orbite circulaire du centre de masse (du centre de segment) dont la vitesse linéaire est $V = n\,a$.

Il en suit qu'on peut choisir l de façon que $V^2 - \frac{2\mu}{\alpha} > 0$, c'est-à-dire pour que la vitesse V soit *hyperbolique*. Par conséquent le cas présent montre que dans le mouvement troublé du type hyperbolique le point mobile peut toujours rester à une distance finie du corps principal.

Je terminerai par une remarque qui n'a pas de rapport direct avec le thème exposé.

On peut trouver des exemples où le mouvement troublé est du type elliptique, mais dans lequel le point mobile s'éloigne infiniment du corps principal.

On long range effects in the motion of artificial satellites

By

Peter Musen

Theoretical Division, Goddard Space Flight Center
National Aeronautics and Space Administration
Greenbelt, Maryland, U.S.A.

Abstract. The long range effects as caused by the Moon and the Sun are of primary importance for establishing the stability of highly eccentric orbits of satellites. For the time being no complete analytical theory exists which can treat such orbits. The use of HALPHEN's method of treating secular planetary effects was suggested for the determination of the long range lunar effects in the motion of artificial satellites.

This choice permits the numerical integration of long range lunar effects over an interval of many years. Numerous examples are computed and plotted. The long range solar effects are treated by averaging the disturbing function over the orbit of the satellite.

The effects in the motion of a 24 hour satellite caused by the ellipticity of Earth's equator are treated using a resonance theory.

This paper contains results obtained by a group working in celestial mechanics in the Theoretical Division at the Goddard Space Flight Center. Consideration is given to the long range effects in the motion of artificial satellites with large eccentricities, large inclinations and large semi-major axes. The effects of the Moon and the Sun and the irregularities of the Earth's gravitational field are considered. In treating the problem of stability of such satellites, the long range effects are of primary importance, but no purely analytical theory considering the long range effects for such extreme elements exists. One has to resort to numerical integration in order to obtain information about the stability of the orbit over a long interval of time and about the life-time of the satellite. The methods based on the use of an un-averaged distrubing function, like COWELL or ENCKE, contain both the short and the long period terms, and, in the case of artificial satellites, require that the interval of integration be much less than that of the period of the satellite. The main long range effects in the elements are produced by the long range terms in the disturbing function and by their "cross actions". The short period terms can produce long range effects through their mutual cross actions in higher approximations, but these are very small (BROWN and SHOOK, 1933).

For these reasons, as well as to diminish the accumulation of round off errors, it is necessary to remove the short period terms from the

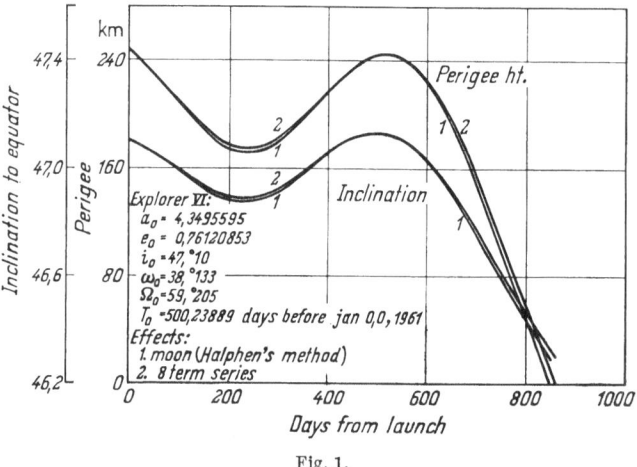

Fig. 1.

disturbing function or from the components of the disturbing force from the very start.

The problem thus formulated does not differ from the problem of determining the secular effects of planets and comets by means of

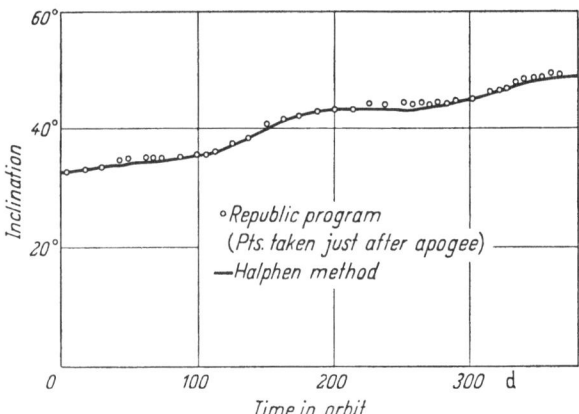

Fig. 2. IMP inclination vs time.

numerical integration using the Gaussian method (1818). With the existence of modern electronic equipment such a solution of the problem has become possible.

The use of HALPHEN's form (1888) of the Gaussian theory was suggested by the author as a practical method for determining the long range lunar effects in the motion of artificial satellites (MUSEN, 1961). The secular changes in the Moon's elements as well as the perturbations in the satellite's elements caused by the Earth's oblateness were included into the programming of the method for the use of IBM 7090.

HALPHEN's method previously was not in use, probably, because of several numerical errors which appear in the original publication. They were all corrected by GORIACHEV (1937), whose name chould be assoc-iated with method as well, and in its present form should justly be called the HALPHEN-GORIACHEV method. The HALPHEN-GORIACHEV method is based on simple operations with matrices. The trigonometrical transformations which are so typical to HILL's method (1901) are absent in the HALPHEN-GORIACHEV method.

The author has suggested (1961) the use of the GOURSAT trans-formation (1881) and of the E-summability process to speed up the convergence of two hypergeometric series which appear in the HALPHEN-GORIACHEV method and to facilitate the numerical computation. All programming for the use of the electronic machine was done by A. SMITH of the Theoretical Division.

A careful comparison of the HALPHEN-GORIACHEV method with some other existing methods was done before it was recommended for use on a large scale. Figure 1 shows the results of a comparsion of the HALPHEN-GORIACHEV method, as applied to the determination of long range lunar effects in the motion of Explorer VI, with the method based on the use of trigonometric series as developed by the author, BAILIE, and UPTON (1961). The life-time prediction agrees with the prediction by KOZAI (1959).

B. SHUTE of the Theoretical Division has applied the HALPHEN-GORIACHEV method extensively to the investigation of long range stability of artificial satellites and compared the results for some satellites with numerical integration using ENCKE's method as programmed by the Republic Aviation Corporation, Farmingdale, New York. Figures 2 and 3 show the results obtained using both methods for the hypothetical satellite IMP ("Interplanetary Monitoring Probe"). This satellite has a semi-major axis of 22 earth radii and an eccentricity equal to .95 initially. The comparison clearly indicates a relative unimportance of short period terms and of their long range cross actions over an extended interval of time. One might expect that the effect of such cross action terms will not exceed approximately 1% to 2% of the perturbations obtained on the basis of the HALPHEN-GORIACHEV method, providing that no sharp commensurability of mean motions does exist.

Figures 4 and 5 show the results, also obtained by B. SHUTE, for IMP over an interval of twelve years. They show that for satellites of IMP-type one might expect large variations of the inclination and the eccen-

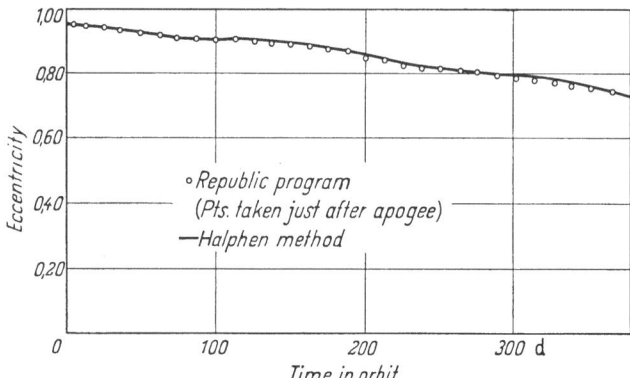

Fig. 3. IMP eccentricity vs time.

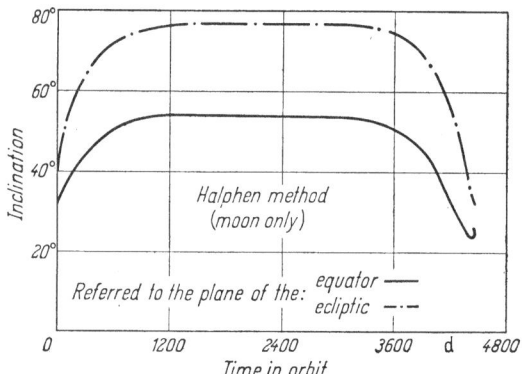

Fig. 4. IMP inclination vs time 10 year study.

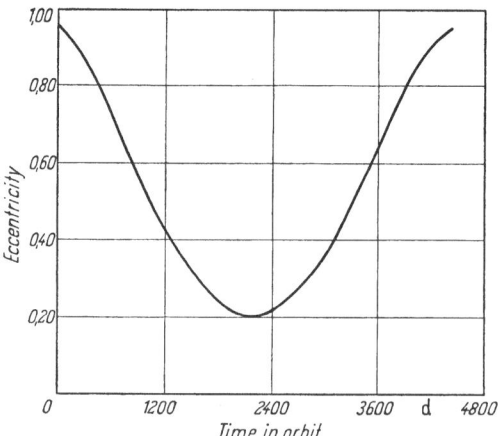

Fig. 5. IMP eccentricity vs time. 10 year study.
HALPHEN method (*Monn only*).

tricity under the influence of the Moon. The life-time of IMP will be approximately 8 years for these initial conditions. Such large "secular"

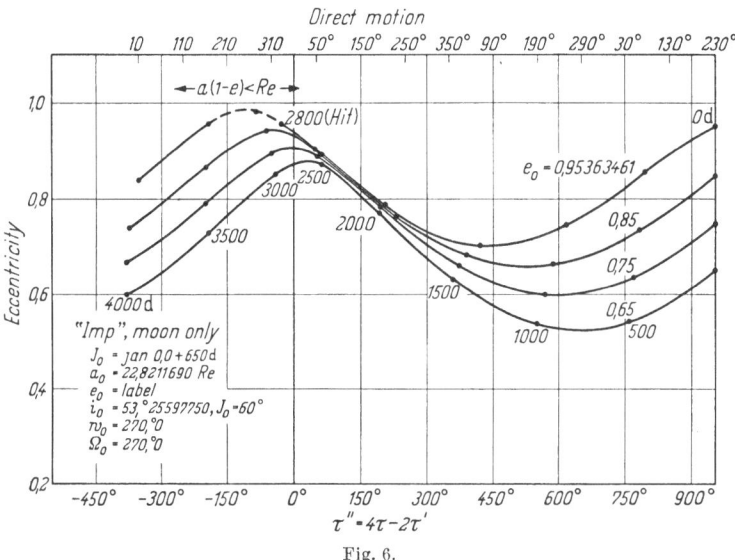

Fig. 6.

changes in the elements might also take place in the case of minor planets and comets, but we never will be able to observe the accumulation of

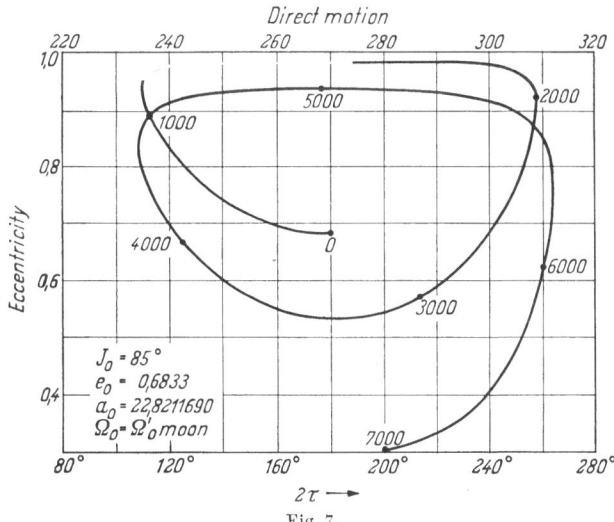

Fig. 7.

such effects directly. Similar events take place in an Earth-Moon-satellite system in a much shorter time. In that fact lies the importance of artificial satellites for controlled experiments in the domain of "secular" perturbations and for sharpening the mathematical tools of celestial mechanics.

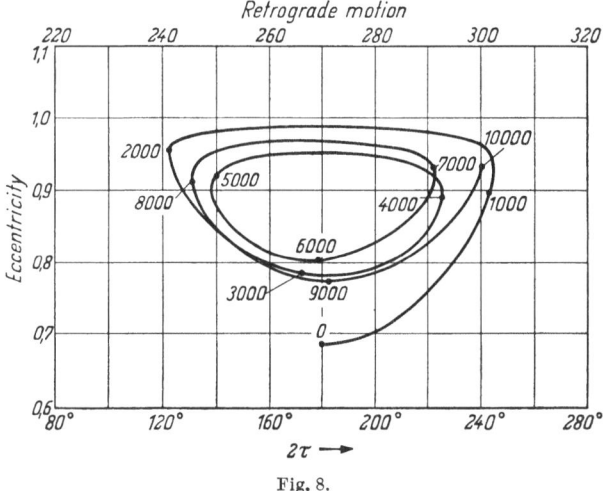

Fig. 8.

The HALPHEN-GORIACHEV method might be helpful in the problem ot the determination of characteristic points in the phase space of orbital elements which are associated with the problem of stability.

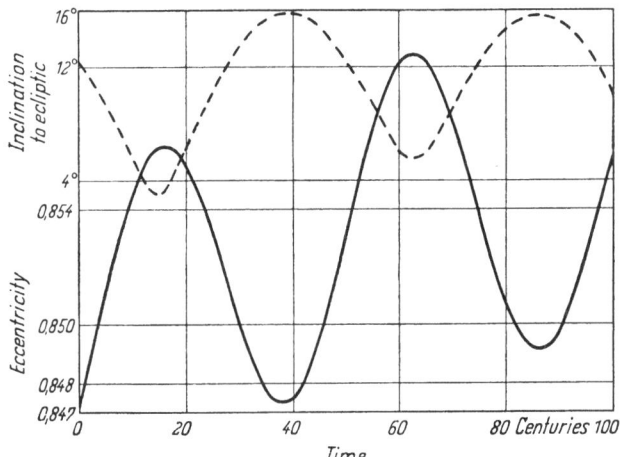

Fig. 9. Secular pertubations of ENCKE's comet.

Let τ be the argument of the perigee and $-\tau'$ be the longitude of the ascending node with respect to the Moon's orbit. Figure 6 is a plot of the eccentricity versus a critical argument $4\tau - 2\tau'$ for some particular values of the initial elements. The existence of a node is evident. Figures 7 and 8 give example of a focus for a direct and for a retrograde motion of an artificial satellite. The case of the retrograde motion shows more stability than the case of the direct motion. This statement can be easily proved analytically for small eccentricities and inclinations

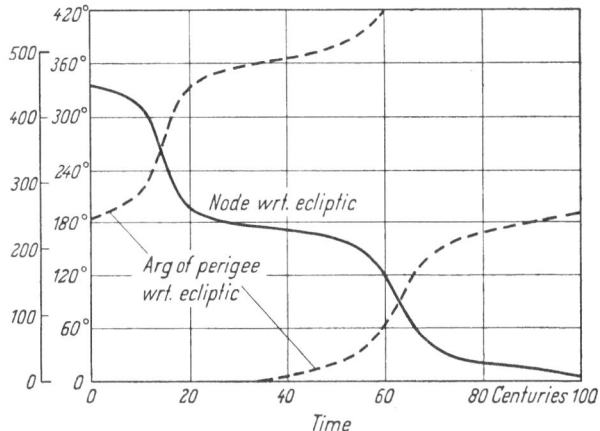

Fig. 10. Secular perturbations of ENCKE's comet.

(HIRAYAMA, 1927). In a more general case the answer can be obtained by systematic investigations based on the HALPHEN-GORIACHEV method.

Finally, the same method can be used to deduce long range variations of elements of comets and minor planets. The variations of the elements of ENCKE's comet were computed as an example (Fig. 9 and 10). The results agree basically with the results obtained by WHIPPLE and HAMID (1950), but the HALPHEN-GORIACHEV method gives a period which is somewhat longer than that given by WHIPPLE.

Let M_0, ω, Ω, i, e, a, n, be the osculating elements of the artificial satellite referred to the Earth's equator and let M_0', ω', Ω', i', e', a', n', m' be the elements and the mass of the Moon referred to a fixed ecliptic and equinox. Put

$$A_1(\alpha) = \begin{bmatrix} +1 & 0 & 0 \\ 0 & +\cos\alpha & -\sin\alpha \\ 0 & +\sin\alpha & +\cos\alpha \end{bmatrix},$$

$$A_3(\alpha) = \begin{bmatrix} +\cos\alpha & -\sin\alpha & 0 \\ +\sin\alpha & +\cos\alpha & 0 \\ 0 & 0 & +1 \end{bmatrix}.$$

The following system of formulae, combined from the system given by GORIACHEV (1937) and from the system given in author's previous work (1961) was used for the actual computations of the long range lunar effects:

$$[\boldsymbol{P},\,\boldsymbol{Q},\,\boldsymbol{R}] = A_3(\Omega) \cdot A_1(i) \cdot A_3(\omega)$$

$$[\boldsymbol{P}',\,\boldsymbol{Q}',\,\boldsymbol{R}'] = A_1(\varepsilon) \cdot A_3(\Omega') \cdot A_1(i') \cdot A_3(\omega')$$

$$\boldsymbol{r} = \boldsymbol{P}\,a(\cos u - e) + \boldsymbol{Q}\,a\sqrt{1 - e^2}\,\sin u$$

$$\varrho = \boldsymbol{P}\,\frac{a}{a'}\,(\cos u - e) + \boldsymbol{Q}\,\frac{a}{a'}\,\sqrt{1 - e^2}\,\sin u + e'\,\boldsymbol{P}'$$

$$\alpha = \varrho \cdot \boldsymbol{P}', \quad \beta = \varrho \cdot \boldsymbol{Q}', \quad \gamma = \varrho \cdot \boldsymbol{R}'$$

$$K_1 = \varrho^2 - 2 + e'^2$$

$$K_2 = (1 - e'^2)\,(1 - \alpha^2) - \beta^2 - (2 - e'^2)\,\gamma^2$$

$$K_3 = \gamma^2(1 - e'^2)$$

$$g_2 = \frac{4}{3}\,(K_1^2 - 3K_2)$$

$$g_3 = \frac{4}{27}\,(2K_1^3 - 9K_1K_2 + 27K_3)$$

$$\xi = \frac{27\,g_3^2}{g_2^3}$$

$$A = \frac{\sqrt{6}\,\sqrt[4]{g_2}}{9g_2^3}\,\frac{144}{\pi}\,\sqrt{\xi}\,\psi'(\xi)$$

$$B = \frac{\sqrt{2}}{\pi g_2\,\sqrt[4]{g_2}}\,\psi(\xi)$$

$$w = \frac{1 - \sqrt{\xi}}{1 + \sqrt{\xi}}$$

$$\psi(\xi) = \left(\frac{2}{1 + \sqrt{\xi}}\right)^{1/6}. \quad \frac{144}{\pi}\,\sqrt{\xi}\,\psi'(\xi) = \left(\frac{2}{1 + \sqrt{\xi}}\right)^{7/6}.$$

$(+2.3870942$	$(-3.7991784$
$-0.0663082\,w$	$+0.3693646\,w$
$+0.0225632\,w^2$	$-0.1556119\,w^2$
$-0.0117691\,w^3$	$+0.0889726\,w^3$
$+0.0073743\,w^4$	$-0.0586828\,w^4$
$-0.0051060\,w^5$	$+0.0419870\,w^5$
$+0.0037250\,w^6$	$-0.0313364\,w^6$
$-0.0027325\,w^7$	$+0.0233758\,w^7$
$+0.0019070\,w^8$	$-0.0165247\,w^8$
$-0.0011936\,w^9$	$+0.0104483\,w^9$
$+0.0006337\,w^{10}$	$-0.0055933\,w^{10}$
$-0.0002710\,w^{11}$	$+0.0024083\,w^{11}$
$+0.0000884\,w^{12}$	$-0.0007898\,w^{12}$
$-0.0000205\,w^{13}$	$+0.0001837\,w^{13}$
$+0.0000030\,w^{14}$	$-0.0000268\,w^{14}$
$-0.0000002\,w^{15})$	$+0.0000018\,w^{15})$

$$K_4 = 9\,K_3 - K_1\,K_2$$

$$K_5 = K_1(K_1\,K_2 - 3\,K_3) - 2\,K_2^2$$

$$a_{11} = K_4(\alpha^2 - 1) + K_5 + \frac{3}{2}\,g_2\,K_3$$

$$a_{22} = K_4(\beta^2 - 1 + e'^2) + K_5 + \frac{3}{2}\,\frac{g_2\,K_3}{1 - e'^2}$$

$$a_{33} = K_4\,\gamma^2 + K_5 + \frac{3}{2}\,g_2[\alpha^2(1 - e'^2) + \beta^2 - (1 - e'^2)]$$

$$a_{12} = a_{21} = K_4\,\alpha\,\beta$$

$$a_{23} = a_{32} = \left(K_4 - \frac{3}{2}\,g_2\right)\beta\,\gamma$$

$$a_{31} = a_{13} = \left[K_4 - \frac{3}{2}\,g_2(1 - e'^2)\right]\gamma\,\alpha$$

$$a_{11}' = \alpha^2 - 1 - \frac{1}{3}\,K_1$$

$$a_{22}' = \beta^2 - 1 + e'^2 - \frac{1}{3}\,K_1$$

$$a_{33}' = \gamma^2 - \frac{1}{3}\,K_1$$

$$a_{12}' = a_{21}' = \alpha\,\beta$$

$$a_{23}' = a_{32}' = \beta\,\gamma$$

$$a_{31}' = a_{13}' = \gamma\,\alpha$$

Then the dyadic $\boldsymbol{\Phi}$ is formed

$$
\begin{aligned}
\boldsymbol{\Phi} = &+ A_{11}\,\boldsymbol{P}'\,\boldsymbol{P}' + A_{12}\,\boldsymbol{P}'\,\boldsymbol{Q}' + A_{13}\,\boldsymbol{P}'\,\boldsymbol{R}' \\
&+ A_{21}\,\boldsymbol{Q}'\,\boldsymbol{P}' + A_{22}\,\boldsymbol{Q}'\,\boldsymbol{Q}' + A_{23}\,\boldsymbol{Q}'\,\boldsymbol{R}' \\
&+ A_{31}\,\boldsymbol{R}'\,\boldsymbol{P}' + A_{32}\,\boldsymbol{R}'\,\boldsymbol{Q}' + A_{33}\,\boldsymbol{R}'\,\boldsymbol{R}' \\
&\quad A_{ij} = a_{ij}\,A + a_{ij}'\,B
\end{aligned}
$$

The disturbing "force" averaged over the orbit of the Moon is

$$F_0 = -\frac{2\,k\,m'}{a'^3}\,\boldsymbol{\Phi}\,\boldsymbol{r}.$$

where k is the Gaussian constant.

S_0, T_0, W_0, the components of the disturbing force in the direction of $\boldsymbol{r}$, $\boldsymbol{R} \times \boldsymbol{r}$ and $\boldsymbol{R}$, are

$$S_0 = -\frac{2\,k\,m'}{a'^3}\,\frac{1}{r}\,\boldsymbol{r}\cdot\boldsymbol{\Phi}\cdot\boldsymbol{r}$$

$$T_0 = -\frac{2\,k\,m'}{a'^3}\,\frac{1}{r}\,(\boldsymbol{R}\times\boldsymbol{r})\cdot\boldsymbol{\Phi}\cdot\boldsymbol{r}$$

$$W_0 = -\frac{2\,k\,m'}{a'^3}\,\boldsymbol{R}\cdot\boldsymbol{\Phi}\cdot\boldsymbol{r}.$$

The equations for the variation of elements take the form:

$$\frac{de}{dt} = \frac{\sqrt{a(1-e^2)}}{2\pi} \int_0^{2\pi} \left[\sqrt{1-e^2}\, S_0 \sin u + T_0 \left(-\frac{3}{2}e + 2\cos u - \frac{1}{2}e\cos 2u \right) \right] du$$

$$\frac{di}{dt} = \sqrt{\frac{a}{1-e^2}}\, \frac{1}{2\pi} \int_0^{2\pi} W_0 \frac{r}{a} \left[(\cos u - e)\cos\omega - \sqrt{1-e^2}\sin u \sin\omega \right] du$$

$$\sin i \frac{d\Omega}{dt} = \sqrt{\frac{a}{1-e^2}}\, \frac{1}{2\pi} \int_0^{2\pi} W_0 \frac{r}{a} \left[(\cos u - e)\sin\omega + \sqrt{1-e^2}\sin u \cos\omega \right] du$$

$$\frac{d\pi}{dt} = \frac{\sqrt{a(1-e^2)}}{e}\, \frac{1}{2\pi} \int_0^{2\pi} \left[-S_0(\cos u - e) + T_0 \left(1 + \frac{r}{a}\frac{1}{1-e^2} \right) \sqrt{1-e^2}\sin u \right] du,$$

where $\pi = \omega + \Omega$.

The solar perturbations present less theoretical and computational difficulties, than do the lunar perturbations. The system of formulae (45'), (46'), (47'), (57) derived in the author's previous work (1961) is adequate to take all significant long period terms into account. This system is in use in the Theoretical Division for the determination of the lifetime and the long range stability of artificial satellites.

Tesseral harmonics in the Earth's potential can also introduce some long range effects, especially if the mean motion of the satellite is commensurable with the angular velocity of Earth's rotation. The most interesting case of long range effects produced by tesseral harmonics is the influence of the ellipticity of Earth's equator on the motion of a 24 hour satellite.

This effect was treated by L. BLITZER, E. M. BOUGHTON, G. KANG and R. M. PAGE (1962) using a linearized system of differential equations of motion of the satellite and by MORANDO (1962) using VON ZEIPEL's method. SEHNAL (1960) treated the same problem under the condition that the commensurability is not very sharp and that it would be possible to develop the integrals of the problem into trigonometric series He made use of the elements of DELAUNAY

$$L, G, H, l, g, h.$$

The author and A. BAILIE (1962) treated the problem using the canonincal elements

$$x_1 = L, \qquad\qquad y_1 = l + g + h - n't$$
$$x_2 = L - G, \qquad y_2 = -g$$
$$x_3 = L - H, \qquad y_3 = -h + n't$$

The disturbing function F was taken in the form

$$F = F_0 + Q_0 \cos 2y_1 + Q_1 \cos(2y_1 + 2y_2) + Q_2 \cos(2y_1 + 4y_2),$$

where F_0 is BROUWER's expression (1959) for the secular part of the disturbing function to which the term $n'H$ is added. BROUWER's expression is transformed to the set x_1, x_2, x_3. The coefficients Q_0, Q_1, Q_2 are:

$$Q_0 = +\frac{3}{8}\frac{\mu A_{22}}{a^3}(1+\cos i)^2\left(1-\frac{5}{2}e^2+\frac{13}{16}e^4-\frac{35}{288}e^6\right)$$

$$Q_1 = +\frac{3}{4}\frac{\mu A_{22}}{a^3}\sin^2 i\left(\frac{9}{4}e^2+\frac{7}{4}e^4+\frac{141}{64}e^6\right)$$

$$Q_2 = +\frac{3}{8}\frac{\mu A_{22}}{a^3}(1-\cos i)^2\left(\frac{1}{24}e^4+\frac{7}{240}e^6\right),$$

where μ is the gravitational constant and A_{22} is the geodetic parameter associated with the Ω_2^2 tesseral harmonic.

Two cases are possible:

1°. A stable case, when the conditions

$$\frac{\partial F_0}{\partial x_1}-\frac{\partial Q_0}{\partial x_1}=0$$

for

$$y_1 = \frac{\pi}{2},\ \frac{3}{2}\pi$$

are satisfield and

2°. An un-stable where

$$\frac{\partial F_0}{\partial x_1}+\frac{\partial Q_0}{\partial x_1}=0$$

for

$$y_1 = 0,\ 2\pi.$$

The problem admits two integrals, the integral of energy

$$F = -C$$

and

$$x_3 = \alpha_3.$$

Let $\alpha_1, \alpha_2, \alpha_3$ be the constants of integration associated with x_1, x_2, x_3 and for the stable case put

$$R_0 = F_0 - Q_0$$

and for the unstable case put

$$R_0 = F_0 + Q_0$$

and let

$$w_{ij} = \frac{\partial^{i+j}}{\partial\alpha_1^i\partial\alpha_2^j}R_0(\alpha_1, \alpha_2, \alpha_3).$$

The case when w_{10} is small is a near resonance case and the integrals are not developable into trigonometric series. BOHLIN's resonance theory (1889) can be used to solve the problem. The complete integral of the Hamiltonian partial differential equation

$$F = -C_0 - C_1 - \cdots,\ \ \text{with}\ \ x_1 = \frac{\partial S}{\partial y_1},\ \ x_2 = \frac{\partial S}{\partial y_2},$$

is developed into series of the form

$$S = \alpha_1 y_1 + \alpha_2 y_2 + \alpha_3 y_3 + \varphi(y_1) + [\varphi_2(y_1) + \psi_2(y_1, y_2)] + \cdots$$

with respect to the parameter $w = w_{10}/w_{20}$. The function ψ_2 is a purely trigonometric function. In the exposition, the value of $Q_0(\alpha_1, \alpha_2, \alpha_3)$ was considered to be of the order of w^2.

The constant of energy is also decomposed into a series in w. This decomposition served to remove possible zero divisors from the higher approximations by the particular choices of $C_1, C_2, \ldots$. The constant C_0 is defined as

$$C_0 = -R_0(\alpha_1, \alpha_2, \alpha_3).$$

For the stable case the solution of the Hamiltonian was obtained in the form

$$S = (\alpha_1 + A_1) y_1 + \alpha_2 y_2 + \alpha_3 y_3 \pm \int \sqrt{A_2 - A_3 \cos^2 y_1} \, dy_1$$

$$+ A_4 \sin 2y_1 + A_5 \sin (2y_1 + 2y_2) + A_6 \sin(2y_1 + 4y_2) + \cdots$$

and for the un-stable case in the form

$$S = (\alpha_1 + B_1) y_1 + \alpha_2 y_2 + \alpha_3 y_3 + \int \sqrt{B_2 + B_3 \sin^2 y_1} \, dy_1$$

$$+ B_4 \sin 2y_1 + B_5 \sin (2y_1 + 2y_2) + B_6 \sin (2y_1 + 4y_2) + \cdots$$

The coefficients A_i and B_i are power series in w with the coefficients depending upon $\alpha_1, \alpha_2, \alpha_3$. They were obtained by the authors with accuracy up to w^2. Determination of the integrals of motion in the form

$$x_i = \frac{\partial S}{\partial y_i} \qquad i = 1, 2, 3$$

$$\frac{\partial C}{\partial \alpha_i} t + \beta_i = \frac{\partial S}{\partial \alpha_i}$$

did not present any difficulty.

For orbits with small eccentricities the terms having the arguments $2y_1 + 2y_2$, $2y_1 + 4y_2$ are very small long period terms. Thus, the main part of the motion of the point $(x_1, x_2, x_3; y_1, y_2, y_3)$ in its phase space in a stable case consists of describing ovals around the libration points in the plane (x_1, y_1). The period of this motion is

$$T = \frac{\pi}{\sqrt{Q_0 \, w_{20}}} \left\{ 1 - w \left[\frac{2}{3\pi} \frac{w_{30}}{w_{20}} + \frac{Q_0'}{Q_0} - \frac{1}{2 w_{20}} \left(\frac{w_{30}}{w_{20}} + 3 \frac{Q_0'}{Q_0} \right) \right] \right\} + O(w^2).$$

In the unstable case the argument y_1 possesses a real secular term, which is not present in the stable case. Its value is

$$\nu_1 = \frac{n_1}{P},$$

where

$$n_1 = - w\, w_{20} + \frac{1}{2}\left(\frac{w_{30}}{w_{20}} + 3\,\frac{Q_0'}{Q_0}\right) w^2$$

$$P = -\frac{1}{3}\, w\, \frac{w_{30}}{w_{20}} + \frac{2\, k'\, K}{\pi}\left(1 - \frac{Q_0'}{Q_0}\right),$$

$$Q_0' = \frac{\partial}{\partial \alpha_1}\, Q_0(\alpha_1,\, \alpha_2,\, \alpha_3),\, \dots$$

K is a complete elliptic integral of the first kind with the modulus

$$k = \sqrt{\frac{B_3}{B_2 + B_3}}$$

and the complementary modulus,

$$k' = \sqrt{1 - k^2}.$$

References

[1] BLITZER, L·, E. M. BOUGHTON, G. KANG and R. M. PAGE: Effect of ellipticity of the equator on 24-hour nearly circular satellite orbits, J. G. R. **67**, 329, (1962).

[2] BOHLIN, K. P.: Über eine neue Annäherungsmethode in der Störungstheorie, Ak. Handl. Bihang, **14** (Afd. 1), Stockholm (1889).

[3] BROUWER, D.: Solution of the problem of artificial satellite theory without drag, Astronom. J., **64**, 378 (1959).

[4] BROWN, E. W., and C. A. SHOOK: Planetary Theory, Cambridge, 159 (1933).

[5] GAUSS, K. F.: Determinatio attractionis quam in punctum quodvis positionis data exerceret planeta, si eius massa, etc., Coll. Works, **III**, 331 (1818).

[6] GORIACHEV, N. N.: On the method of Halphen of the computation of secular perturbations (Russ.), University of Tomsk., 1—115 (1937).

[7] GOURSAT, E.: Ann. Scient. École Norm. Sup. (2), **10**, 3—142 (1881).

[8] HALPHEN, G. H.: Traite des fonction élliptiques, **II**, Paris (1888).

[9] HILL, G. W.: Secular perturbations of the planets, Amer. J. Math. **23**, 317—336 (1901).

[10] HIRAYAMA, K.: Proc. Imp. Acad. Jap., **3**, 9 (1927).

[11] KOZAI, Y.: Private communication (1959).

[12] MORANDO, B.: Libration d'un satellite de 24 h., C. R. Acad. Sci., Paris du 22 Jan., 635 (1962).

[13] MUSEN, P.: On the long period lunisolar effect in the motion of the artificial satellite, J. G. R., **66**, 1659 (1961).

[14] MUSEN, P., and A. BAILIE: On the motion of a 24-hour satellite, J. G. R., **67**, 1123 (1962).

[15] MUSEN, P., A. BAILIE, and E. UPTON: Development of the lunar and solar perturbations in the motion of an artificial satellite, NASA TN D-494, January (1961).

[16] SEHNAL, L.: The influence of the equatorial ellipticity of the earth gravitational field on the motion of a close satellite, Bull. Astron. Inst. Czechoslovakia, **11** (3), 90—93 (1960).

[17] WHIPPLE, F. L., and S. E. D. HAMID: Harv. Abstract, **9**, 248 (1950).

Analytical study of resonance caused by solar radiation pressure[1]

By

Dirk Brouwer

Yale University Observatory, New Haven, Connecticut, U.S.A.

Abstract. A qualitative analytical treatment of the solar radiation effect on artificial satellite motion is presented. It is shown that a development in powers of the time for polar orbits in the resonance region confirms the results obtained by numerical integration.

Let F be the Hamiltonian of the equations in the DELAUNAY variables of the drag-free problem of artificial satellite motion. The effect of solar radiation pressure may be allowed for by modifying F by the addition of

$$\Delta F = -\beta \, \frac{x\,x' + y\,y' + z\,z'}{r'^{\,3}} \, .$$

The unprimed quantities refer to the satellite, the primed quantities to the sun. The coordinates are geocentric, referred to the mean equator and mean equinox of date. The factor β, proportional to the intensity of the solar radiation and the cross section of the satellite, is treated as a constant.

The expansion of ΔF in terms of the mean anomalies yields as the leading terms

$$\Delta F = \frac{3}{2} \, \frac{\beta\,a\,e}{a'^{\,2}} \left[C\,C' \cos(g + h - \lambda') + S\,S' \cos(g - h - \lambda') + \right.$$
$$+ \frac{1}{2} \sin I \sin \varepsilon \cos(g - \lambda') +$$
$$+ C\,S' \cos(g + h + \lambda') + S\,C' \cos(g - h + \lambda') -$$
$$\left. - \frac{1}{2} \sin I \sin \varepsilon \cos(g + \lambda') \right],$$

which was given by MUSEN (1960). For the sake of brevity,

$$C = \cos^2 \frac{I}{2}, \qquad S = \sin^2 \frac{I}{2}, \qquad C' = \cos^2 \frac{\varepsilon}{2}, \qquad S' = \sin^2 \frac{\varepsilon}{2},$$

ε = obliquity of the ecliptic,
λ' = sun's mean longitude = $\nu\,t$ + const.

---

[1] This work was supported by a contract with the National Aeronautics and Space Administration.

Only terms independent of the mean anomaly of the satellite are of interest.

This paper is concerned with a qualitative discussion of the character of the motion in a resonance region. The DELAUNAY variables

$$L = \sqrt{\mu\, a} \qquad\qquad l = \text{mean anomaly}$$

$$G = L\sqrt{1 - e^2} \qquad g = \text{argument of perigee}$$

$$H = G\cos I \qquad\quad h = \text{longitude of ascending node}$$

will be used, in which μ is the constant of gravitation times the earth's mass. It may be noted that μ and β have the same dimension.

The following discussion will be limited to polar orbits ($I \approx 90°$). For such orbits the mean motion of g is negative. Hence the semi-major axis can be chosen such that the mean motion of $g + \lambda'$ is approximately zero. A simplifying circumstance for polar orbits is that the mean motion of h is zero for $I = 90°$. Consequently the last three terms of $\varDelta F$ have arguments with approximately zero mean motion. The first three terms can be eliminated in the same manner as the periodic oblateness terms, and the principal features of the motion in the resonance region in the vicinity of $I = 90°$ can be obtained by limiting $\varDelta F$ to the last three terms.

For the treatment of first-order effects it is permissible to add $\varDelta F$ to the Hamiltonian that results after the elimination of the periodic oblateness terms. By introducing $g + \lambda'$ as the second angular variable instead of g, the time is eliminated from F. This is accomplished by adding $-\nu G$ to the original Hamiltonian.

Finally, since l is absent from the Hamiltonian, L is a constant, and it will be permissible to use G/L, H/L as the variables. The only change required is division of F by L.

Since L is a constant, the term $\frac{1}{2}\,\mu^2/L^2$ in the original Hamiltonian may be omitted, and the Hamiltonian (divided by L) may be written

$$F^* = -\nu\,\frac{G}{L} + j_2\,n_0\,\frac{L^3}{G^3}\left(-\frac{1}{4} + \frac{3}{4}\,\frac{H^2}{G^2}\right) +$$

$$+ j_2^2\,n_0\left(\frac{15}{128}\,\frac{L^5}{G^5} + \frac{3}{32}\,\frac{L^6}{G^6} - \frac{15}{128}\,\frac{L^7}{G^7}\right) +$$

$$+ j_4\,n_0\left(\frac{27}{128}\,\frac{L^5}{G^5} - \frac{45}{128}\,\frac{L^7}{G^7}\right) + \varDelta F/L.$$

In polar orbits H/G is small, zero for $I = 90°$. Terms factored by H^2/G^2 have been retained in the part of F^* that has j_2 as a factor; they have been omitted in the second-order parts. The symbols j_2, j_4, n_0

were defined by

$$j_2 = \frac{\mu^2 J_2 R^2}{L^4}, \qquad j_4 = \frac{\mu^4 J_4 R^4}{L^8}, \qquad n_0 = \frac{\mu^2}{L^3},$$

R being the equatorial radius of the earth.

The canonical transformation from the set

$$L, G, H, l, g + \lambda', h$$

to

$$L, L - G, H, l + g + \lambda', -g - \lambda', h$$

leaves the original Hamiltonian unchanged.

Omitting the equations for L, $l + g + \lambda'$, new variables

$$x_1 = \frac{L - G}{L} = 1 - \sqrt{1 - e^2}, \qquad y_1 = -g - \lambda',$$

and finally

$$\xi_1 = \sqrt{2x_1} \cos y_1, \qquad \eta_1 = \sqrt{2x_1} \sin y_1$$

$$x_2 = \frac{H}{L}, \qquad\qquad y_2 = h$$

are introduced. The equations are

$$\frac{d\xi_1}{dt} = \frac{\partial F^*}{\partial \eta_1}, \qquad \frac{d\eta_1}{dt} = -\frac{\partial F^*}{\partial \xi_1}$$

$$\frac{dx_2}{dt} = \frac{\partial F^*}{\partial y_2}, \qquad \frac{dy_2}{dt} = -\frac{\partial F^*}{\partial x_2}$$

with

$$F^* = \nu x_1 + j_2 n_0 \left[-\frac{1}{4}(1 - x_1)^{-3} + \frac{3}{4} x_2^2 (1 - x_1)^{-5} \right] +$$

$$+ j_2^2 n_0 \left[\frac{15}{128}(1 - x_1)^{-5} + \frac{3}{32}(1 - x_1)^{-6} - \frac{15}{128}(1 - x_1)^{-7} \right] +$$

$$+ j_4 n_0 \left[\frac{27}{128}(1 - x_1)^{-5} - \frac{45}{128}(1 - x_1)^{-7} \right] +$$

$$+ \left[(A_0 + A_1 \cos y_2)\, \xi_1 + B \sin y_2\, \eta_1 \right],$$

in which

$$x_1 = \frac{1}{2}(\xi_1^2 + \eta_1^2)$$

$$A_0 = -\frac{\beta}{\mu} \frac{a^2}{a'^2} n_0 \sqrt{1 - \frac{x_1}{2}} \left[1 - x_2^2 (1 - x_1)^{-2} \right] \sin \varepsilon$$

$$A_1 = +\frac{\beta}{\mu} \frac{a^2}{a'^2} n_0 \sqrt{1 - \frac{x_1}{2}} \left[1 - x_2 (1 - x_1)^{-1} \cos 2\varepsilon \right]$$

$$B = -\frac{\beta}{\mu} \frac{a^2}{a'^2} n_0 \sqrt{1 - \frac{x_1}{2}} \left[\cos 2\varepsilon - x_2 (1 - x_1)^{-1} \right].$$

A first approximation to the solution for small values of x_1 and x_2 is obtained by expanding F^* in powers of x_1, x_2, retaining only the first power of x_1, and putting $x_2 = 0$. The equations are then reduced to the simple form

$$\frac{d\xi_1}{dt} = \varkappa \eta_1 + B \sin y_2,$$

$$\frac{d\eta_1}{dt} = -\varkappa \xi_1 - A_0 - A_1 \cos y_2,$$

in which

$$\varkappa = \nu - \frac{3}{4} j_2 n_0 \left[1 - \frac{7}{16} j_2 + \frac{15}{8} \frac{j_4}{j_2} \right]$$

and A_0, A_1, B, y_2 are treated as constants.

The case $\varkappa = 0$, with

$$n_0 = 17.04337 \, (R/a_0)^{3/2} \text{ rev./day},$$
$$\nu = 0.00273,$$
$$J_2 = 1.0823 \times 10^{-3},$$
$$J_4 = 1.8 \times 10^{-6},$$
$$R = 6378.165 \text{ km}$$

requires

$$\frac{a_0}{R} = 1.5880.$$

The general solution of these equations is

$$\xi_1 = -\frac{A_0 + A_1 \cos y_2}{\varkappa} + p \cos \varkappa t + q \sin \varkappa t,$$

$$\eta_1 = -\frac{B \sin y_2}{\varkappa} + q \cos \varkappa t - p \sin \varkappa t,$$

p and q being constants of integration. If $\xi_1 = \xi_{10}$, $\eta_1 = \eta_{10}$ for $t = 0$, the solution becomes

$$\xi_1 = \xi_{10} \cos \varkappa t + \eta_{10} \sin \varkappa t + \frac{A_0 + A_1 \cos y_2}{\varkappa} (\cos \varkappa t - 1) + \frac{B \sin y_2}{\varkappa} \sin \varkappa t,$$

$$\eta_1 = \eta_{10} \cos \varkappa t - \xi_{10} \sin \varkappa t - \frac{A_0 + A_1 \cos y_2}{\varkappa} \sin \varkappa t + \frac{B \sin y_2}{\varkappa} (\cos \varkappa t - 1).$$

For $\varkappa \to 0$ this form of the solution becomes illusory, but in an expansion in powers of t only positive powers of $\varkappa$ appear. The expressions become

$$\xi_1 = \xi_{10} + (B \sin y_2 + \varkappa \eta_{10}) \left(t - \frac{1}{6} \varkappa^2 t^3 + \cdots \right) +$$

$$+ (A_0 + A_1 \cos y_2 + \varkappa \xi_{10}) \left(-\frac{1}{2} \varkappa t^2 + \frac{1}{24} \varkappa^3 t^4 + \cdots \right),$$

$$\eta_1 = \eta_{10} - (A_0 + A_1 \cos y_2 + \varkappa \xi_{10}) \left(t - \frac{1}{6} \varkappa^2 t^3 + \cdots \right) +$$

$$+ (B \sin y_2 + \varkappa \eta_{10}) \left(-\frac{1}{2} \varkappa t^2 + \frac{1}{24} \varkappa^3 t^4 + \cdots \right).$$

For the simple case $\xi_{10} = \eta_{10} = 0$ it is found that

$$(\xi_1^2 + \eta_1^2)^{1/2} = e\left(1 + \frac{1}{4}e^2 + \cdots\right)$$
$$= [(A_0 + A_1 \cos y_2)^2 + (B \sin y_2)^2]^{1/2}\left(t - \frac{\varkappa^2}{24}t^3 \ldots\right),$$

so that, for small values of $\varkappa$, the eccentricity increases almost uniformly with the time.

On the other hand, if $\varkappa$ is sufficiently different from zero, the tri-gonometric form the solution is applicable. It is then found that the eccentricity has a long-period oscillation.

These results are in general agreement with the results obtained by numerical integration by SHAPIRO and JONES (1961). The significant oscillations of shorter period present in the numerical integration results are caused by the first three terms in ΔF the arguments of which have a period of one-half year.

In a problem such as that of the motion of the dipoles of the West Ford project, the main concern is to establish the class of orbits in the resonance region for which the increase of the eccentricity and the corre-sponding diminution of the perigee height yield short life times. A development in powers of t is not only justified, but has considerable merit. The principal advantage is that even for $\varkappa = 0$ the solution presents nothing unusual.

In order to proceed with the more general solution, x_2, y_2 are obtained by integrating the equations

$$\frac{dx_2}{dt} = -A_1 \xi_1 \sin y_2 + B \eta_1 \cos y_2$$
$$\frac{dy_2}{dt} = -\frac{3}{2}j_2 n_0 x_2 - \left[\frac{\partial A_0}{\partial x_2} + \frac{\partial A_1}{\partial x_2}\cos y_2\right]\xi_1 - \frac{\partial B}{\partial x_2}\eta_1 \sin y_2.$$

In the right-hand members the first approximation results for ξ_1, η_1 are substituted, while A_0, A_1, B, y_2 are still treated as constants and x_{20}, y_{20} are introduced as constants of integration. If $\xi_{10} = \eta_{10} = 0$, the right-hand member of the first equation has no constant term, and the solution for x_2 will have the form

$$x_2 = x_{20} + x_{22}t^2 \ldots$$

The developments given are sufficient to indicate the procedure by which a solution in powers of t may be obtained for polar orbits in the resonance region.

Essentially the procedure may be followed for the treatment of resonance in gravitational minor perturbations in artificial satellite motion. The developments of ΔF for lunar and solar perturbations have been explored by MUSEN (1961a, b). The motion of a 24-hour satellite (MUSEN and BAILIE 1962) is a further example of the same type of problem.

Although the main object of this paper is to present a method for the analytical treatment of the problem, it may be remarked that the equations obtained are suitable also for exploration by the method of numerical integration. If this method is used, allowance can also be made for the effect of the earth's shadow. However, since $L =$ constant, the equations cannot yield the change in the semi-major axis caused by the earth's shadow.

References

MUSEN, P. 1960: J. Geophys. Research **65** (5) 1391—1396. — 1961a: ibid **66** (6) 1659—1666. — 1961b: ibid **66** (9) 2797—2806.

MUSEN, P., and ANN E. BAILIE 1962: ibid **67** (3) 1123—1132.

SHAPIRO, I. I., and H. M. JONES 1961: Science (October 6) **134**, 973—979.

Sur l'évolution des systèmes planétaires

Par

Albert M. Moltchanov

Institut Steklov de Mathématiques Moscou, U.S.S.R.

Introduction

Il sera question dans ce rapport de la possibilité d'étudier le comportement des systèmes planétaires au long de nombreuses fois la durée d'une révolution autour d'un astre central. Il y sera démontré que l'on peut exprimer l'équation définissant les lentes variations des paramètres de l'orbite, sans intervention d'aucune variable rapide, déterminant le mouvement orbital. Il en résulte que les orbites planes et circulaires apparaissent stationnaires, dans le terme principal. Ce théorème fait prévoir que même un système primitivement non plan doit le devenir par le processus d'évolution. Mais des facteurs dissipatifs, indispensables à l'évolution, n'entrent pas, à proprement parler, dans les équations de la mécanique céleste. D'où la réflexion suivante: déjà, dans les limites de la mécanique céleste, on voit qu'une décomposition du système est possible, lorsqu'une partie des planètes quitte le système en question et se rattache à des orbites hyperboliques. Ainsi, la possibilité d'une évolution (il est vrai, aux dépens de la décomposition) est déjà contenue dans les équations de la mécanique céleste. De même, il faut prendre note que, sur des périodes longues, même avec des facteurs dissipatifs petits (par exemple, sur le Soleil, des phénomènes de marées), des phénomènes minimes peuvent jouer un rôle essentiel.

De là, on peut concevoir la perspective suivante: les équations de la Mécanique céleste déterminent des états stationnaires du système planétaire — les orbites planes circulaires. Toutefois, pour obtenir de tels états, il est indispensable d'avoir des facteurs dissipatifs petits. Un intéressant problème se trouve posé, celui de l'énumération systématique de tous les phénomènes dissipatifs. Bien entendu, il se peut qu'ils soient tous trop petits pour assurer l'obtention d'un régime stationnaire, durant la durée d'existence du système solaire. Il restera alors une possibilité alternative et mettre: le système solaire est plan, étant donné qu'il l'était depuis le moment de son apparition.

Comme nous venons déjà de le dire, au cours de l'étude de l'évolution, on peut se dispenser de faire les vérifications des équations fondamentales si l'on examine les systèmes pouvant se décomposer. Cette question est intéressante en elle-même, car elle peut trouver des applications dans le problème relatif à la décomposition des associations stellaires.

§ 1. Théorème sur la division des mouvements

Tous les énoncés qui suivent sont basés sur le théorème relatif à la théorie des oscillations dont l'expression est indiquée ci-dessous. En raison du manque de place, la démonstration n'en sera pas donnée.

Etudions le système des équations différentielles, comprenant le petit paramètre ε:

$$\frac{dJ}{dt} = \varepsilon F(J, \Phi, \varepsilon),$$
$$\frac{d\Phi}{dt} = \omega(J) + \varepsilon \Omega(J, \Phi, \varepsilon). \quad (1)$$

Ici $\Phi = (\Phi_1, \ldots, \Phi_l)$ sont variables de phase (les fonctions F et Ω sont périodiques par rapport à Φ_i avec une période 2π), $J = (J_1, \ldots, J_m)$ est le système des intégrales premières du mouvement de non-perturbation.

Le théorème se rapportant à la division des mouvements, pour un tel système, se formule de la façon suivante:
il existe un changement de variables:

$$J = I + \varepsilon P(I, \Phi, \varepsilon), \quad (2)$$

tel que l'équation en I s'exprime sans variables de phase:

$$\frac{dI}{dt} = \varepsilon G(I, \varepsilon). \quad (3)$$

A l'heure actuelle, l'auteur ne connaît pas dans sa totalité la démonstration de ce théorème, elle est seulement limitée à quelques suppositions. La principale limitation (excepté les suppositions concernant l'uniformité, peu importantes) se trouve dans le fait que pour l'instant le théorème n'est démontré que pour une zone environnant un point *non-singulier*. De façon plus précise, examinons la valeur moyenne de la fonction F lorsque $\varepsilon = 0$:

$$F_0(I) = \frac{1}{(2\pi)^l} \int_0^{2\pi} \ldots \int_0^{2\pi} F(I, \Phi, 0) \, d\Phi_1 \ldots d\Phi_l. \quad (4)$$

Le théorème est démontré pour n'importe quel point I tel que $F_0(I) \neq 0$. Il est vrai que, dans ce cas-là, il nous est permis de démontrer en outre — et ceci est très important pour les applications — que le terme principal de l'équation de l'évolution coïncide avec $F_0(I)$:

$$G(I, 0) = F_0(I). \quad (5)$$

Remárquons que la formule (4) montre pourquoi la méthode habituelle, utilisée pour former la moyenne, ne peut être appliquée aux oscillations fréquentes. En effet, en général, on examine la moyenne de la trajectoire, ce qui selon le théorème de l'ergodicité ne coincide avec la moyenne spatiale que pour les points où les fréquences ne peuvent être mesurées. C'est pourquoi la moyenne temporelle, au contraire de la moyenne spatiale (4), constitue la fonction discontinue I dans tous les points I où les fréquences sont mesurées, c'est-à-dire sur un ensemble dense partout.

En ce qui concerne la fonction P dans la formule (2), l'auteur, jusqu'à présent, n'a pu en établir qu'une évaluation très rudimentaire:

$$|P(I, \Phi, \varepsilon)| \leq \frac{c}{\sqrt{\varepsilon}} \, . \qquad (6)$$

Une semblable évaluation est raisonnable aux environs d'un point stationnaire de résonance I, et doit être remplacée simplement par une limite P dans n'importe quel autre point.

L'évaluation (6) montre même que le problème des plus petits dénominateurs, dans la théorie des oscillations (en particulier, dans la mécanique céleste), apparaît à cause d'une forme malencontreuse de la théorie des perturbations — désagrégation en série, par degrés du petit paramètre ε.

Cette désagrégation a une signification dans le cadre de délais limités, alors que dans des délais $t \sim 1/\varepsilon$, ou davantage encore $t \sim 1/\varepsilon^2$ le processus de la désagrégation en série par degrés de ε équivant à une désagrégation en une série de fonctions ayant une singularité importante. Si l'on utilisait la méthode rationelle des approximations successives, aucun plus petit dénominateur n'apparaîtrait alors. On l'aperçoit très clairement si l'on écrit l'équation aux dérivées partielles qui détermine la fonction P:

$$\frac{\partial P}{\partial \Phi}[\omega(I + \varepsilon P) + \varepsilon \Omega] + \varepsilon \frac{\partial P}{\partial I} G(I, \varepsilon) = F(I + \varepsilon P, \Phi, \varepsilon) - G(I, \varepsilon).$$

Si l'on élimine tous les termes comprenant ε, on obtient un premier rapprochement avec la théorie classique des perturbations, où apparaît le problème des plus petits dénominateurs. Mais, il suffit de remplacer dans le terme contenant $\partial P/\partial I$ le coefficient $G(I, \varepsilon)$ par $F_0(I)$ conformément à (5) pour faire disparaître le problème des plus petits dénominateurs. Il en ressort, en somme, que l'évaluation (6) est très rudimentaire.

Présentons encore une remarque de caractère terminologique.

Le système (3) est tout naturellement désigné comme «évolutif» par rapport au système (1), étant donné qu'il décrit des changements lents, se produisant à une vitesse ε. Le sens de la formule (2) se résume en ceci qu'elle montre comment se comportent les intégrales J du mouvement de non-perturbation, lorsqu'une perturbation est mise en jeu. On voit alors que, lors du fonctionnement doux des paramètres I qui évoluent (pour des périodes $t \sim 1/\varepsilon$), il se superpose une vibration de moindre amplitude $\varepsilon\,P$, mais de haute fréquence, car il entre en P des variables rapides Φ.

Il faut souligner que l'application du théorème sur la division des mouvements conduit à étudier le système (1) à partir du système (3), qui est de degré moins élevé puisqu'il n'y entre pas de Φ à phases variables. Si l'on introduit un temps « lent » $\tau = \varepsilon\,t$, alors le système d'évolution (3) peut être mis sous la forme suivante, après avoir supprimé l'égalité (5):

$$\frac{dI}{d\tau} = F_0(I) + \varepsilon\,G_1(I,\varepsilon). \tag{7}$$

Ce système, comme celui sur lequel il est basé, comprend un petit paramètre ε. Son comportement fondamental est représenté par le terme $F_0(I)$, et des périodes de l'ordre de $\tau \sim 1$ y correspondent à $t \sim 1/\varepsilon$.

Ecrivons l'équation de « non-perturbation » pour le système (7):

$$\frac{dI}{d\tau} = F_0(I). \tag{8}$$

Le système (8) peut avoir, en général, des points stationnaires stables. Soit I_0 un tel point. Alors, n'importe quelle trajectoire issue d'un point assez voisin de I_0 tendra avec l'écoulement du temps à venir en I_0. Lorsqu'en particulier le régime devient stationnaire, il y a une évolution du système.

Ensuite, il peut se produire qu'une partie seulement des paramètres composant I tende vers la limite, alors que l'autre partie, ou même tous les paramètres, accomplissent un mouvement presque périodique. En prenant pour nouvelles variables le système des intégrales premières et les variables de phase de l'équation (8), nous mettrons dans ce cas le système (7) sous la forme (1), toutefois avec un nombre réduit de variables.

Le théorème sur la division des mouvements nous permet, de cette façon, en principe à la suite d'un nombre donné de démarches, soit de trouver un régime stationnaire vers lequel le système tend à évoluer, soit d'en décrire le mouvement sous forme de superposition de mouvements presque périodiques, qui se produisent avec des vitesses brusquement différentes $1, \varepsilon, \varepsilon^2, \ldots$, correspondant à l'échelle des temps $1, 1/\varepsilon, 1/\varepsilon^2, \ldots$.

Signalons que le nombre des démarches dans ce processus n'est pas supérieur à la dimension du système, étant donné qu'à chaque démarche une variable sort du jeu, soit pour le compte de la tendance vers la limite, soit pour le compte d'un terme moyen.

Il est important de remarquer que la nature de l'équation de l'évolution est déterminée par le terme $F_0(I)$, dont la recherche n'exige pas de connaître les changements de variables (2). Remarquons de même que, pour le théorème sur la division des mouvements, il faut seulement que les variables Φ forment un ensemble d'ergodicité en présence de presque tous les points I. Dans ce cas plus général, il suffit de remplacer la formule (4) par sa généralisation naturelle :

$$F_0(I) = \int\limits_{\mathfrak{M}(I)} F(I, \Phi, 0)\, d\mu(\Phi), \tag{9}$$

où $\mu(\Phi)$ indique une mesure invariante sur un ensemble d'ergodicité $\mathfrak{M}(I)$.

Si, toutefois, la première démarche ne met pas en évidence les paramètres qui subissent des évolutions, il serait nécessaire, avant d'effectuer la démarche suivante, de trouver un changement des variables, ce qui exige de résoudre l'équation aux dérivées partielles. Une économie de travail particulièrement notable est procurée par la remarque suivante : un changement des variables n'a pas besoin d'être recherché avec trop de précision, il suffit de le trouver avec une précision de l'ordre ε^2 pour la première démarche, de l'ordre de ε^3 pour la seconde, et ainsi de suite.

§ 2. Problème concernant le système planétaire

Il est aisé de vérifier que, selon une notation adimensionnelle, les équations de système « n » des planètes peuvent être exprimées, dans le système héliocentrique de coordonnées, de la façon suivante :

$$\left. \begin{aligned} \frac{d\bar{u}_i}{dt} &= -\frac{\bar{r}_i}{r_i^3} + \varepsilon \sum_{\varkappa \neq i} \Theta_\varkappa \left[\frac{r_\varkappa - \bar{r}_i}{|r_\varkappa - \bar{r}_i|^3} - \frac{\bar{r}_\varkappa}{r_\varkappa^3} + \frac{\bar{r}_i}{r_i^3} \right], \\ \frac{d\bar{r}_i}{dt} &= \bar{u}_i. \end{aligned} \right\} \tag{10}$$

Ici, le petit paramètre désigné par ε, soit :

$$\varepsilon = \frac{m_1 + \cdots + m_n}{m_0 + m_1 + \cdots + m_n}, \ 0 \tag{11}$$

représente le rapport de la masse de toutes les planètes à la masse entière du système (m_0 — masse du corps central), alors que

$$\Theta_\varkappa = \frac{m_\varkappa}{m_1 + \cdots + m_n} \tag{12}$$

représente la part $\varkappa$ d'une planète dans la masse de l'ensemble des planètes.

Remarquons que, pour le système solaire :

$$\varepsilon = 1{,}34 \cdot 10^{-3},$$

et que, pour le système des satellites de Jupiter :

$$\varepsilon = 4{,}6 \cdot 10^{-5}.$$

Afin de pouvoir appliquer le théorème § 1. au système (10), il faut exprimer celui-ci au moyen de nouvelles variables. Pour celles-ci, il est commode de choisir des moments spécifiques de la quantité de mouvement des planètes, vecteurs dirigés vers le péricentre de l'orbite. Ayant complété le système des intégrales premières du mouvement de non-perturbation à l'aide de variables de phase, nous obtiendrons la possibilité de ramener le système (10) à la forme (1).

A titre d'exemple, voici les formules de passage aux nouvelles variables (une lettre non surlignée désigne le module du vecteur correspondant):

$$
\left.
\begin{aligned}
\bar{L}_i &= \bar{r}_i \times \bar{u}_i, \\
\bar{a}_i &= \left(u_i^2 - \frac{1}{r_i}\right)\bar{r}_i - (\bar{r}_i, \bar{u}_i)\,\bar{u}_i, \\
\cos\varphi_i &= \frac{1}{r_i\,a_i}(\bar{r}_i, \bar{a}_i).
\end{aligned}
\right\}
\tag{13}
$$

Il faut signaler que $\bar{L}_i$ et $\bar{a}_i$ sont corrélés par $(\bar{L}_i, \bar{a}_i) = 0$. C'est pourquoi les dimensions $\bar{L}_i$, $\bar{a}_i$, φ_i, pour chaque indice i, constituent six (et non sept!) grandeurs indépendantes, ce qui permet d'exprimer, grâce à elles et réciproquement, six grandeurs $\bar{r}_i$, $\bar{u}_i$. La transformation inverse de (13) s'effectue par les formules:

$$
\left.
\begin{aligned}
\bar{r}_i &= r_i[\cos\varphi_i\,\bar{\alpha}_i + \sin\varphi_i\,\bar{\beta}_i], \\
\bar{u}_i &= \frac{1}{L_i}[-\sin\varphi_i\,\bar{\alpha}_i + (a_i + \cos\varphi_i)\,\bar{\beta}_i],
\end{aligned}
\right\}
\tag{14}
$$

où figurent les notations suivantes:

$$
\left.
\begin{aligned}
r_i &= \frac{L_i^2}{1 + a_i\cos\varphi_i}, \\
\bar{\alpha}_i &= \frac{1}{a_i}\,\bar{a}_i, \\
\bar{\beta}_i &= \frac{1}{L_i\,a_i}(\bar{L}_i \times \bar{a}_i).
\end{aligned}
\right\}
\tag{15}
$$

Il n'y aurait aucun intérêt à exprimer tout le système des équations dans les nouvelles variables, puisque seules les équations des évolutions nous intéressent. C'est pourquoi nous ne ferons la transformation que pour les grandeurs $\bar{L}_i$, $\bar{a}_i$, qui constituent d'ailleurs le vecteur J dans le cas présent. On peut vérifier que ces équations sont de la forme:

$$
\frac{d\bar{L}_i}{dt} = \varepsilon(\bar{r} \times \bar{F}_i),
\tag{16}
$$

$$
\frac{d\bar{a}_i}{dt} = \varepsilon[2(\bar{u}_i, \bar{F}_i)\,\bar{r}_i - (\bar{r}_i, \bar{F}_i)\,\bar{u}_i - (\bar{r}_i, \bar{u}_i)\,\bar{F}_i].
\tag{17}
$$

Nous introduirons alors les notations suivantes:

$$
\left.
\begin{aligned}
\bar{F}_i &= \sum_{\varkappa \neq i} \Theta_\varkappa \bar{F}_{\varkappa i}, \\
\bar{F}_{\varkappa i} &= \frac{\bar{r}_\varkappa - \bar{r}_i}{|\bar{r}_\varkappa - \bar{r}_i|^3} - \frac{\bar{r}_\varkappa}{\bar{r}_\varkappa^3} + \frac{\bar{r}_i}{\bar{r}_i^3},
\end{aligned}
\right\}
\tag{18}
$$

où il est sous-entendu que les grandeurs $\bar{r}_i$ et $\bar{u}_i$ doivent être exprimées en $\bar{L}_i$, $\bar{a}_i$, φ_i suivant les formules (14).

Notre problème est celui de l'étude du comportement du système durant les longues périodes. Précisons cette expression trop vague de « longues périodes ». Ainsi qu'il a été dit au § 1., l'existence d'un petit paramètre engendre différentes échelles de périodes. On peut parler de périodes de l'ordre d'une seule rotation. Dans de telles périodes, les corrections de l'ordre de ε sont petites, et négligeables. Pour des périodes de l'ordre de $1/\varepsilon$, le comportement du système est précisément défini par les petits termes de l'ordre de ε, mais les termes de l'ordre de ε^2 et davantage sont susceptibles d'être négligés. Si l'on prend comme exemple le système solaire, pour lequel $\varepsilon \sim 10^{-3}$, les périodes de l'ordre de $1/\varepsilon$ concernent des milliers de révolutions de Jupiter autour du Soleil, soit des dizaines de milliers d'années. Vraisemblablement, lorsque l'on examinera pour ce cas des périodes de l'ordre de $1/\varepsilon$ et $1/\varepsilon^2$, il sera judicieux de se limiter aux équations de la mécanique céleste. Mais, déjà, des périodes de l'ordre de $1/\varepsilon^3$, comparables à la durée de l'existence du système solaire, et par conséquent à l'approximation de la Mécanique céleste, peuvent être notoirement insuffisantes. Il est intéressant, en particulier, de noter que les corrections relativistes concernant le mouvement de Jupiter deviennent essentielles pour des périodes de l'ordre de $1/\varepsilon^3$. N'importe quel facteur qui n'a pas été retenu dans l'établissement des équations de la mécanique céleste (étendue des planètes, force des marées, champs magnétiques, etc.) doit être pris en compte dans le cadre de la théorie des perturbations, car ces facteurs y deviennent essentiels. Il est nécessaire de remarquer que le théorème du § 1. permet de tenir compte de tous ces facteurs indépendamment de leur origine, à condition de connaître les formules exprimant les forces correspondantes en fonction de la position et de la vitesse des planètes.

Tout en ne tenant pas compte des termes qu'on aura à évaluer, dans des problèmes à venir concernant la théorie des perturbations, la première démarche doit se tenir dans les limitations de la mécanique céleste. Le problème étudié dans le présent paragraphe consiste à déterminer quelques caractéristiques de l'équation du premier degré de l'évolution, selon la théorie des perturbations. On a vu au § 1. que, pour l'établissement de l'équation de l'évolution, il faut opérer la moyenne du membre de droite du système de départ, en fonction d'une mesure invariante. On peut vérifier que, dans notre cas, une mesure invariante est fournie par les différentielles d'anomalies moyennes. Si l'on préfère utiliser les anomalies réelles, on obtient la formule suivante :

$$d\mu = r_1^2 \, r_2^2 \ldots r_n^2 \, d\varphi_1 \, d\varphi_2 \ldots d\varphi_n. \tag{19}$$

Le premier résultat qui mérite d'être signalé se trouve établi lorsque, en présence d'équations exprimant les moyennes de (16) et (17) en des grandeurs $\overline{F}_{ki}$, on peut ne retenir que le premier terme, les deux autres étant de moyenne nulle. Ceci est particulièrement manifeste pour le terme $\bar{r}_\varkappa/\bar{r}_\varkappa^3$, étant donné qu'après l'avoir multiplié par $\bar{r}_\varkappa^2 \, d\varphi_\varkappa$ on obtient $(\bar{r}_\varkappa/r_\varkappa) \, d\varphi_\varkappa$. Dans des intégrations par période, cette expresion fournit un résultat nul, comme on l'observe aisément dans la formule (14). Il est un peu plus difficile de vérifier que le terme $\bar{r}_i/\bar{r}_i^3$ peut être également tenu pour nul. Le calcul en est omis ici, en raison de son caractère élémentaire.

En tenant compte de cette remarque, le système d'évolution peut être exprimé de façon analogue au système (16), (17). Il faut toutefois noter que, dans les membres de droite, on peut substituer l'expression:

$$\bar{F}^*_{\varkappa i} = \frac{\bar{r}_\varkappa - \bar{r}_i}{|\bar{r}_\varkappa - \bar{r}_i|^3} \, . \tag{20}$$

Ensuite, ces membres de droite doivent être calculés en moyenne, pour les variables $\varphi_1, \ldots, \varphi_n$, avec un poids déterminé par la formule (19).

Puis, une fois les équations établies, on peut démontrer le très important théorème suivant:

Si tous les moments $\bar{L}_i$ sont parallèles, alors que les excentricités sont nulles, alors les membres de droite des équations des évolutions du premier ordre sont également nuls.

Cette affirmation se démontre aussi par quelques calculs dont la simplicité rend ici leur présentation superflue.

Il est facile de comprendre que le théorème précédent affirme le fait évolutionnaire important que voici:

Les systèmes planétaires plans et circulaires apparaissent stationnaires au premier ordre en ε.

En conclusion, nous remarquerons que, sur des machines à calculer perfectionnées, on peut aujourd'hui faire intégrer sans difficulté les équations évolutionnaires pour le cas d'un système de quelques planètes.

Les calculs, dans le cas de deux planètes, ont été effectués par Mademoiselle VALENTINA NICOLAEVNA IVANOV. L'angle formé entre les plans des premiers orbites est assez grand $\left(> \dfrac{\pi}{6}\right)$, et les excentricités assez importantes (~ 0.3). Il s'avère que, même dans ce cas, le système est très stable. Mademoiselle V. N. IVANOV a très aimablement fourni à l'auteur les calculs rassemblés dans les dix graphiques ci-joints. Si, pour unité de mesure, on choisit un ordre de grandeur caractéristique pour le système solaire, les calculs englobent une durée en temps réel supérieure à trois millions d'années.

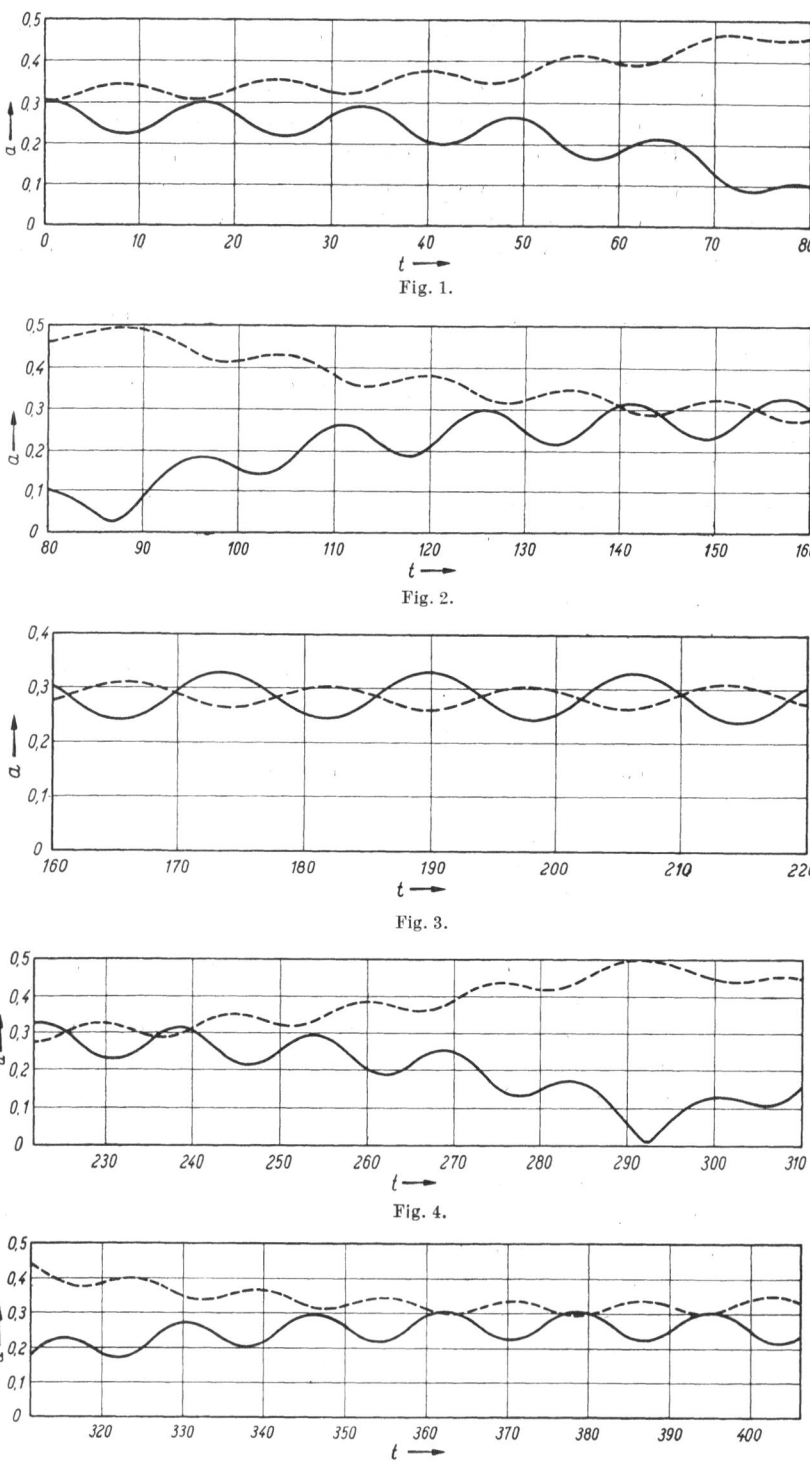

Fig. 1.

Fig. 2.

Fig. 3.

Fig. 4.

Fig. 5.

Fig. 1—5. Variation de l'éxcentricité en fonction du temps.

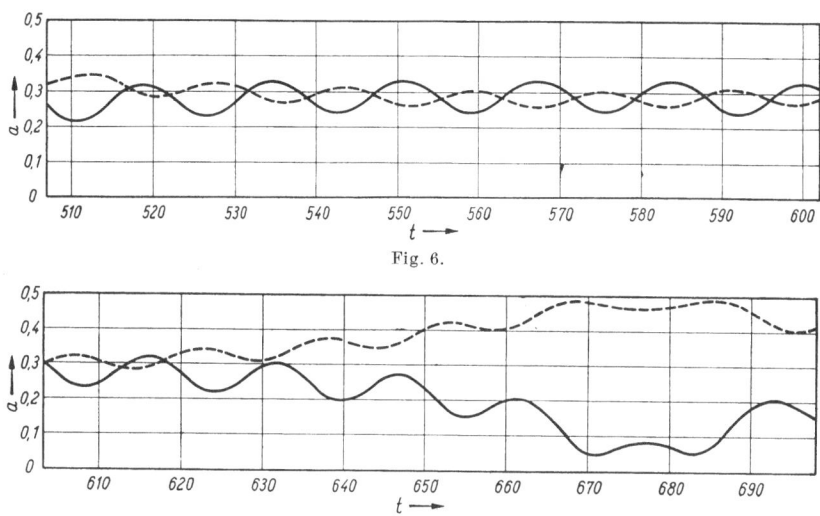

Fig. 6.

Fig. 7.

Fig. 6 et 7. Variation de l'éxcentricité en fonction du temps.

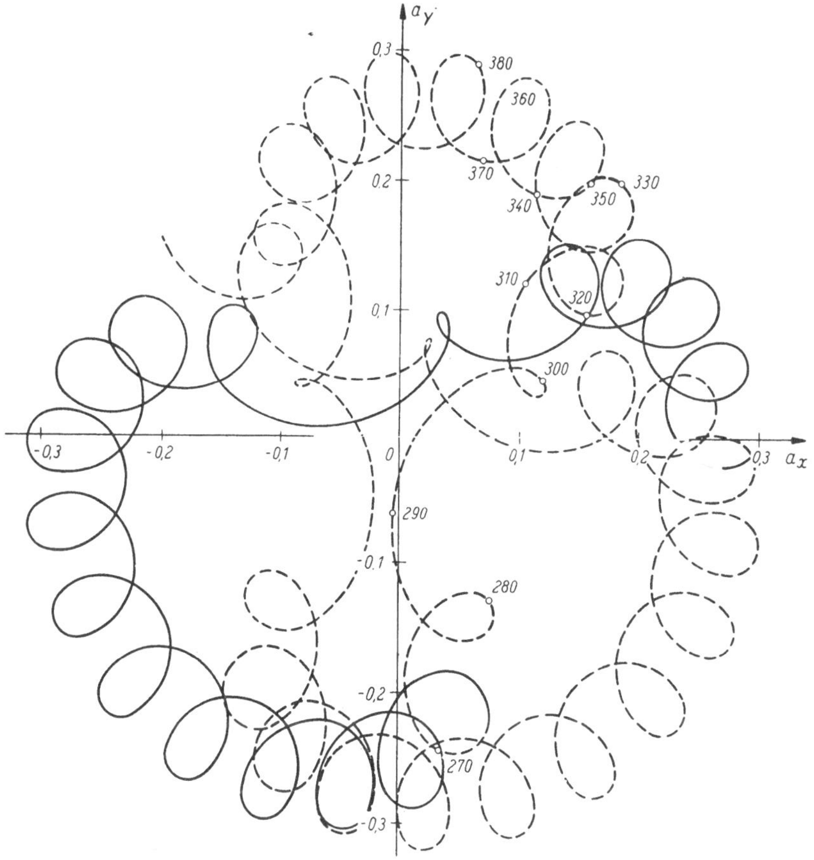

Fig. 8. Projection du vecteur de LAPLACE de ,,Jupiter'' sur le plan invariant du système.

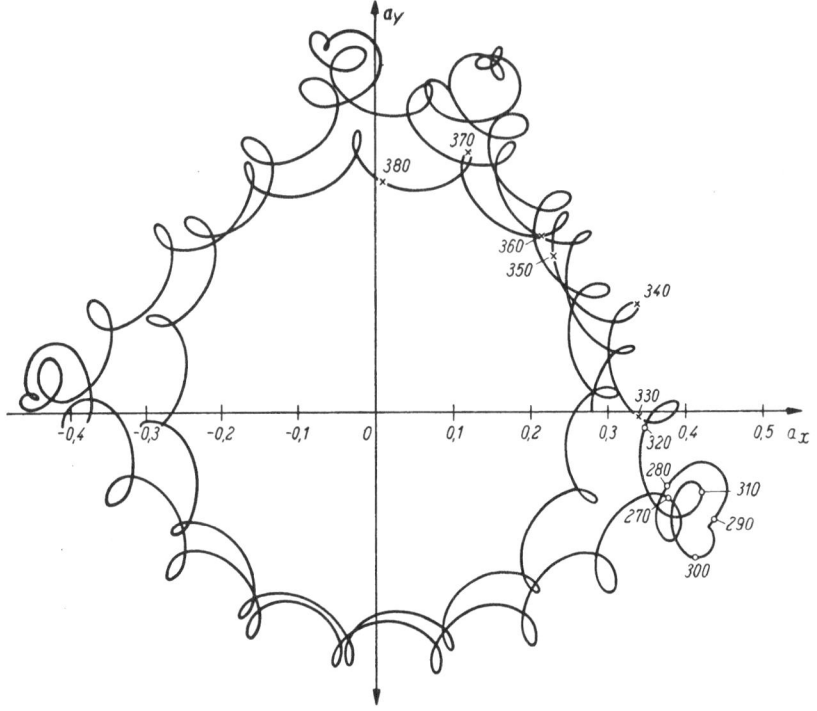

Fig. 9. Projection du vecteur de LAPLACE de ,,Saturne'' sur le plan invariant du système.

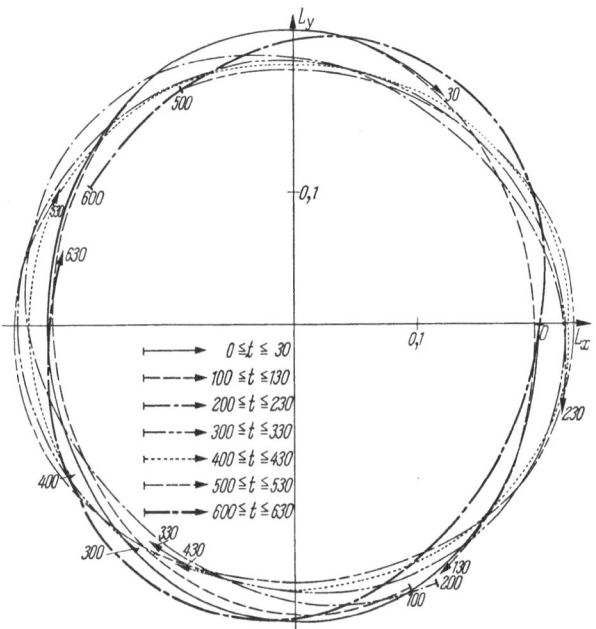

Fig. 10. Projection du moment cinétique de ,,Jupiter'' sur le plan invariant du système.

A comparison of astronomical and ballistic traditions in orbit correction

By

Samuel Herrick

University of California, Los Angeles, Calif., U.S.A.

A. Introduction

In space navigation the confluence of traditions from sea and air navigation, from aerodynamics and ballistics, from guidance and control theory, and from celestial mechanics, especially in the determination of orbits, is productive both of cross-fertilization and of conflict. One area having particular interest is that of differential formulae, which are useful in the mathematical differential correction of an approximate orbit into better agreement with observation, in the physical or thrust differential correction of a non-rendezvous trajectory into one that intercepts its target, in guidance, in error analysis, and in optimization.

So many are the problems to which differential formulae may be applied, and so varied are the basic equations and parameters that may be used to solve them, that a comparison and analysis of the methods from differing traditions should be severely limited in a number of ways, at least preliminarily. Accordingly, I propose to limit my discussion to a comparison of astronomical differential correction methods with the "adjoint method" that has come into space navigation from ballistics, to problems in observational differential correction and in rendezvous or targeting, to residuals or miss-components in r and $\dot{r}$ the dynamically centered position and velocity vectors at time t, and to the correction of the "elements" or parameters, r_0, $\dot{r}_0$, the position and velocity vectors at the "epoch", t_0. The discussion can readily be extended to observational residuals in such quantities as α, δ, ϱ, the topocentric right ascension, declination, distance (or "range"), or in the rates of change thereof, or restricted to miss-components in two dimensions, by simple analytic formulae; or the "elements" r_0, $\dot{r}_0$ can be replaced by or expressed in terms of others.

Let us specify that the "conjectured" or assumed values of the adopted parameters are specified by r_{0C} and $\dot{r}_{0C}$, that $\varDelta r_0$, $\varDelta \dot{r}_0$ are the desired corrections to the conjectured values, and that

$$r_0 = r_{0C} + \varDelta r_0, \quad \dot{r}_0 = \dot{r}_{0C} + \varDelta \dot{r}_0 \tag{1}$$

4*

are the unknown values that are the "objective" of the differential correction process, whether it be concerned with observation or rendezvous.

Let us further specify that r, $\dot{r}$ and r_C, $\dot{r}_C$ designate "observed" and "computed" values of r, $\dot{r}$ at time t. The former represent observed quantities in observational differential correction, or the position of the target in rendezvous differential correction, known quantities in either circumstance, but functionally related to the unknown objective r_0, $\dot{r}_0$ by

$$\left. \begin{aligned} r &= r(x_0, y_0, z_0, \dot{x}_0, \dot{y}_0, \dot{z}_0, t - t_0) \\ \dot{r} &= \dot{r}(x_0, y_0, z_0, \dot{x}_0, \dot{y}_0, \dot{z}_0, t - t_0). \end{aligned} \right\} \tag{2}$$

The latter, r_C, is the result of an actual calculation or "representation" from the conjectured r_{0C}, $\dot{r}_{0C}$, whether by two-body formulae or by integrations inclusive of perturbations:

$$\left. \begin{aligned} r_C &= r(x_{0C}, y_{0C}, z_{0C}, \dot{x}_{0C}, \dot{y}_{0C}, \dot{z}_{0C}, t - t_0) \\ \dot{r}_C &= \dot{r}(x_{0C}, y_{0C}, z_{0C}, \dot{x}_{0C}, \dot{y}_{0C}, \dot{z}_{0C}, t - t_0). \end{aligned} \right\} \tag{3}$$

Some or all of the components of the "residual" vectors

$$\Delta r = r - r_C, \quad \Delta \dot{r} = \dot{r} - \dot{r}_C \tag{4}$$

accordingly are known quantities in the basic differential correction formulae, which may be derived from Eqs. (1)–(4), excluding higher-order terms, as follows:

$$\left. \begin{aligned} \Delta r &= \frac{\partial r}{\partial x_0}\Delta x_0 + \frac{\partial r}{\partial y_0}\Delta y_0 + \frac{\partial r}{\partial z_0}\Delta z_0 + \frac{\partial r}{\partial \dot{x}_0}\Delta \dot{x}_0 + \frac{\partial r}{\partial \dot{y}_0}\Delta \dot{y}_0 + \frac{\partial r}{\partial \dot{z}_0}\Delta \dot{z}_0 \\ \Delta \dot{r} &= \frac{\partial \dot{r}}{\partial x_0}\Delta x_0 + \frac{\partial \dot{r}}{\partial y_0}\Delta y_0 + \frac{\partial \dot{r}}{\partial z_0}\Delta z_0 + \frac{\partial \dot{r}}{\partial \dot{x}_0}\Delta \dot{x}_0 + \frac{\partial \dot{r}}{\partial \dot{y}_0}\Delta \dot{y}_0 + \frac{\partial \dot{r}}{\partial \dot{z}_0}\Delta \dot{z}_0 \end{aligned} \right\} \tag{5}$$

and which are to be solved, usually by a numerical matrix-inversion, for the components of Δr_0 and $\Delta \dot{r}_0$, for use in the correction of r_{0C} and $\dot{r}_{0C}$ by Eq. (1). It is possible to avoid the numerical inversion of these equations only in systems in which all six of the components of Δr and $\Delta \dot{r}$ are obtained by observation or measurement; in such circumstances r, $\dot{r}$ and r_0, $\dot{r}_0$ are simply interchanged in Eq. (5). The usefulness of such systems has been widely misunderstood and exaggerated.

In passing, we note that the terms "objective" and "conjectured" are selected so that O–C residuals in the components of r — "observed minus computed" in the traditional astronomical sense — will lead to O–C ("objective minus computed") corrections to the elements $\dot{r}_{0C}$ and $\dot{r}_{0C}$.

The links between the known components of Δr, $\Delta \dot{r}$ and the unkown components of Δr_0, $\Delta \dot{r}_0$ are evidently the 36 partial differential coefficients, $\partial r/\partial x_0$, $\partial r/\partial y_0$, ... $\partial \dot{r}/\partial \dot{z}_0$, and the key to the differential correction is in their evaluation. It may be achieved in several ways, of which I list the following:

(a) By analytic differentiation of the "representation" formulae schematically indicated by Eq. (3). In this process we usually take into account only two-body terms, perhaps supplemented by secular terms from the perturbations, but neglecting changes in, and the differentiation of, most of the perturbation terms.

(b) By variant calculations (e. g., by COWELL's method of special perturbations) in which the conjectured values of the elements or parameters (r_{0C} and $\dot{r}_{0C}$) are altered one by one to produce changes in the calculated values, r_C, $\dot{r}_C$ from which the partial differential coefficients are obtained. This process may include all of the effect of changes in the perturbation terms, but would be equivalent to process (a) if the perturbation terms were not recalculated. These variant calculations are often referred to as "perturbed calculations" by the modern engineer, extending the astronomical concept of "perturbations" to arbitrary displacements or variants unconnected with the force field, but rather appropriately in view of process (c).

(c) By variant calculations specifically utilizing ENCKE's method for integrating special perturbations—with "engineering perturbations" (as mentioned above) instead of astronomical ones. These integrations lead to differences directly proportional to the partial differential coefficients, rather than gross values that must be differenced, as in COWELL's method.

(d) By the "adjoint method", in which linearized variants are integrated backwards to give the partial differential coefficients directly.

(e) By a linearized version of process (c), which utilizes some of the same quantities as process (d), but in a forward integration.

These methods, with the exception of the first, we shall compare formally, and then as practical solutions to specific problems.

B. Variant or "Perturbed" calculation of the partial differential coefficients

For the specified observed, known, or intermediate variables r, $\dot{r}$ and for the specified elements, initial conditions, or intermediate parameters r_0 and $\dot{r}_0$ of Eqs. (1)–(5), the variant calculation of the partial differential coefficients is simple and direct. If we vary x_{0C} by an arbitrary amount $\Delta_1 x_0$, leaving y_{0C}, z_{0C}, $\dot{x}_{0C}$, $\dot{y}_{0C}$, $\dot{z}_{0C}$ unchanged, we may calculate the corresponding residuals,

$$\left. \begin{aligned} \Delta_1 r &= r(x_{0C} + \Delta_1 x_0, y_{0C}, z_{0C}, \dot{x}_{0C}, \dot{y}_{0C}, \dot{z}_{0C}, t - t_0) - r_C \\ \Delta_1 \dot{r} &= \dot{r}(x_{0C} + \Delta_1 x_0, y_{0C}, z_{0C}, \dot{x}_{0C}, \dot{y}_{0C}, \dot{z}_{0C}, t - t_0) - \dot{r}_C \end{aligned} \right\} \quad (6)$$

and since Eqs. (5) reduce to

$$\Delta_1 r = \frac{\partial r}{\partial x_0} \Delta_1 x_0 \qquad \Delta_1 \dot{r} = \frac{\partial r}{\partial x_0} \Delta_1 x_0 \qquad (7)$$

we obtain approximations to the partial differential coefficients,

$$\frac{\partial r}{\partial x_0} = \frac{\Delta_1 r}{\Delta_1 x_0} \qquad \frac{\partial \dot{r}}{\partial x_0} = \frac{\Delta_1 \dot{r}}{\Delta_1 x_0}. \qquad (8)$$

Similarly a second variant calculation, with y_{0C} adjusted by an arbitrary $\Delta_2 y_0$, but with x_{0C}, z_{0C}, $\dot{x}_{0C}$, $\dot{y}_{0C}$, $\dot{z}_{0C}$ unaltered from the original "representation" or calculation of r_C, will lead to

$$\frac{\partial r}{\partial y_0} = \frac{\Delta_2 r}{\Delta_2 y_0} \qquad \frac{\partial \dot{r}}{\partial y_0} = \frac{\Delta_2 \dot{r}}{\Delta_2 y_0} \qquad (9)$$

and like arbitrary variants in the other four elements, resulting in like calculated residuals, will yield

$$\left.\begin{array}{ll} \dfrac{\partial r}{\partial z_0} = \dfrac{\Delta_3 r}{\Delta_3 z_0} & \dfrac{\partial \dot{r}}{\partial z_0} = \dfrac{\Delta_3 \dot{r}}{\Delta_3 z_0} \\[2mm] \dfrac{\partial r}{\partial \dot{x}_0} = \dfrac{\Delta_4 r}{\Delta_4 \dot{x}_0} & \dfrac{\partial \dot{r}}{\partial \dot{x}_0} = \dfrac{\Delta_4 \dot{r}}{\Delta_4 \dot{x}_0} \\[2mm] \dfrac{\partial r}{\partial \dot{y}_0} = \dfrac{\Delta_5 r}{\Delta_5 \dot{y}_0} & \dfrac{\partial \dot{r}}{\partial \dot{y}_0} = \dfrac{\Delta_5 \dot{r}}{\Delta_5 \dot{y}_0} \\[2mm] \dfrac{\partial r}{\partial \dot{z}_0} = \dfrac{\Delta_6 r}{\Delta_6 \dot{z}_0} & \dfrac{\partial \dot{r}}{\partial \dot{z}_0} = \dfrac{\Delta_6 \dot{r}}{\Delta_6 \dot{z}_0} \end{array}\right\} \qquad (10)$$

The calculation of r_C, $\dot{r}_C$ and of the residuals $\Delta_1 r$, $\Delta_1 \dot{r}$ as indicated by Eq. (6) may be undertaken in a variety of ways, but for purposes of comparison with the sections that follow we shall suppose that they are determined from COWELL integrations of the acceleration equations,

$$\ddot{r} = -\frac{\mu r}{r^3} + \dot{r}^{\backslash}, \qquad (11)$$

where μ is a constant involving the constant of gravitation and some function of the masses in the problem; where the first term represents the "two-body" accelerations; and where $\dot{r}^{\backslash}$ represents the perturbative accelerations, or the effects of all forces other than the central two-body gravitational one, of whatever character. In COWELL's method, of course, it is unnecessary to distinguish between the two terms of Eq. (11), but in the variant calculations it may sometimes be possible and convenient to recalculate only the $-\mu r/r^3$ term, leaving the perturbations unchanged as with the usual analytic formulation.

C. Variant calculation by Encke's method

ENCKE's method can be used effectively in the integration of the partial differential coefficients of Eq. (5), or, strictly, of the variants of r, $\dot{r}$ that appear in Eqs. (8)–(10).

A rather common misconception of ENCKE's method is that it is limited to the use of fixed two-body reference orbits. In fact, it can be used with perturbed or varying reference orbits, including those integrated numerically by any method. The analytical basis of a perturbed integration might even be something other than the two-body problem, although we shall accept that limitation, for simplicity, in what follows. The inclusion of perturbation terms in the reference orbit, however, is a necessity to the variant calculation method.

Accordingly we develop the basic ENCKE formula here inclusive of perturbation terms. Given a reference orbit that satisfies, and enables us to obtain a reference position, r_e, that satisfies [cf. Eq. (11)]

and the fact that

$$\ddot{r}_e = -\frac{\mu\, r_e}{r_e^3} + \dot{r}_e \qquad (12)$$

$$r = r_e + \varrho, \qquad (13)$$

where the components of ϱ are ξ, η, ζ, according to the usual ENCKE notation. Then the basic ENCKE equation may be written

$$\ddot{\varrho} = \frac{\mu}{r_e^3}\left[r_e - r\left(\frac{r_e}{r}\right)^3\right] + \dot{r} - \dot{r}_e \qquad (14)$$

or

$$\ddot{\varrho} = \frac{\mu}{r_e^3}[f\, q\, r - \varrho] + \dot{r} - \dot{r}_e, \qquad (15)$$

where

and

$$q = \frac{1}{r_e^2}\left(r_e + \frac{1}{2}\varrho\right)\cdot\varrho \qquad (16)$$

$$f\, q = 1 - (1 + 2q)^{-3/2} = 3q - \frac{3\cdot 5}{1\cdot 2}\, q^2 + \frac{3\cdot 5\cdot 7}{1\cdot 2\cdot 3}\, q^3 - \cdots. \qquad (17)$$

It is in details that ENCKE's method is limited to the two-body problem, especially in Eqs. (15)–(17), but not in philosophy or in practical extensibility.

It is usual to use an osculating reference orbit with ENCKE's method, so that we integrate Eqs. (14) or (15) with initial conditions

$$\varrho_0 = 0 \qquad \dot{\varrho}_0 = 0. \qquad (18)$$

For simplicity we may think of the initial calculation as having been carried through from this start, perhaps with a fixed two-body reference orbit. For the variant calculations, however, the reference orbit will be the result of the initial calculation. Then, if in the first variant calculation we set

$$\xi_0 = \varDelta_1 x_0, \qquad \eta_0 = \zeta_0 = \dot{\xi}_0 = \dot{\eta}_0 = \dot{\zeta}_0 = 0 \qquad (19)$$

the result of the integration will be

$$\varrho = \varDelta_1 r, \qquad \dot{\varrho} = \varDelta_1 \dot{r} \qquad (20)$$

for use in Eq. (8). In the second variant calculations we set

$$\eta_0 = \varDelta_2 y_0, \quad \xi_0 = \zeta_0 = \dot{\xi}_0 = \dot{\eta}_0 = \dot{\zeta}_0 = 0 \tag{21}$$

to obtain

$$\varrho = \varDelta_2 r, \quad \dot{\varrho} = \varDelta_2 \dot{r} \tag{22}$$

for use in Eq. (9). Four more variants yield the data for Eq. (10).

As with Eq. (11) it may be unnecessary to recalculate the perturbation terms, i. e., in the notation of this section it may be possible to set

$$\dot{r}^\backslash = \dot{r}_e^\rangle. \tag{23}$$

D. Partial differential coefficients by "Adjoint" techniques

The "adjoint method" can be developed with varying degrees of generality and elegance. We shall restrict the generality to the degree necessary to encompass Eqs. (5), and the elegance to certain concepts from matrix theory without the efficient notations that require specialized knowledge of that subject.

It is convenient, however, to replace the notations r and $\dot{r}$ by

$$\begin{rcases} x = x_1, \quad y = x_2, \quad z = x_3, \\ \dot{x} = x_4, \quad \dot{y} = x_5, \quad \dot{z} = x_6. \end{rcases} \tag{24}$$

Since $\ddot{r}$ is a function of r and $\dot{r}$, and since $\dot{r}$ likewise is such a function, at least formally, we may write

$$\varDelta \dot{x}_i = \sum_j \frac{\partial \dot{x}_i}{\partial x_j} \varDelta x_j. \tag{25}$$

The adjoint to this equation is

$$\dot{\lambda}_j = -\sum_i \lambda_i \frac{\partial \dot{x}_i}{\partial x_j}, \tag{26}$$

where λ_i is a multiplier unspecified for the present. Assuming that the derivatives are with respect to t, or perhaps $\tau = k'(t - t_0)$ as in other orbit problems, where k' is an arbitrary constant, which is here particularized to unity,

$$\frac{d}{dt}\left(\sum_j \lambda_j \varDelta x_j\right) = \sum_j \lambda_j \varDelta \dot{x}_j + \sum_j \dot{\lambda}_j \varDelta x_j = \sum_i \lambda_i \varDelta \dot{x}_i + \sum_j \dot{\lambda}_j \varDelta x_j. \tag{27}$$

Introducing Eqs. (25) and (26) into (27), we obtain

$$\frac{d}{dt}\left(\sum_j \lambda_j \varDelta x_j\right) = \sum_i \sum_j \lambda_i \frac{\partial \dot{x}_i}{\partial x_j} \varDelta x_j - \sum_j \sum_i \lambda_i \frac{\partial \dot{x}_i}{\partial x_j} \varDelta x_j = 0. \tag{28}$$

Thus $\sum_j \lambda_j \varDelta x_j$ is constant, or, specifically

$$\sum_j \lambda_j \varDelta x_j = \sum_j \lambda_{j0} \varDelta x_{j0}. \tag{29}$$

Accordingly, if we integrate a set of six λ_j by means of Eqs. (26), starting from "terminal" conditions at t rather than from "initial" conditions at t_0, as follows,

$$\lambda_1 = \frac{\partial x_1}{\partial x_1} = \frac{\partial x}{\partial x} = 1 \qquad \lambda_4 = \frac{\partial x_1}{\partial x_4} = \frac{\partial x}{\partial \dot x} = 0$$

$$\lambda_2 = \frac{\partial x_1}{\partial x_2} = \frac{\partial x}{\partial y} = 0 \qquad \lambda_5 = \frac{\partial x_1}{\partial x_5} = \frac{\partial x}{\partial \dot y} = 0 \qquad (30)$$

$$\lambda_3 = \frac{\partial x_1}{\partial x_3} = \frac{\partial x}{\partial z} = 0 \qquad \lambda_6 = \frac{\partial x_1}{\partial x_6} = \frac{\partial x}{\partial \dot z} = 0$$

we shall obtain, at the integrals at t_0,

$$\lambda_{10} = \frac{\partial x}{\partial x_0}, \qquad \lambda_{20} = \frac{\partial x}{\partial y_0}, \qquad \lambda_{30} = \frac{\partial z}{\partial z_0}$$

$$\lambda_{40} = \frac{\partial x}{\partial \dot x_0}, \qquad \lambda_{50} = \frac{\partial x}{\partial \dot y_0}, \qquad \lambda_{60} = \frac{\partial x}{\partial \dot z_0} \qquad (31)$$

and shall find, in fact, that Eq. (29) is identical with the x Eq. of (5). The partial differential coefficients in the y, z and the $\dot x$, $\dot y$, $\dot z$ equations, of course, are obtained with five additional and similar sets of six integrations of Eqs. (26), starting from the following terminal conditions at time t:

$$\lambda_2 = 1, \qquad \lambda_1 = \lambda_3 = \lambda_4 = \lambda_5 = \lambda_6 = 0$$

$$\lambda_3 = 1, \qquad \lambda_1 = \lambda_2 = \lambda_4 = \lambda_5 = \lambda_6 = 0 \qquad (32)$$

$$\lambda_4 = 1, \qquad \lambda_1 = \lambda_2 = \lambda_3 = \lambda_5 = \lambda_6 = 0$$

$$\lambda_5 = 1, \qquad \lambda_1 = \lambda_2 = \lambda_3 = \lambda_4 = \lambda_6 = 0 \qquad (33)$$

$$\lambda_6 = 1, \qquad \lambda_1 = \lambda_2 = \lambda_3 = \lambda_4 = \lambda_5 = 0$$

As an additional and very useful capacity of the adjoint method, we may note that it can generate the partial differential coefficients for any quantity ψ_i associated with x, y, z (or, for that matter, with $\dot x$, $\dot y$, $\dot z$), such as ϱ, α, δ, or such as miss-components, simply by adopting as terminal conditions at t, the values,

$$\lambda_1 = \frac{\partial \psi_i}{\partial x}, \qquad \lambda_2 = \frac{\partial \psi_i}{\partial y}, \qquad \lambda_3 = \frac{\partial \psi_i}{\partial z},$$

$$\lambda_4 = \frac{\partial \psi_i}{\partial \dot x}, \qquad \lambda_5 = \frac{\partial \psi_i}{\partial \dot y}, \qquad \lambda_6 = \frac{\partial \psi_i}{\partial \dot z}. \qquad (34)$$

The results of the integration will be, of course,

$$\lambda_{10} = \frac{\partial \psi_i}{\partial x_0}, \dots \qquad \lambda_{60} = \frac{\partial \psi_i}{\partial \dot z_0}. \qquad (35)$$

For all of the proposed integrations we may now rewrite Eqs. (26) exclusive of the terms in which the partial differential coefficients are zero, making the special assumption, in order to simplify our discussion, that the changes in the perturbation terms of Eq. (11) may be neglected as second-order, as follows:

$$
\begin{aligned}
\dot\lambda_1 &= -\lambda_4 \frac{\partial \ddot x}{\partial x} - \lambda_5 \frac{\partial \ddot y}{\partial x} - \lambda_6 \frac{\partial \ddot z}{\partial x} \\[2mm]
\dot\lambda_2 &= -\lambda_4 \frac{\partial \ddot x}{\partial y} - \lambda_5 \frac{\partial \ddot y}{\partial y} - \lambda_6 \frac{\partial \ddot z}{\partial y} \\[2mm]
\dot\lambda_3 &= -\lambda_4 \frac{\partial \ddot x}{\partial z} - \lambda_5 \frac{\partial \ddot y}{\partial z} - \lambda_6 \frac{\partial \ddot z}{\partial z} \\[2mm]
\dot\lambda_4 &= -\lambda_1, \qquad \dot\lambda_5 = -\lambda_2, \qquad \dot\lambda_6 = -\lambda_3
\end{aligned}
\tag{36}
$$

where

$$
\begin{aligned}
\frac{\partial \ddot x}{\partial x} &= 3\frac{\mu x^2}{r^5} - \frac{\mu}{r^3}, &
\frac{\partial \ddot y}{\partial x} &= 3\frac{\mu x y}{r^5}, &
\frac{\partial \ddot z}{\partial x} &= 3\frac{\mu x z}{r^5} \\[2mm]
\frac{\partial \ddot x}{\partial y} &= 3\frac{\mu x y}{r^5}, &
\frac{\partial \ddot y}{\partial y} &= 3\frac{\mu y^2}{r^5} - \frac{\mu}{r^3}, &
\frac{\partial \ddot z}{\partial y} &= 3\frac{\mu y z}{r^5} \\[2mm]
\frac{\partial \ddot x}{\partial z} &= 3\frac{\mu x z}{r^5}, &
\frac{\partial \ddot y}{\partial z} &= 3\frac{\mu y z}{r^5}, &
\frac{\partial \ddot z}{\partial z} &= 3\frac{\mu z^2}{r^5} - \frac{\mu}{r^3}
\end{aligned}
\tag{37}
$$

These coefficients are computed for each step in the numerical integrations, but it is to be noted that they are identical for each set of integrations, i. e., for the sets of terminal conditions specified by Eqs. (30), (32), (33), (34). Accordingly it will be advantageous if either: 1. all of the desired integrations can be carried through simultaneously, or 2. the partial differential coefficients specified by Eqs. (37) can be stored for each date. If the latter possibility can be realized, it will provide a convenience also for the integration of partial differential coefficients for more than one terminal time t.

E. The relationship of the adjoint method to Encke's method

Linearization of ENCKE's method, as used with variant calculations, reveals its close relationship to the adjoint method. If we omit second-order and higher-order terms from Eqs. (16) and (17), they reduce to

$$
f q = 3 q = \frac{3}{r_e^3} (\boldsymbol{r} \cdot \boldsymbol{\varrho})
\tag{38}
$$

and Eqs. (15) may then be written

$$
\ddot{\boldsymbol{\varrho}} = \frac{\partial \ddot{\boldsymbol{r}}}{\partial x}\, \xi + \frac{\partial \ddot{\boldsymbol{r}}}{\partial y}\, \eta + \frac{\partial \ddot{\boldsymbol{r}}}{\partial z}\, \zeta
\tag{39}
$$

where the components of $\partial \ddot{\boldsymbol{r}}/\partial x$, $\partial \ddot{\boldsymbol{r}}/\partial y$, $\partial \ddot{\boldsymbol{r}}/\partial z$ are exactly those that enter into Eqs. (36), and are given, if again we neglect $\dot r^\setminus - \dot r_e^\setminus$, by Eq. (37).

To appreciate better the significance and form of Eq. (39), we may recollect that $\varrho, \xi, \eta, \zeta$, traditional ENCKE notations, might well be replaced by $\varDelta \boldsymbol{r}, \varDelta x, \varDelta y, \varDelta z$, as in Eqs. (19)–(22). We may carry the substitution still further, to the point where it becomes evident that we are actually integrating the partial differential coefficients instead of the differences from which they may be obtained through Eqs. (8), (9), (10), (19–(22), etc. For example, the integration of ξ for Eqs. (8), (20), from initial conditions as given by Eq. (19), may be expressed formally as

$$
\left.
\begin{aligned}
\xi &= \xi_0 + \int_{t_0}^{t}\int \left(\frac{\partial \ddot{x}}{\partial x} \xi + \frac{\partial \ddot{x}}{\partial y} \eta + \frac{\partial \ddot{x}}{\partial z} \zeta \right) dt^2 \\
\dot{\xi} &= \qquad \int_{t_0}^{t} \left(\frac{\partial \ddot{x}}{\partial x} \xi + \frac{\partial \ddot{x}}{\partial y} \eta + \frac{\partial \ddot{x}}{\partial z} \zeta \right) dt .
\end{aligned}
\right\} \tag{40}
$$

Dividing these equations through by ξ_0 and taking account of Eqs. (8), (9), (10), (19)–(22), etc., we obtain

$$
\left.
\begin{aligned}
\frac{\partial x}{\partial x_0} &= 1 + \int_{t_0}^{t}\int \left(\frac{\partial \ddot{x}}{\partial x} \frac{\partial x}{\partial x_0} + \frac{\partial \ddot{x}}{\partial y} \frac{\partial y}{\partial x_0} + \frac{\partial \ddot{x}}{\partial z} \frac{\partial z}{\partial x_0} \right) dt^2 \\
\frac{\partial \dot{x}}{\partial x_0} &= \qquad \int_{t_0}^{t} \left(\frac{\partial \ddot{x}}{\partial x} \frac{\partial x}{\partial x_0} + \frac{\partial \ddot{x}}{\partial y} \frac{\partial y}{\partial x_0} + \frac{\partial \ddot{x}}{\partial z} \frac{\partial z}{\partial x_0} \right) dt .
\end{aligned}
\right\} \tag{41}
$$

For comparison we may introduce the adjoint-method Eqs. (30) or (31) into selected Eqs. from (36), and indicate the necessary integrations by

$$
\frac{\partial x}{\partial x_0} = 1 - \int_{t}^{t_0} \left(\frac{\partial \ddot{x}}{\partial x} \frac{\partial x}{\partial \dot{x}_0} + \frac{\partial \ddot{y}}{\partial x} \frac{\partial x}{\partial \dot{y}_0} + \frac{\partial \ddot{z}}{\partial x} \frac{\partial x}{\partial \dot{z}_0} \right) dt \tag{42}
$$

where

$$
\left.
\begin{aligned}
\frac{\partial x}{\partial \dot{x}_0} &= - \int_{t}^{t_0} \frac{\partial x}{\partial x_0} dt \\
\frac{\partial x}{\partial \dot{y}_0} &= - \int_{t}^{t_0} \frac{\partial z}{\partial y_0} dt \\
\frac{\partial x}{\partial \dot{z}_0} &= - \int_{t}^{t_0} \frac{\partial x}{\partial z_0} dt .
\end{aligned}
\right\} \tag{43}
$$

[Some careful thought about signs would be necessary if Eqs. (42) and (43) were actually to be used in calculation.]

It will be noted that the equivalent of the double integration of Eqs. (41) is implicit in Eqs. (42) and (43). Accordingly we find, as a first comparison, that the full set of 36 partial differential coefficients can be obtained by the 36 *single* integrations implied by the adjoint-method

Eqs. (30)–(33), (36), or (42), (43), etc., whereas 18 *double* integrations are necessary to produce the same set by the linearized ENCKE-method Eqs. (39) and (19)–(22), etc., or (41), etc. (In many problems, where the Gaussian summation integration procedures are so much more effective than other techniques, double integration is often preferable to single integration.) Comparisons will be made, in the sections that follow, for problems that require less than the full set of 36 partial differential coefficients.

The question of linearization vs. non-linearization has already been touched upon by some of the remarks at the and of section F. With modern automatic calculating equipment it is necessary to generate or store the data used to calculate the perturbation terms required for the initial integration of the calculated positions, whether by COWELL's method and Eq. (11) or by ENCKE's method and Eq. (15). Exactly the same data would be used for the variant calculations by the COWELL and ENCKE methods, if the perturbations terms were to be recalculated; if they were not to be recalculated, no reference data at all would be needed by COWELL's method, and only the two-body or positional reference data would be needed by ENCKE's method, unlinearised. For the adjoint method or the linearised form of ENCKE's method, it would be necessary additionally to store or generate, for each step of the integration, the partial differential coefficients of Eqs. (37).

The adequacy of linearization, as opposed to unlinearized perturbation techniques such as those of COWELL and ENCKE, can be considered only in the light of a particular problem. It is assuredly a question that must be considered for each problem; one cannot turn unguardedly to the linearized version here developed for ENCKE's method, or to the equivalently linearized adjoint method.

Shortly after the foregoing was written I was privileged to hear an enlightening lecture, on May 14, 1962, by Dr. OLIVER K. SMITH of Space Technology Laboratories, who had developed what I have here termed the "linearized ENCKE" method as a "variational orbit" method. His s ources were: Poincaré, *Méthodes Nouvelles*; McShane, Kelley, Reno, *Exterior Ballistics*; Coddington, Levinson, *Ordinary Differential Equations*; with some reference to G. W. HILL's well known variational orbit for the moon. It is evident that I should have taken note, in my opening sentence, not only of the differing traditions of different fields, but also of the differing traditions within celestial mechanics itself. After Dr. SMITH's lecture, the discussion developed also the fact that Dr. R. M. L. BAKER, JR., had independently approached the equations of this section by linearizing the ENCKE method.

Dr. SMITH indicated that he preferred the carrying of 42 single or 21 double integrations simultaneously to the storage of data for the six sets of linearized integrations that may be separated from the original set of 6 single or 3 double non-linearized integrations. This possibility led him to reject the adjoint method for observational differential correction (section F).

Of the two approaches to the "linearized-ENCKE-variational-orbit" equations it seems fair to state that the linearization of the "variational orbit" approach takes place too early in the development, so that it is not immediately apparent what one should do when linearization is inadequate. The linearization of the older ENCKE method, on the other hand, is clearly unnecessary, and one reverts to ENCKE's method itself, with simpler programming though perhaps more calculation, when linearization is inadequate. ENCKE's method does not require that the 21 double integrations be performed simultaneously; they may be performed in 7 separate sets of 3 each.

F. Observational differential correction

The adaptability of the various techniques for obtaining the partial differential coefficients of Eqs. (5) to the demands of observational differential correction is a matter of considerable interest and of some misconceptions.

First of all we note that we have a problem involving a number of t's, the dates of the various observations, but only one t_0, the epoch of the elements to be corrected. Accordingly it is preferable, generally, to start at t_0 and carry the integrals to the first of the t's, and on through it and the others to the last of the t's. This procedure is a feature, of course, of the variant-calculation techniques, whether COWELL or ENCKE or other, and whether linearized or unlinearized. It is also possible to use the analytic differential formulae for any t.

In order to meet this challenge the adjoint method must be supplemented, probably by a large storage capacity if the calculating equipment is automatic, to make the step-by-step values of partial differential coefficients readily available to each set of integrations.

If there is only one kind of observation of sufficient accuracy to be valuable, e. g., range, ϱ, the adjoint method has a compensating advantage, indicated by Eqs. (34), (35). It is that only six single integrations are necessary to obtain the partial differential coefficients for one ϱ and one t. If there are only six ϱ's, and storage of the reusable partial differential coefficients (37) is possible, the necessary 36 single integrations would compare favorably with the 18 double integrations required by the COWELL and ENCKE methods. But each additional ϱ would require 6 additional single integrations, whereas the 18 double integrations would cover the additional ϱ's as well as the initial 6.

G. Interception or targeting

Rendezvous, or homing on a target, we shall consider here only for interception at a specified position and time, r and t, and for impulsive correction of the velocity r_0 at the epoch t_0. That is to say, we shall resist so far as possible the temptation to extend to interception at a variable time, or to the interesting questions associated with continuous guidance and thrust over a finite length of time, and with terminal guidance or the "docking" maneuver.

The problem of physically correcting an orbit to intercept a target and the problem of mathematically correcting an approximation to an intercepting design or reference orbit into exact interception are formulated in exactly the same manner.

It is sometimes assumed that the mathematical interception problem consists in finding by analytical formulae, though necessarily by successive approximations, the orbit that passes through two positions, r_0 and r, and satisfies the time interval, $t - t_0$. The well known two-body solutions to this problem, however, are only approximations to reality, very much as orbits achieved by thrust are only approximations to orbits designed to intercept a target. A two-body orbit so obtained will give us an $\dot{r}_0$ to serve with r_0 as a set of elements or initial conditions, or an $\dot{r}$ to serve in the same way with r. But if we integrate inclusive of perturbations from the former set, we shall find that the orbit no longer satisfies r, or, if we integrate from the latter set, the orbit no longer satisfies r_0. It must be differentially corrected to satisfy both ends, if it is necessary for it to do so.

Be it physical or mathematical, then, the differential correction of an orbit to interception may be solved formally in several illuminating ways, of which we shall discuss and compare four, together with the advantages and disadvantages of the various ways of obtaining the necessary partial differential coefficients:

Method (a). Elements taken to be $r_{0C} = r_0$ and $\dot{r}_{0C}$; Δr obtained by integration. Since $\Delta r_0 = 0$, and since $\Delta \dot{r}$ is not of interest, Eqs. (5) reduce to

$$\Delta r = \frac{\partial r}{\partial \dot{x}_0} \Delta \dot{x}_0 + \frac{\partial r}{\partial \dot{y}_0} \Delta \dot{y}_0 + \frac{\partial r}{\partial \dot{z}_0} \Delta \dot{z}_0. \tag{44}$$

The nine partial differential coefficients require nine double integrations by COWELL's method, by ENCKE's method, or by the linearized form of the latter. The adjoint method, however, still requires the eighteen single integrations of sections D and E, since the partials with respect to x_0, y_0, z_0 must still be obtained as well as those with respect to $\dot{x}_0$, $\dot{y}_0$, $\dot{z}_0$. The calculation required by an analytic formulation may also be noted as halved.

It is evident that the method adopted for the determination of the partial differential coefficients of Eqs. (44) has no effect upon the form of the equations themselves. In all circumstances they must be inverted to yield the desired components of $\Delta \dot{\boldsymbol{r}}_0$ from the known components of $\Delta \boldsymbol{r}$. The three following methods address themselves to the problem of avoiding the 3×3 matrix inversion of Eq. (44). Method (b) unsuccessfully explores the effect of integrating from $\boldsymbol{r}$, $\dot{\boldsymbol{r}}$ instead of $\boldsymbol{r}_0$, $\dot{\boldsymbol{r}}_0$.

Method (b). Elements taken to be $\boldsymbol{r}_C = \boldsymbol{r}$ and $\dot{\boldsymbol{r}}_C$; $\dot{\boldsymbol{r}}_{0C}$ and $\Delta \boldsymbol{r}_0$ obtained by integration. Since $\Delta \boldsymbol{r} = 0$, Eqs. (5) or (44) are replaced by

$$\Delta \dot{\boldsymbol{r}}_0 = \frac{\partial \dot{\boldsymbol{r}}_0}{\partial \dot{x}} \Delta \dot{x} + \frac{\partial \dot{\boldsymbol{r}}_0}{\partial \dot{y}} \Delta \dot{y} + \frac{\partial \dot{\boldsymbol{r}}_0}{\partial \dot{z}} \Delta \dot{z}, \qquad (45)$$

$$\Delta \boldsymbol{r}_0 = \frac{\partial \boldsymbol{r}_0}{\partial \dot{x}} \Delta \dot{x} + \frac{\partial \boldsymbol{r}_0}{\partial \dot{y}} \Delta \dot{y} + \frac{\partial \boldsymbol{r}_0}{\partial \dot{z}} \Delta \dot{z}. \qquad (46)$$

Eq. (45) does in fact make possible the direct calculation of $\Delta \dot{x}_0$, $\Delta \dot{y}_0$, $\Delta \dot{z}_0$, but unfortunately only with quantities, $\Delta \dot{x}$, $\Delta \dot{y}$, $\Delta \dot{z}$, that must be obtained by inverting Eqs. (46), since the only known residuals are the components of $\Delta \boldsymbol{r}_0$. The number of needed partial differential coefficients has been increased to 18 without enabling us to avoid a 3×3 matrix inversion.

In passing, one may note that it is possible to devise a method that is even more stupid than Method (b) — one that requires 36 partial differential coefficients but still does not avoid a 3×3 matrix inversion.

Method (c). Elements taken to be $\boldsymbol{r}_{0C} = \boldsymbol{r}_0$ and $\dot{\boldsymbol{r}}_{0C}$; $\Delta \boldsymbol{r}$ obtained by integration; $\Delta \boldsymbol{r}_0 = 0$; all as in Method (a). But Eqs. (44) are replaced by

$$\Delta \dot{\boldsymbol{r}}_0 = \frac{\partial \dot{\boldsymbol{r}}_0}{\partial x} \Delta x + \frac{\partial \dot{\boldsymbol{r}}_0}{\partial y} \Delta y + \frac{\partial \dot{\boldsymbol{r}}_0}{\partial z} \Delta z. \qquad (47)$$

These equations provide for the direct calculation of $\Delta \dot{\boldsymbol{r}}_0$ from $\Delta \boldsymbol{r}$, but the required partial differential coefficients cannot be obtained by numerical means, whether variant-calculation or adjoint, except by the 3×3 matrix inversion of Eqs. (44) that we are trying to avoid. They can be obtained, however, by analytic differentiation of one of the sets of two-body formulae for determining an orbit from $\boldsymbol{r}_0$, $\boldsymbol{r}$, and $t - t_0$, when the changes in the perturbations are no greater than the neglected second-order terms of Eqs. (47). A possibly more useful solution, however, is to be found in Method (d).

Method (d). Elements taken to be $\boldsymbol{r}_C = \boldsymbol{r}$ and $\dot{\boldsymbol{r}}_C$; $\dot{\boldsymbol{r}}_{0C}$ and $\Delta \boldsymbol{r}_0$ obtained by integration; $\Delta \boldsymbol{r} = 0$; all as in Method (b). But the differential formulae required for direct calculation of $\Delta \dot{\boldsymbol{r}}_0$ are

$$\Delta \boldsymbol{r}_0 = \frac{\partial \dot{\boldsymbol{r}}_0}{\partial x_0} \Delta x_0 + \frac{\partial \dot{\boldsymbol{r}}_0}{\partial y_0} \Delta y_0 + \frac{\partial \dot{\boldsymbol{r}}_0}{\partial z_0} \Delta z_0. \qquad (48)$$

The indicated partial differential coefficients, like those of Method (c), can be obtained directly only by analytic differentiation. Comparing Methods (c) and (d) as they are specified by Eqs. (47) and (48), we note that the latter are in the form desired by basic continuous guidance procedures.

Both of Methods (c) and (d), accordingly, direct our attention pointedly to the several available sets of formulae for determining a two-body orbit that satisfies r_0, r, and $t - t_0$. The discussion of these formulae, and of analytic differential coefficients in general, we reserve to another paper.

To JERRY L. ALDERMAN and DARIL W. HAHN of the Boeing Aerospace Division, and to THEODORE I. FINE of the Lockheed Astrodynamics Research Center, I want to acknowledge my indebtedness for my introduction to the adjoint method. To them should go the credit for whatever excellences there are in the presentation of the method in Section. D.

For support in the studies that led to this paper, I am indebted to the Office of the Chief Scientist, Lockheed-California Company; to the Space Flight Research Section, Aerospace Division, Boeing Company; and to the Mechanics Division, Air Force Office of Scientific Research Air Research and Development Command.

References

BLISS, G. A.: Mathematics for Exterior Ballistics. Wiley: New York (1944).

BRYSON, A. E., et al.: Determination of the lift or drag program that minimizes re-entry heating ... IAS 61-6, 29th Annual Meeting, January 23—25 (1961).

CODDINGTON, E. A., and N. LEVINSON: Theory of Ordinary Differential Equations. New York: McGraw-Hill (1955).

DUNN, J. C.: Green's function for space-trajectory perturbation analysis. J. Astronaut. Sci., 8, No. 4, Winter (1961).

DUNN, J. C., and C. GIANETTO: Lunar-trajectory perturbation analysis—some computational results via the adjoint method. ARS 2072-61.

FINE, T. I.: Theory and application of adjoint differential equations to astrodynamics. Lockheed Res. 15742, March (1962).

HAHN, D. W.: Trajectory sensitivity coefficients by the adjoint method. Boeing 2-5770/3-050, July (1961).

HERRICK, S.: Step-by-step integration of $\ddot{x} = f(x, y, z, t)$ without a "corrector." MTAC, 5, 61—67, April, 1951; UCLA Astr. Pap., 1, no. 12, 123—130.

JUROVICS, S. A.: The adjoint method and its application to trajectory optimization. ARS 2064-61, Sp. Flt. Report to the Nation, October (1961).

MCSHANE, E. J., J. L. KELLEY, and F. V. RENO: Exterior Ballistics. Denver, Univ. of Denver Press (1953). Chaps. 8, 9.

POINCARÉ, H.: Les Méthodes Nouvelles de la Mécanique Céleste. Paris; Gauthier, Villars, 1892—99; New York; Dover Catalog, 1, chap. 4 (1957).

SMITH, F. T.: A discussion of a midcourse guidance technique for space vehicles, Rand RM-2581, October (1960).

The potential of the earth derived from satellite motions

By

Yoshihide Kozai[1]

Smithsonian Astrophysical Observatory Cambridge, Mass., U.S.A.

Recently I determined the coefficients of zonal spherical harmonics up to ninth order in the potential of the earth by analyzing motions of artificial satellites (KOZAI, 1962a).

I used many observational materials by several authors. However, in solving equations of condition I assigned the largest weights to the data reduced by myself from precisely reduced BAKER-NUNN observations, which have an accuracy of four seconds of arc.

In section 1 of this paper, I explain my method of reduction of observations in detail, using Satellite 1959 Eta as an example.

In section 2, I report new values of the coefficients of the zonal spherical harmonics; and in section 3, I summarize my studies on tesseral harmonics of the potential.

1. Reduction of observations

At the Smithsonian Astrophysical Observatory the Differential Orbit Improvement program (DOI) is available for the reduction of observations of artificial satellites. This program, initiated by G. VEIS and C. MOORE (1960), can reduce every kind of observation except DOPPLER. In the DOI program short-periodic perturbations of the first order due to the oblateness of the earth are taken out of each observation by KOZAI's formulas (1959b), and KEPLER's orbital elements and their secular changes are improved by the method of least squares. Therefore, the orbital elements derived by the DOI are in a certain sense, their mean values, and they are not affected by any bias in the distribution of observations in orbit. Of the orbital elements the inclination i and the argument of perigee ω are referred to the equator of date; and the longitude of the ascending node Ω is measured from the mean equinox at 1950.0. By adopting this coordinate system, we very much reduce effects due to the motion of earth's equator (KOZAI 1960).

We derive the mean orbital elements at a regular interval of time for each satellite and publish them in the Special Report series of the Smith-

---

[1] Now at Tokyo Astronomical Observatory Mitaka, Tokyo, Japan.

sonian Astrophysical Observatory. Usually they are based on field-reduced BAKER-NUNN and MINITRACK observations having an accuracy of several minutes of arc. Although these orbital elements are determined with less than five significant figures, we have been able, by using these data, to determine rather accurate values for the coefficients of the zonal spherical harmonics in the potential of the earth (KOZAI 1961 a).

Films taken and roughly reduced at each station of the BAKER-NUNN camera are sent to the Cambridge Headquarters, where the satellite positions on these films are measured to an accuracy of four seconds of

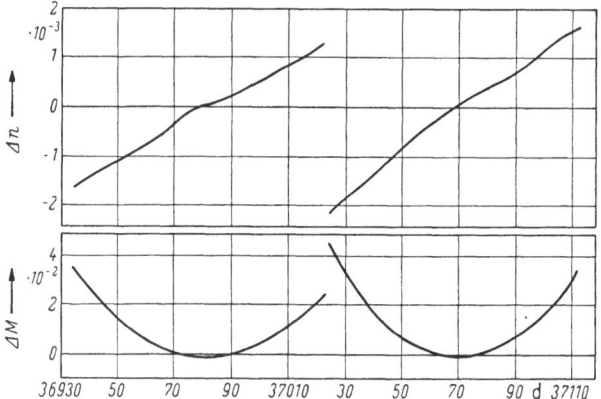

Fig. 1. Air-drag effects in mean motion and mean anomaly (Days Modified JULIAN DAYS).

arc. The mean orbital elements of the satellites are then redetermined more accurately by these precisely reduced observations.

In Figure 1 through 5, variations of the mean orbital elements based on precisely reduced observations of Satellite 1959 Eta (Vanguard III) during 178 days from January 1.0 to June 27.0, 1960 are given.

The period of the observations is divided into two periods, each of which covers 88 days; the mean epoch of the first is February 14.0; and that of the second, May 14.0, 1960.

The perturbations due to the air drag are evaluated from the variations of the anomalistic mean motion n (revolutions per day) and the mean anomaly M (revolutions). The solar-radiation pressure effects, which are computed by my program on the IBM-7090 (KOZAI 1961 b), and the long-periodic terms in the mean anomaly have been taken out of the observed values of the mean motion and the mean anomaly.

The values plotted in Figure 1 are not the actual ones. Rather they are differences of the observed values and the mean motions at the mean epochs for the mean motion. For the mean anomaly, they are those of the observed values and the ones computed by using the mean anomalies and the mean motions at the mean epochs; that is, for the first period

$$\begin{aligned} \varDelta n &= n - 11.06540, \\ \varDelta M &= M - (0.8768 + 11.065402t); \end{aligned} \Bigg\} \tag{1}$$

and for the second period

$$\begin{aligned} \varDelta n &= n - 11.06886, \\ \varDelta M &= M - (0.8958 + 11.068856t). \end{aligned} \Bigg\} \tag{2}$$

Therefore, the variations of the mean motion and the mean anomaly plotted in Figure 1 may be entirely due to air-drag effects.

Since the disturbing force of the air drag is not constant, the curve of $\varDelta n$ deviates considerably from a straight line. From this curve we find that on the average the air-drag changes the semimajor axis and the mean motion by 2×10^{-7} during one revolution of the satellite.

From $\varDelta n$ and $\varDelta M$ for each epoch we can evaluate the air-drag perturbations in the eccentricity, the argument of perigee, and the longitude of the ascending node by assuming that the perigee distance of the satellite is not perturbed by the air drag. The following formulas are used:

$$\begin{aligned} \varDelta e &= -\frac{2}{3}(1 - e)\varDelta n/n = -0.0491\,\varDelta n, \\ \varDelta \omega &= \frac{1}{3}\frac{\dot\omega}{n}\frac{7 - e}{1 + e}\varDelta M = 0.°844\,\varDelta M, \\ \varDelta \Omega &= \frac{1}{3}\frac{\dot\Omega}{n}\frac{7 - e}{1 + e}\varDelta M = -0.°567\,\varDelta M, \end{aligned} \Bigg\} \tag{3}$$

where $\dot\omega$ and $\dot\Omega$ are the secular motions of the argument of perigee and the node, respectively.

Amplitudes of long-periodic terms due to even harmonics in the potential can be computed by approximate values of the even harmonics (KOZAI 1959b). The long-periodic terms are expressed as

$$\begin{aligned} \delta e &= -0.23 \times 10^{-4}\cos 2\omega, \\ \delta i &= 0.°40 \times 10^{-3}\cos 2\omega, \\ \delta \omega &= 0.°67 \times 10^{-2}\sin 2\omega, \\ \delta \Omega &= 0.°75 \times 10^{-3}\sin 2\omega. \end{aligned} \Bigg\} \tag{4}$$

In the upper halves of Figures 2 through 5 the observed values of four orbital elements are plotted every two days. In Figures 4 and 5 the variations due to the secular motions have been subtracted from the observed values by the following expressions:
For the first period,

$$\begin{aligned} \Omega &- (106.°1189 - 3.°273988t), \\ \omega &- (138.°4484 + 4.°874325t); \end{aligned} \Bigg\} \tag{5}$$

5*

for the second period,

$$\Omega - (171.°3949 - 3.°275719t), \\ \omega - (217.°2524 + 4.°876874t). \Big\}$$ \hfill (6)

The lunisolar gravitational perturbations of long periods are computed by my program (Kozai 1959 a) and are shown by solid lines in the fig-

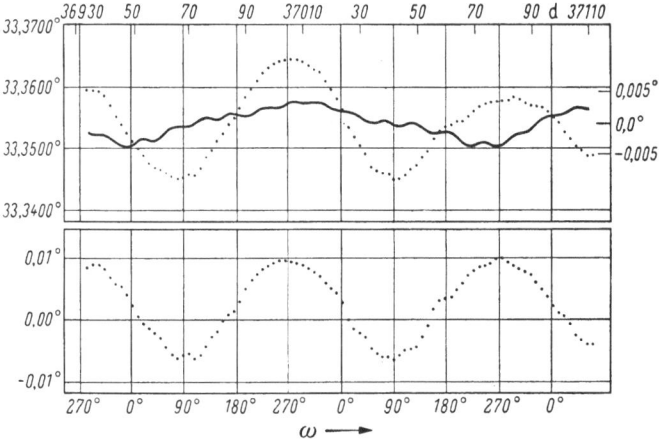

Fig. 2. Variations of inclination. Lunisolar perturbation is expressed by a solid line with a scale on the right side.

ures. The solar-radiation pressure effects based on area-mass ratio (A/M) 0.173 (cgs) are too small to be plotted except for the longitude of the ascending node in Figure 4, where they are expressed by a broken line.

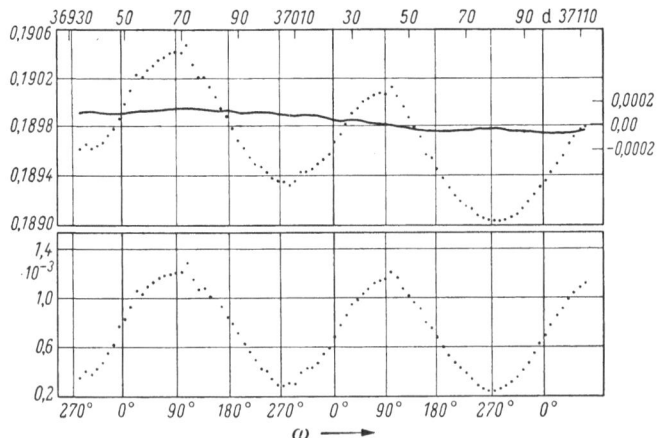

Fig. 3. Variations of eccentricity. Lunisolar perturbation is expressed by a solid line.

Roughly speaking, for this satellite the effect of the sun during each revolution of the satellite is 0.6×10^{-7}; that of the moon, 1.3×10^{-7};

and that of the solar-radiation pressure, 0.796×10^{-8}. The effects of the lunisolar perturbation during each revolution are proportional to $1/n^2$ (n being the mean motion); and that of the radiation pressure, to

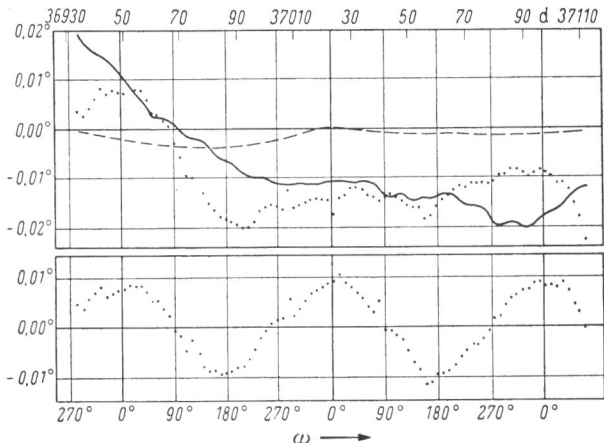

Fig. 4. Variations of longitude of ascending node. Lunisolar perturbation is expressed by a solid line and solar radiation effect by a broken line.

$(A/M)\, a^2$. Thus, for the Echo satellite ($A/M = 100$), the effects of the solar-radiation pressure are very large.

After subtracting the air-drag, lunisolar, solar-radiation pressure perturbations, and the long-periodic terms given in Eqs. (4), we have

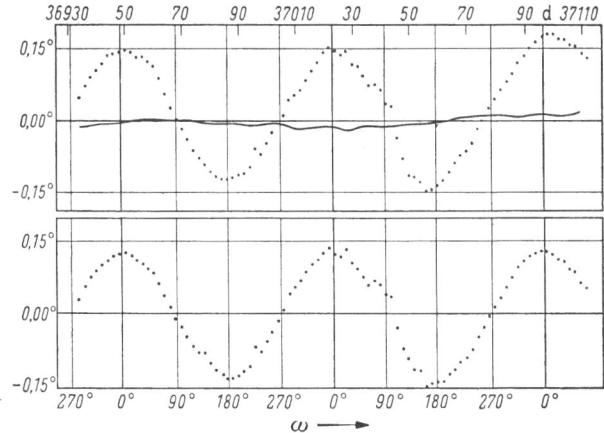

Fig. 5. Variations of argument of perigee. Lunisolar perturbation is expressed by a solid line.

the corrected values of the orbital elements plotted in the lower halves of Figures 2 through 5. The corrected variations may be regarded as those due to the odd-order harmonics in the potential. We can see that the observed real variations for the inclination and the longitude of the ascending node are greatly distorted by the lunisolar perturbations. Therefore, if the lunisolar perturbations had not been taken into consideration, we would have derived quite different values for the amplitudes of the long-periodic terms and the secular motions from the observations.

By fitting the corrected observed values by $\sin \omega$ or $\cos \omega$ plus a secular term, we can derive the following set of expressions for each period:

Epoch	Feb. 14.0, 1960	May 14.0, 1960
n	11.06540 ± 1	11.06886 ± 1
i	$33.^{\circ}35428 \pm 6$	$33.^{\circ}35376 \pm 10$
	$-(0.^{\circ}763 \pm 9) \times 10^{-2} \sin \omega$	$-(0.^{\circ}765 \pm 14) \times 10^{-2} \sin \omega$
e	0.189755 ± 3	0.189498 ± 3
	$+(0.459 \pm 5) \times 10^{-3} \sin \omega$	$+(0.438 \pm 4) \times 10^{-3} \sin \omega$
Ω	$106.^{\circ}1189 \pm 2$	$171.^{\circ}3949 \pm 2$
	$-(3.^{\circ}273988 \pm 7) t$	$-(3.^{\circ}275719 \pm 8) t$
	$+(0.^{\circ}89 \pm 2) \times 10^{-2} \cos \omega$	$+(0.^{\circ}92 \pm 2) \times 10^{-2} \cos \omega$
ω	$138.^{\circ}4484 \pm 10$	$217.^{\circ}2524 \pm 11$
	$+(4.^{\circ}87433 \pm 4) t$	$+(4.^{\circ}87687 \pm 5) t$
	$+(0.^{\circ}1267 \pm 12) \cos \omega$	$+(0.^{\circ}1326 \pm 17) \cos \omega$

From this table we can evaluate the accuracy of the observed orbital elements. The significant figures of the mean orbital elements are near six, except that for the argument of perigee, which is five. The inclination is very well determined because of the geographical distribution of the BAKER-NUNN cameras; there is no camera between latitudes $-16°$ and $+12°$, but there are many cameras near the apices of this satellite.

The differences in the mean values of the eccentricity and the inclination for the two epochs can be explained by the solar-radiation and air-drag effects.

However, the amplitudes of $\sin \omega$ and $\cos \omega$ in the argument of perigee and the eccentricity for the second period seem to be affected by systematic errors.

The standard errors may be reduced by consideration of tesseral harmonics in the potential, which are not yet known accurately and may have rather large effects on the orbital elements based on poorly distributed optical observations.

The other sources of errors may be in the values of the equatorial radius of the earth GM, and the coordinates of the observing stations adopted in the DOI program.

2. Zonal spherical harmonics

The observed amplitudes of the long-periodic term with argument ω and the secular motions are expressed by polynomials of J_n, where J_n

is defined in the formula of the potential by

$$U = \frac{GM}{r}\left\{1 - \sum_{n=2}^{\infty} J_n \left(\frac{R}{r}\right)^n P_n(\sin\beta)\right\}. \tag{7}$$

Here, R is the equatorial radius of the earth and P_n is the LEGENDRE polynomial of the n-th order.

By analytical expressions of the secular motions and the long-periodic terms given by KOZAI (1959b) and KAULA (1961), we can furnish the equations of condition to solve the values of J_n from the observed quantities. When we do not need more than three significant figures, 0.10822×10^{-2} is adopted as the value of J_2. The adopted values for R and GM are

$$\left. \begin{array}{l} R = 6.378165 \times 10^8 \text{ cm}, \\ GM = 3.986032 \times 10^{20} \text{ cm}^3/\text{sec}^2. \end{array} \right\} \tag{8}$$

By including the spherical harmonics up to ninth order, we can write the equations of conditions for the first period as follows:
For the amplitudes,

$$\left. \begin{array}{l} e = -190.5 J_3 + 84.2 J_5 + 50.7 J_7 - 71.4 J_9 = (0.459 \pm 5) \times 10^{-3}, \\ i = 3264 J_3 - 1443 J_5 - 869 J_7 + 1223 J_9 = -(0.°763 \pm 9) \times 10^{-2}, \\ \omega = -54711 J_3 + 41712 J_5 + 10806 J_7 - 39020 J_9 = 0.°1267 \pm 12, \\ \Omega = -5937 J_3 - 11177 J_5 + 14653 J_7 - 299 J_9 = (0.°89 \pm 2) \times 10^{-2}; \end{array} \right\} \tag{9}$$

for the secular motions,

$$\left. \begin{array}{l} \dot{\omega} = 4499.379 J_2 - 1267.4 J_4 - 2537 J_6 + 2117 J_8 + 0.°001901 + \\ \quad + 0.°000568 = 4.°87433 \pm 4, \\ \dot{\Omega} = -3020.487 J_2 + 2268.5 J_4 - 95 J_6 - 1036 J_8 + 0.°000221 - \\ \quad - 0.°000396 = -3.°273988 \pm 7. \end{array} \right\} \tag{10}$$

In the equations for the secular motions the last two terms on the left-hand sides are J_2^2-terms and lunisolar secular perturbations, respectively.

In my recent report (KOZAI 1962), using 40 equations of the amplitudes and 40 equations of the secular motions of 13 satellites, I derived the values of J_n up to the ninth order as follows:

$$\left. \begin{array}{ll} J_2 = 1082.48 \times 10^{-6}, & J_3 = -2.566 \times 10^{-6}, \\ \quad \pm 0.04 & \quad \pm 0.012 \\ J_4 = -1.84 \times 10^{-6}, & J_5 = -0.063 \times 10^{-6}, \\ \quad \pm 0.09 & \quad \pm 0.019 \\ J_6 = 0.39 \times 10^{-6}, & J_7 = -0.469 \times 10^{-6}, \\ \quad \pm 0.09 & \quad \pm 0.021 \\ J_8 = -0.02 \times 10^{-6}, & J_9 = 0.114 \times 10^{-6}. \\ \quad \pm 0.07 & \quad \pm 0.025 \end{array} \right\} \tag{11}$$

3. Tesseral harmonics

The determination of coefficients of tesseral and sectorial harmonics
in the potential from the orbits of the satellites is much more difficult
compared with that of the zonal harmonics, partly because the periods
of the principal perturbations are not long, near one day and its integral
fractions. Because of the length of the periods the amplitudes of the
perturbations are not increased so much; therefore, to determine the
constants associated with the tesseral harmonics, we must have observations that are well distributed even in one day.

Since the periods of the orbital variations due to the tesseral and
sectorial harmonics are known from theory, the amplitudes and
the phase angles of the principal variations are determined by
the DOI.

I used precisely reduced BAKER-NUNN observations of 1958 $\delta 2$,
1959 $\delta 1$, and 1959 Eta to determine the tesseral and sectorial harmonics
up to the fourth order (KOZAI 1961 c). Some of my results are not
reliable because the optical BAKER-NUNN observations are badly distributed; here I give only data for the second sectorial harmonics:

$$J_{2,2} = (2.3 \pm 0.5) \times 10^{-6}, \quad \lambda_{2,2} = (37° \pm 9°) \, W, \qquad (12)$$

where $J_{2,2}$ is the coefficient of the associated LEGENDRE function $P_2^{(2)}$
$(\sin\beta)$, and $\lambda_{2,2}$ is the phase angle. The standard errors given here seem
to be smaller than the actual deviations.

Recently IZSAK (1961) revised the DOI program by introducing a
new coordinate system, in which one axis is perpendicular to the apparent
topocentric path of the observer. When the residuals in this component
are used, the air drag may have very little effect in the determination of
the tesseral harmonics even if we use observations covering rather long
intervals of time. Also we want to include some of the lunar perturbations
and the second-order short-periodic perturbations of the oblateness
(KOZAI, unpublished) in the DOI to reduce systematic errors.

Therefore, we can expect that good determination of the tesseral
harmonics will be achieved in the future by precisely reduced BAKER-
NUNN observations.

References

IZSAK, I. G. (1961): Differential orbit improvement with the use of rotated resi-
    duals. Smithsonian Astrophys. Obs., Special Report No. 73.
KAULA, W. M. (1961): Analysis of gravitational and geometric aspects of geodetic
    utilization of satellites. Geophys. J., **5**, 104—133.
KOZAI, Y. (1959)a: On the effects of the sun and the moon upon the motion of
    a close earth satellite. Smithsonian Astrophys. Obs., Special Report No. 22,
    pp. 7—10.

(1959)b: The motion of a close earth satellite. Astronom. J., **64**, 367—377.

(1960): Effect of precession and nutation on the orbital elements of a close earth satellite. Astronom. J., **65**, 621—623.

(1961)a: The gravitational field of the earth derived from motions of three satellites. Astronom. J. **66**, 8—10.

(1961)b: Effects of solar radiation pressure on the motion of an artificial satellite. Smithsonian Astrophys. Obs., Special Report No. 56, 25—33.

(1961)c: Tesseral harmonics of the gravitational potential of the earth as derived from satellite motions. Astronom. J., **66**, 355—358.

(1962): Numerical results from orbits. Paper presented at the Symposium on the Use of Artificial Satellites for Geodesy, Washington, D. C.

VEIS, G., and C. H. MOORE (1960): The Smithsonian Astrophysical Observatory differential improvement program. Seminar Proceedings, Tracking Programs and Orbit Determination, Jet Propulsion Laboratory, 195

Perturbations in the motion of a satellite due to the second zonal harmonic of the Earth's potential

By

Iu. V. Batrakov

Institute for Theoretical Astronomy

Leningrad, U.S.S.R.

Abstract. The secular perturbations of the second order and long periodic perturbations of the first order in the orbital elements of Earth's satellites due to the second zonal harmonic of the gravitational potential of the Earth are calculated up to the terms of order e_0^2. The short periodic perturbations are given up to the terms with e_0^5. The results of this paper and those of BROUWER's appear to be identical with the assumed accuracy provided the difference in constants of integration is taken into account.

Due to the asphericity of the Earth the motion of an Earth satellite is greatly influenced by different perturbations, the largest of which being the perturbations from the second zonal harmonic in the Earth's gravitational potential. The accuracy with which these perturbations must be known depends on the character of the problem under consideration. To determine the orbital elements of a satellite from observations within rather a short interval of time one must have only the short periodic and secular perturbations of the first order with respect to the coefficient of the second zonal harmonic. If the coefficients of the harmonics in the Earth's potential must be determined these perturbations are inadequate and not only the perturbations from other harmonics are to be taken into consideration but also the perturbations due to the second zonal harmonics must be known with more accuracy. In particular for determining the coefficients of zonal harmonics one must have the long periodic perturbations of the first order and the secular perturbations of the second order with respect to the coefficient of the second zonal harmonic.

The analytical expressions for the perturbations as series in powers of eccentricity published before (PROSKURIN, BATRAKOV, 1959, 1960) were intended for the solution of the first problem: determining elements from observations in the short interval of time. In this paper are presented the results for additional secular as well as long periodic perturbations which together with the perturbations from other harmonics are to be known to determine the coefficients of the Earth's potential development by the motion of Earth's satellites. The results of our calculations have been compared with those of BROUWER's (1959).

The potential of the Earth's attraction when the second zonal harmonic only is taken into account is

$$V = \frac{f\,m}{r}\left\{1 + c_{20}\left(\frac{r_0}{r}\right)^2 P_{20}(\sin\varphi)\right\},\tag{1}$$

where $f\,m$ is the product of the gravitational constant and the Farth's mass, c_{20} is a dimensionless parameter, r_0 is a constant having the dimension of the length, φ is the latitude, $P_{20}(\sin\varphi) = \frac{3}{2}\sin^2\varphi - \frac{1}{2}$ is the LEGENDRE polynomial. The constant by JEFFREYS J is connected with C_{20} by means of equality $c_{20} = -\frac{2}{3}J$.

The LAGRANGE equations for osculating elements are

$$\begin{aligned}
\frac{da}{dt} &= \frac{2}{n\,a}\frac{\partial R}{\partial M}, \\[2mm]
\frac{de}{dt} &= \frac{1 - e^2}{n\,a^2\,e}\frac{\partial R}{\partial M} - \frac{\sqrt{1 - e^2}}{h\,a^2\,e}\frac{\partial R}{\partial\omega}, \\[2mm]
\sin i\,\frac{di}{dt} &= \frac{\cos i}{n\,a^2\sqrt{1 - e^2}}\frac{\partial R}{\partial\omega} - \frac{1}{n\,a^2\sqrt{1 - e^2}}\frac{\partial R}{\partial\Omega}, \\[2mm]
\sin i\,\frac{d\Omega}{dt} &= \frac{1}{n\,a^2\sqrt{1 - e^2}}\frac{\partial R}{\partial i}, \\[2mm]
\frac{d\omega}{dt} &= -\frac{\cot i}{n\,a^2\sqrt{1 - e^2}}\frac{\partial R}{\partial i} + \frac{\sqrt{1 - e^2}}{n\,a^2\,e}\frac{\partial R}{\partial e}, \\[2mm]
\frac{dM}{dt} &= n - \frac{1 - e^2}{n\,a^2\,e}\frac{\partial R}{\partial e} - \frac{2}{n\,a}\frac{\partial R}{\partial a}.
\end{aligned}\right\}\tag{2}$$

At the right-hand side of the equation for the mean anomaly the term of zero order with respect to C_{20} is presented, so the perturbations of M of some order can be obtained only after the perturbations of the semimajor axis of the same order are known.

The secular and short periodic perturbations of the first order due to the second zonal harmonic of the Earth's potential, which can be found in (PROSKURIN, BATRAKOV, 1960) are of the form

$$\begin{aligned}
\frac{\delta_1 a}{a_0} &= \sum a_{ik}\cos(iM + k\omega), \\[2mm]
\delta_1 e &= \sum e_{ik}\cos(iM + k\omega), \\[2mm]
\delta_1 I &= \sum I_{ik}\cos(iM + k\omega), \\[2mm]
\delta_1 \Omega &= \Omega' t + \sum \Omega_{ik}\sin(iM + k\omega), \\[2mm]
\delta_1 \pi &= \pi' t + \sum \pi_{ik}\sin(iM + k\omega), \\[2mm]
\delta_1 \varepsilon &= \varepsilon' t + \sum \varepsilon_{ik}\sin(iM + k\omega),
\end{aligned}\right\}\tag{3}$$

where I is the inclination of the orbit to the equatorial plane, ω is the argument of perigee, M is the mean anomaly, $\pi = \omega + \Omega$ is the perigee longitude, $\varepsilon = M_0 + \omega + \Omega$ is the mean longitude at the epoch. Coefficients $a_{ik}, \ldots, \varepsilon_{ik}$ are the power series in eccentricity, calculated up to the terms with e^5. The finite expressions for Ω', π', ε' are given. The perturbations of ω and M can be received from the perturbations of π and ε using formulae

$$
\left.
\begin{aligned}
\delta_1\, \omega &= \delta_1\, \pi - \delta_1\, \Omega, \\
\delta_1\, M &= \delta_1\, \varepsilon - \delta_1\, \pi + \int \delta_1\, n\, dt, \\
\delta_1\, n &= -\frac{3}{2}\, n_0 \frac{\delta_1 a}{a_0}.
\end{aligned}
\right\}
\tag{4}
$$

We need the secular perturbations of the second order and long periodic perturbations of the first order with respect to C_{20}. For this purpose the right-hand sides of Eqs. (2) are to be developed in terms of $\frac{\delta_1 a}{a}$, $\delta_1 e, \ldots \delta_1 M$, only the constant and long periodic terms of order C_{20}^2 being taken into account in these developments. It must be remembered that the long periodic perturbations of the first order of angular elements Ω, ω, M are also caused by the coefficients of the secular perturbations of the first order Ω', ω', M_0. For determining these long periodic perturbations one needs long periodic perturbations of a, e, I.

The matter is more complicated in the case of the perturbations of the mean anomaly. For the complete determination of the long periodic terms of the first order in this element one must know the second order long periodic perturbations of the semimajor axis the calculation of which directly from differential equations requiring the very troublesome third approximation. This circumstance was marked in Kozai's paper (1959). Nowever this difficulty can be avoided using the energy integral which exists in this problem

$$
\frac{f\,m}{2\,a} + R = C, \qquad R = f\,m\,C_{20} \frac{r_0^2}{r^3}\, P_{20}\,(\sin\varphi).
\tag{5}
$$

Taking into account the above mentioned consideration the long periodic perturbations of the semimajor axis can be derived at the second approximation.

The secular perturbations of the second order and long periodic perturbations of the first order due to the second zonal harmonic of the Earth's potential calculated here together with short periodic and secular perturbations of the first order from the paper (Proskurin, Batrakov, 1960), transformed with the help of formulae (4), are given beneath.

The coefficients of the second order secular perturbations are accurate to e_0^2 through. The secular perturbations are denoted with square brackets, the long periodic ones with braces and the short periodic ones with round brackets. The osculating elements are calculated with the formulae

$$E = E_0 + [\delta_1 E] + [\delta_2 E] + \{\delta_1 E\} + (\delta_1 E), \qquad (6)$$

where E_0 are the constants of the given theory a_0, e_0, I_0, Ω_0, ω_0. The nonperturbed mean motion n_0 is connected with a_0 by the formula of the third KEPLER's low $n_0^2 a_0^3 = f m$. The perturbed mean anomaly is calculated with

$$M = M_0 + n_0(t - t_0) + [\delta_1 M] + [\delta_2 M] + \{\delta_1 M\} + (\delta_1 M). \qquad (7)$$

Here the following abbreviations are used $\lambda = \sin I_0$, $\alpha = \dfrac{r_0}{a_0}$, besides all zero indices in e_0, n_0, a_0 are neglected for brevity which mustn't give rise to misunderstanding.

1. Secular perturbations

$$
\left.
\begin{aligned}
& [\delta_1 a] = [\delta_2 a] = [\delta_1 e] = [\delta_2 e] = [\delta_1 I] = [\delta_2 I] = 0, \\[4pt]
& [\delta_1 \Omega] = \Omega' = \frac{3}{2} C_{20} \alpha^2 \sqrt{1 - \lambda^2}\,(1 - e^2)^{-2}\, n(t - t_0), \\[4pt]
& [\delta_2 \Omega] = \frac{3}{32} C_{20}^2 \alpha^4 \sqrt{1 - \lambda^2}[-60 + 76\lambda^2 + e^2(-232 + 281\lambda^2)]\,n(t - t_0), \\[4pt]
& [\delta_1 \omega] = \omega' = -\frac{3}{4} C_{20} \alpha^2(4 - 5\lambda^2)\,(1 - e^2)^{-2}\, n(t - t_0), \\[4pt]
& [\delta_2 \omega] = \frac{3}{128} C_{20}^2 \alpha^4[576 - 1352\lambda^2 + 790\lambda^4 + e^2(2264 - 2590\lambda^2 + \\
& \qquad\qquad + 2935\lambda^4)]\,n(t - t_0), \\[4pt]
& [\delta_1 M] = M_0' = -\frac{3}{4} C_{20} \alpha^2(2 - 3\lambda^2)\,(1 - e^2)^{-\frac{3}{2}}\, n(t - t_0), \\[4pt]
& [\delta_2 M] = \frac{3}{128} C_{20}^2 \alpha^4[288 - 752\lambda^2 + 574\lambda^4 + e^2(1304 - 3392\lambda^2 + \\
& \qquad\qquad + 2779\lambda^4)]\,n(t - t_0).
\end{aligned}
\right\} \quad (8)
$$

2. Long periodic perturbations[1]

$$\{\delta_1 a\} = 0,$$

$$\{\delta_1 e\} = -\frac{1}{8} C_{20} \alpha^2 e\, \frac{13 - 15\lambda^2}{4 - 5\lambda^2}\cos 2\omega,$$

$$\{\delta_1 I\} = \frac{1}{8} C_{20} \alpha^2 e^2\, \frac{13 - 15\lambda^2}{4 - 5\lambda^2}\, \lambda \sqrt{1 - \lambda^2}\cos 2\omega,$$

$$\{\delta_1 \Omega\} = -\frac{1}{8} C_{20} \alpha^2 e^2\, \frac{-52 + 120\lambda^2 - 75\lambda^4}{(4 - 5\lambda^2)^2}\, \sqrt{1 - \lambda^2}\sin 2\omega,$$

[1] In the arguments of the long and short periodic perturbations ω and M must be calculated with the first order secular perturbations only,

$$\omega = \omega_0 + [\delta_1 \omega], \qquad M = M_0 + n_0(t - t_0) + [\delta_1 M].$$

$$\{\delta_1\omega\} = \frac{1}{16}\,C_{20}\times$$

$$\times\,\alpha^2\,\frac{104\,\lambda^2 - 250\,\lambda^4 + 150\,\lambda^6 + e^2(-104 + 620\,\lambda^2 - 1055\,\lambda^4 + 550\,\lambda^6)}{(4 - 5\,\lambda^2)^2}\,\sin 2\omega,$$

$$\{\delta_1 M\} = -\frac{1}{16}\,C_{20}\,\alpha^2\,\frac{26\,\lambda^2 - 30\,\lambda^4 + e^2(17\,\lambda^2 - 20\,\lambda^4)}{4 - 5\,\lambda^2}\,\sin 2\omega.$$

3. Short periodic perturbations

$$\frac{(\delta_1 a)}{a_0} = -\frac{3}{2}\,C_{20}\,\alpha^2(2 - 3\lambda^2)\left[\left(\ +\frac{9}{8}\,e^3 + \frac{87}{64}\,e^5\right)\cos M\ +\right.$$

$$+\frac{3}{2}\left(e^2 + \frac{7}{9}\,e^4\right)\cos 2M + \frac{53}{24}\left(e^3 + \frac{393}{848}\,e^5\right)\cos 3M\ +$$

$$+\frac{77}{24}\,e^4\cos 4M + \frac{591}{128}\,e^5\cos 5M\right] - \frac{3}{2}\,C_{20}\,\alpha^2\,\lambda^2\ \times$$

$$\times\left[-\frac{1}{2}\left(e - \frac{1}{8}\,e^3 + \frac{5}{192}\,e^5\right)\cos(M + 2\omega) + \frac{1}{48}\left(e^3 + \frac{11}{16}\,e^5\right)\times\right.$$

$$\times\,\cos(M - 2\omega) + \left(1 - \frac{5}{2}\,e^2 + \frac{13}{16}\,e^4\right)\cos(2M + 2\omega)\ +$$

$$+\frac{1}{24}\,e^4\cos(2M - 2\omega) + \frac{7}{2}\left(e - \frac{123}{56}\,e^3 + \frac{489}{448}\,e^5\right)\times$$

$$\times\,\cos(3M + 2\omega) + \frac{81}{1280}\,e^5\cos(3M - 2\omega)\ +$$

$$+\frac{17}{2}\left(e^2 - \frac{115}{51}\,e^4\right)\cos(4M + 2\omega) + \frac{845}{48}\left(e^3 - \frac{6505}{2704}\,e^5\right)\times$$

$$\times\,\cos 5(M + 2\omega) + \frac{533}{16}\,e^4\cos(6M + 2\omega)\ +$$

$$\left. +\,\frac{228347}{3840}\,e^5\cos(7M + 2\omega)\right], \tag{9}$$

$$(\delta_1 e) = -\frac{3}{4}\,C_{20}\,\alpha^2(2 - 3\lambda^2)\left[\left(1 + \frac{1}{8}\,e^2 + \frac{15}{64}\,e^4\right)\cos M\ +\right.$$

$$+\frac{3}{2}\left(e - \frac{2}{9}\,e^3 + \frac{29}{144}\,e^5\right)\cos 2M + \frac{53}{24}\left(e^2 - \frac{455}{848}\,e^4\right)\cos 3M\ +$$

$$+\frac{77}{24}\left(e^3 - \frac{641}{770}\,e^5\right)\cos 4M + \frac{591}{128}\,e^4\cos 5M + \frac{3167}{480}\,e^5\cos 6M\right]-$$

$$-\frac{3}{4}\,C_{20}\,\alpha^2\,\lambda^2\left[\frac{1}{2}\left(1 - \frac{1}{8}\,e^2 - \frac{43}{192}\,e^4\right)\cos(M + 2\omega)\ +\right.$$

$$+\frac{1}{16}\left(e^2 + \frac{1}{48}\,e^4\right)\cos(M - 2\omega) - \frac{1}{2}\left(e - \frac{11}{4}\,e^3 + \frac{21}{16}\,e^5\right)\times$$

$$\times\,\cos(2M + 2\omega) + \frac{1}{12}\left(e^3 - \frac{1}{20}\,e^5\right)\cos(2M - 2\omega)\ +$$

$$+\frac{7}{6}\left(1 - \frac{235}{56}\,e^2 + \frac{2569}{448}\,e^4\right)\cos(3M + 2\omega) + \frac{27}{256}\,e^4\ \times$$

$$\times \cos(3M - 2\omega) + \frac{17}{4}\left(e - \frac{383}{102}e^3 + \frac{254}{51}e^5\right)\cos(4M + 2\omega) +$$

$$+ \frac{2}{15}e^5 \cos(4M - 2\omega) + \frac{169}{16}\left(e^2 - \frac{30331}{8112}e^4\right)\cos(5M + 2\omega) +$$

$$+ \frac{533}{24}\left(e^3 - \frac{40979}{10660}e^5\right)\cos(6M + 2\omega) + \frac{32621}{768}e^4 \cos(7M + 2\omega) +$$

$$+ \frac{73369}{960}e^5 \cos(8M + 2\omega)\Big],$$

$$(\delta_1 I) = \frac{1}{2}\cot I \frac{(\delta_1 a)}{a} - \cot I \frac{e}{1 - e^2}(\delta_1 e),$$

$$(\delta_1 \Omega) = \frac{3}{2}C_{20}\alpha^2 \sqrt{1 - \lambda^2}\left[3\left(e + \frac{13}{8}e^3 + \frac{147}{64}e^5\right)\sin M +\right.$$

$$+ \frac{9}{4}\left(e^2 + \frac{23}{18}e^4\right)\sin 2M + \frac{53}{24}\left(e^3 + \frac{817}{848}e^5\right)\sin 3M +$$

$$+ \frac{77}{32}e^4 \sin 4M + \frac{1773}{640}e^5 \sin 5M + \frac{1}{2}\left(e + \frac{3}{8}e^3 + \frac{65}{192}e^5\right)\times$$

$$\times \sin(M + 2\omega) - \frac{1}{48}\left(e^3 + \frac{19}{16}e^5\right)\sin(M - 2\omega) -$$

$$- \frac{1}{2}\left(1 - 2e^2 - \frac{1}{16}e^4\right)\sin(2M + 2\omega) - \frac{1}{48}e^4 \sin(2M - 2\omega) -$$

$$- \frac{7}{6}\left(e - \frac{95}{56}e^3 + \frac{165}{448}e^5\right)\sin(3M + 2\omega) -$$

$$- \frac{27}{1280}e^5 \sin(3M - 2\omega) - \frac{17}{8}\left(e^2 - \frac{179}{102}e^4\right)\sin(4M + 2\omega) -$$

$$- \frac{169}{48}\left(e^3 - \frac{5153}{2704}e^5\right)\sin(5M + 2\omega) - \frac{533}{96}e^4 \sin(6M + 2\omega) -$$

$$- \frac{32621}{3840}e^5 \sin(7M + 2\omega)\Big],$$

$$e(\delta_1 \omega) = -\frac{3}{4}C_{20}\alpha^2(2 - 3\lambda^2)\left[\left(1 + \frac{23}{8}e^2 + \frac{319}{64}e^4\right)\sin M +\right.$$

$$+ \frac{3}{2}\left(e + \frac{19}{18}e^3 + \frac{293}{144}e^5\right)\sin 2M + \frac{53}{24}\left(e^2 + \frac{231}{848}e^4\right)\sin 3M +$$

$$+ \frac{77}{24}\left(e^3 - \frac{383}{1540}e^5\right)\sin 4M + \frac{591}{128}e^4 \sin 5M +$$

$$+ \frac{3167}{480}e^5 \sin 6M\Big] -$$

$$- \frac{3}{2}C_{20}\alpha^2 \lambda^2\left[-\frac{1}{4}\left(1 - \frac{7}{8}e^2 + \frac{37}{192}e^4\right)\sin(M + 2\omega) +\right.$$

$$+ \frac{1}{32}\left(e^2 + \frac{31}{48}e^4\right)\sin(M - 2\omega) - \frac{5}{4}\left(e - \frac{23}{20}e^3 + \frac{83}{240}e^5\right)\times$$

$$\times \sin(2M + 2\omega) + \frac{1}{24}\left(e^3 + \frac{11}{20}e^5\right)\sin(2M - 2\omega) +$$

$$+ \frac{7}{12}\left(1 - \frac{397}{56} e^2 + \frac{3865}{448} e^4\right) \sin(3M + 2\omega) + \frac{27}{512} e^4 \times$$

$$\times \sin(3M - 2\omega) + \frac{17}{8}\left(e - \frac{511}{102} e^3 + \frac{334}{51} e^5\right) \sin(4M + 2\omega) +$$

$$+ \frac{1}{15} e^5 \sin(4M - 2\omega) + \frac{169}{32}\left(e^2 - \frac{36581}{8112} e^4\right) \sin(5M + 2\omega) +$$

$$+ \frac{533}{48}\left(e^3 - \frac{46811}{10660} e^5\right) \sin(6M + 2\omega) + \frac{32621}{1536} e^4 \times$$

$$\times \sin(7M + 2\omega) + \frac{73369}{1920} e^5 \sin(8M + 2\omega)\Big] -$$

$$- \frac{3}{2} C_{20}\, \alpha^2 (1 - \lambda^2)\left[3\left(e^2 + \frac{13}{8} e^4\right)\sin M + \frac{9}{4}\left(e^3 + \frac{23}{18} e^5\right) \times\right.$$

$$\times \sin 2M + \frac{53}{24} e^4 \sin 3M + \frac{77}{32} e^5 \sin 4M +$$

$$+ \frac{1}{2}\left(e^2 + \frac{3}{8} e^4\right)\sin(M + 2\omega) - \frac{1}{48} e^4 \sin(M - 2\omega) -$$

$$- \frac{1}{2}\left(e - 2e^3 - \frac{1}{16} e^5\right)\sin(2M + 2\omega) - \frac{1}{48} e^5 \times$$

$$\times \sin(2M - 2\omega) - \frac{7}{6}\left(e^2 - \frac{95}{56} e^4\right)\sin(3M + 2\omega) -$$

$$- \frac{17}{8}\left(e^3 - \frac{179}{102} e^5\right)\sin(4M + 2\omega) - \frac{169}{48} e^4 \times$$

$$\left.\times \sin(5M + 2\omega) - \frac{533}{96} e^5 \sin(6M + 2\omega)\right],$$

$$(\delta_1 M) + (\delta_1 \omega) = -\frac{3}{2} C_{20}\, \alpha^2 (2 - 3\lambda^2)\left[\frac{7}{4}\left(e + \frac{79}{56} e^3 + \frac{895}{448} e^5\right)\sin M +\right.$$

$$+ \frac{3}{2}\left(e^2 + \frac{131}{144} e^4\right)\sin 2M + \frac{53}{32}\left(e^3 + \frac{1229}{2544} e^5\right)\sin 3M +$$

$$\left.+ \frac{385}{192} e^4 \sin 4M + \frac{6501}{2560} e^5 \sin 5M\right] -$$

$$- \frac{3}{2} C_{20}\, \alpha^2 \lambda^2\left[-\frac{7}{8}\left(e - \frac{11}{56} e^3 + \frac{7}{192} e^5\right)\sin(M + 2\omega) +\right.$$

$$+ \frac{3}{64}\left(e^3 + \frac{109}{144} e^5\right)\sin(M - 2\omega) + \frac{3}{4}\left(1 - \frac{10}{3} e^2 + \frac{25}{16} e^4\right) \times$$

$$\times \sin(2M + 2\omega) + \frac{5}{96} e^4 \sin(2M - 2\omega) +$$

$$+ \frac{49}{24}\left(e - \frac{1121}{392} e^3 + \frac{6061}{3136} e^5\right)\sin(3M + 2\omega) + \frac{297}{5120} e^5 \times$$

$$\times \sin(3M - 2\omega) + \frac{17}{4}\left(e^2 - \frac{2351}{816} e^4\right)\sin(4M + 2\omega) +$$

$$+ \frac{507}{64}\left(e^3 - \frac{73583}{24336} e^5\right)\sin(5M + 2\omega) + \frac{2665}{192} e^4 \sin(6M + 2\omega) +$$

$$+ \frac{358\,831}{15\,360} e^5 \sin(7M + 2\omega) \Bigg] -$$

$$- \frac{3}{2} C_{20} \alpha^2 (1 - \lambda^2) \Bigg[3 \left(e + \frac{13}{8} e^3 + \frac{147}{64} e^5 \right) \sin M +$$

$$+ \frac{9}{4} \left(e^2 + \frac{23}{18} e^4 \right) \sin 2M + \frac{53}{24} \left(e^3 + \frac{817}{848} e^5 \right) \sin 3M + \frac{77}{32} e^4 \times$$

$$\times \sin 4M + \frac{1773}{640} e^5 \sin 5M + \frac{1}{2} \left(e + \frac{3}{8} e^3 + \frac{65}{192} e^5 \right) \times$$

$$\times \sin(M + 2\omega) - \frac{1}{48} \left(e^3 + \frac{19}{16} e^5 \right) \sin(M - 2\omega) -$$

$$- \frac{1}{2} \left(1 - 2e^2 - \frac{1}{16} e^4 \right) \sin(2M + 2\omega) - \frac{1}{48} e^4 \sin(2M - 2\omega) -$$

$$- \frac{7}{6} \left(e - \frac{95}{56} e^3 + \frac{165}{448} e^5 \right) \sin(3M + 2\omega) - \frac{27}{1280} e^5 \times$$

$$\times \sin(3M - 2\omega) - \frac{17}{8} \left(e^2 - \frac{179}{102} e^4 \right) \sin(4M + 2\omega) -$$

$$- \frac{169}{48} \left(e^3 - \frac{5153}{2704} e^5 \right) \sin(5M + 2\omega) - \frac{533}{96} e^4 \times$$

$$\times \sin(6M + 2\omega) - \frac{32\,621}{3840} e^5 \sin(7M + 2\omega) \Bigg].$$

The expressions received for secular and long periodic perturbations have been compared with the related expressions by BROUWER (1959). The constants of integration of our theory up to the quantities of order C_{20} through are identical with the constants of BROUWER's theory. The difference can be presented only in the terms of the second order and can influence only the mean anomaly. The secular terms of the both theories are to be identical in all elements but M. In order to compare the long periodic terms one must develop short periodic terms of BROUWER's theory with the true anomaly as an argument to FOURIER series in multiples of the mean anomaly. The constant terms of these developments which are periodic functions of the argument of perigee are to be added to the long periodic terms of BROUWER's theory. The so received long periodic parts with assumed accuracy must coincide with our long periodic terms.

The secular terms of the mean anomaly in both theories are to be different by two causes: firstly, because the secular terms of the mean anomaly in BROUWER's theory were derived under the inexplicit supposition that the arguments of the short periodic perturbations of the semimajor axis depend on the perturbed mean anomaly and the argument of perigee and, secondly, because BROUWER defines the nonperturbed mean motion n_0 through L_0 which is the mean value of element $L = \sqrt{f\,m\,a}$ with respect to the true anomaly. There exists the following relation between the nonperturbed mean motions of BROUWER's theory and our theory

$$n_{0\,(\text{BROUWER})} = n_{0\,(\text{our})} \left(1 + \frac{3}{2}\left[\delta_1\,\omega\,\frac{\partial}{\partial\omega}\,\frac{\delta_1\,a}{a_0}\,+\right.\right.$$
$$\left.\left.+\,\delta_1\,M\,\frac{\partial}{\partial M}\,\frac{\delta_1\,a}{a_0}\right] + \frac{3}{8}\left[\frac{\delta_1^2\,a}{a_0^2}\right]\right),$$

where the square brackets denote the constant terms of the expressions within, and $\delta_1\,\omega$, $\delta_1\,a$, $\delta_1\,M$ are the short periodic parts of the corresponding elements with the mean anomaly as an argument.

The calculations using formulae (9) give

$$n_0\left[\delta_1\,\omega\,\frac{\partial}{\partial\omega}\,\frac{\delta_1\,a}{a_0} + \delta_1\,M\,\frac{\partial}{\partial M}\,\frac{\delta_1\,a}{a_0}\right]$$
$$= \frac{9}{8}\,C_{20}^2\,\alpha^4\,n_0\left\{2 - 5\lambda^2 + \frac{47}{12}\,\lambda^4 + e\left(10 - \frac{155}{6}\,\lambda^2 + \frac{45}{2}\,\lambda^4\right)\right\},$$

$$n_0\left[\frac{\delta_1^2\,a}{a_0^2}\right] = \frac{9}{8}\,C_{20}^2\,\alpha^4\,n_0\left\{\lambda^4 + e^2\left(4 - 12\,\lambda^2 + \frac{33}{2}\,\lambda^4\right)\right\}.$$

The comparison of our results with those of BROUWER's taking into account the above mentioned considerations has shown these results to coincide with assumed accuracy.

More detailed development of the method and the results of this paper including the perturbations from the other zonal harmonics will be published in Bulletin of the Institute of Theoretical Astronomy (U.S.S.R).

References

[1] PROSKURIN, V. F., and YU. V. BATRAKOV (1959): Iskusstvennye sputniki Zemli, N 3 (in russian).
[2] PROSKURIN, V. F., and YU. V. BATRAKOV (1960): Bulleten Instituta Teoreticheskoy Astronomii AN SSSR, 7, N 7 (in russian).
[3] BROUWER, D. (1959): Astronom. J. 64, N 9.
[4] KOZAI, Y. (1959): Astronom. J. 64, N 9.

A Pegasus computer programme for the improvement of the orbital parameters of an earth-satellite

By

R. H. Merson

Royal Aircraft Establishment, Farnborough, England

Abstract. The dynamic model adopted for the motion of a satellite includes the first order short-period and secular perturbations of the orbital elements due to earth's oblateness and other perturbations of the elements are represented by polynomials in time up to the third degree.

The values of the six basic elements at a given time, together with such polynomial coefficients as are required, are determined from a set of observations by a differential correction technique. In this procedure the observation times and angles are weighted according to a priori estimates of their accuracy. In addition to the estimates of the orbital parameters, estimates of their variances and covariances are also determined.

1. Introduction

During the early weeks in the lifetime of *Sputnik 2* (November, 1957) a *Pegasus* digital computer programme was prepared for analysing a set of kinetheodolite or theodolite observations made from one station during a single transit of the satellite [1]. This programme required prior knowledge of the semi-major axis (or period) and the eccentricity (or perigee height) of the satellite and was restricted in its accuracy and usage, although reasonably accurate tables of orbital elements of *Sputnik 2* [2], *Sputnik 3* [3] and *Sputnik 3 Rocket* [4] were obtained with its aid.

For various reasons it was decided, in the summer of 1960, to start developing a more sophisticated approach to the orbit improvement problem. KING-HELE was asking for more accurate orbits in connection with his air density studies and the author was interested in the work of improving the knowledge of the earth's gravitational field. In the latter case, it is essential to use the same basic theory of satellite motion in determining the orbital elements at a given time and in combining these results to study their long-period changes. For this reason, the author developed a theory of the effect of the earth's gravitational field on the motion of a satellite [5] and this was used as the basis for the dynamical model in the new computer programme.

The problem of orbit determination on a computer can be broken down into three main parts. First, the punching and storage of observational data; second, the selection and preliminary processing of suitable observations in the vicinity of an epoch at which the orbit elements are required; and finally, the use of these observations to improve the orbital elements by an iterative procedure.

In the case of the *R.A.E. Pegasus* computer, a Mk.I type with a 5192—word drum and magnetic tape backing store, it was decided to store data (station coordinates and observations) on magnetic tape, in other words, to form a "magnetic tape library" for each satellite. The programme for initiating and updating a magnetic tape library was prepared by R. J. TAYLER. The programme for selecting specified observations from the library tape, processing them in a manner to be explained later, and storing the processed data on a second magnetic tape, was prepared by R. H. GOODING. The orbit improvement programme was written by R. H. GOODING and the author. The first successful run of the programme was made in January, 1962 on a set of observations of Sputnik 2. Sufficient results have been obtained from the analysis of Sputnik 2 observations to show that the previously published elements [2] lie within the quoted standard deviations.

2. The magnetic tape library

The magnetic tape library is in three sections. First, copies of the *library update* and *select/process* programmes; second, a list of stations with their coordinates; and third, a list of observational data. Originally, all this information is punched on paper tape.

For a specified station, the data punched consists of a station serial number, a spheroid letter, and the longitude, latitude and height (above the spheroid) of the station. Various spheroids, including those of AIRY, CLARKE and HAYFORD may be used. On input to the computer, a block of 8 items is formed, comprising the station number, the three cartesian coordinates, the sine and cosine of the latitude and the sine and cosine of the longitude. This block is stored on the magnetic tape.

For a specified observation, the data punched consists of eight items:

Observation No.	Density factor
Station No.	Date and time
Burst No.	Azimuth or right ascension
Code index	Elevation or declination.

Three of these items require some explanation. Two or more observations from one station are considered to belong to a 'burst' if it is known that their time errors are highly correlated, e.g. observations made during one transit.

The code index used is the same as that used by the Smithsonian Institution Astronomical Observatory [6]. It is a six decimal-digit number in which the first digit indicates the precision of the time measurement, the second and third together indicate the precision of the angular measurements, the fourth indicates whether the coordinate system is R.A./Dec. or azimuth/elevation and whether or not refraction corrections have been applied, the fifth digit indicates the epoch of the equator and equinox and the sixth digit indicates the type of instrumentation.

The density factor is the ratio of measured air density to standard air density, for use in refraction correction. So far, however, no such measurements appear to have been made and the density factor is set to unity.

On input to the computer, the eight data items are scaled and stored on the magnetic tape.

As well as storing new data, it is also possible to correct data already on the tape.

3. The programme for selecting and processing observations

3.1 Selection

In order to use the programme for selecting and processing observations it is necessary to have a parameter tape on which are punched estimates of the orbital parameters (defined in section 4 below) at some particular ascending node and a list of the observations to be used in improving these parameters. The computer scans the magnetic tape library for these observations and transfers the blocks of data to a second magnetic tape, the "selected data tape", at the same time storing on the drum of the computer the station numbers corresponding to these observations. The library tape is then scanned for the required station data, which is transferred to the drum.

3.2 Processing

In the processing procedure, the data for each observation is combined with the corresponding station data to form a single block of information. The main steps in the procedure are the transformation of station coordinates to space axes, the application of refraction corrections where required, the transformation of angular coordinates to right ascension and declination and the calculation of weights to be allotted to the observations.

The system of space axes used is the true equator and equinox of date.

3.2.1 Apparent sidereal time. In order to transform the observations to space axes, the apparent sidereal time is required. Apparent sideral time is obtained from mean sidereal time by the addition of the equation of the equinoxes (the "nutation in right ascension") [7]. Denoting by t'_s the mean sidereal time in days and by t the observation time in days U.T.,

$$t'_s = 0.2760837847 + 1.0027379093\,(t - 1957\ \text{JAN}\ 0.0)$$

to the accuracy required.

The equation of the equinoxes is $\Delta\psi\cos\varepsilon$, where $\Delta\psi$ is the nutation in longitude and ε the obliquity of the ecliptic.

Retaining only the two main terms in $\Delta\psi$, which should be adequate for the present purpose,

$$\Delta\psi = -17\overset{''}{.}2\sin\Omega - 1\overset{''}{.}3\sin 2L,$$

where

$$\Omega = 0.6576742222\,2 - 0.0001470940\,5\,(t - 1957\ \text{JAN}\ 0.0)$$

and

$$2L = 0.5521990444\,4 + 0.0054758185\,8\,(t - 1957\ \text{JAN}\ 0.0),$$

the unit here being one revolution (360°).

The obliquity of the ecliptic is taken to be constant in the equation of the equinoxes:

$$\varepsilon = 23\overset{\circ}{.}4444.$$

Finally, the apparent sidereal time t_s is given by

$$t_s = t'_s + \Delta\psi\cos\varepsilon,$$

where, in common with the Ref. [7], unnecessary extra symbols are not here introduced to represent the conversion from angular to time units.

In section 3.2.5 following, the nutation in obliquity is required, and this is given by

$$\Delta\varepsilon = 9\overset{''}{.}2\cos\Omega + 0\overset{''}{.}6\cos 2L.$$

3.2.2. Transformation of station coordinates. If X_0, Y_0, Z_0 are the geocentric cartesian coordinates of the station in earth axes, Z_0 being towards the North pole and X_0 towards the Greenwich meridian, the cartesian coordinates in space axes are

$$X = X_0\cos t_s - Y_0\sin t_s,$$

$$Y = X_0\sin t_s + Y_0\sin t_s,$$

$$Z = Z_0.$$

3.2.3 Refraction corrections. The refraction corrections are based on formulae given by GARFINKEL [8] and BROWN [9] (who refers to GARFINKEL's work).

There are two cases, according to whether total refraction is required or merely parallactic refraction.

For total refraction, if E' is the observed elevation and E the corrected elevation, E is calculated by the sequence of formulae:

$$8.4 \tan 2\eta = \cot E'$$

where
$$E_\infty = E' - (a \tan\eta + b \tan^3\eta + c \tan^5\eta + d \tan^7\eta),$$

$$a = 956.''513, \quad b = 657.''638, \quad c = 179.''109, \quad d = 248.''314,$$

$$(1 + h\, R_s^{-1}) \cos\gamma = 1.000276454 \cos E',$$

$$\Theta = \gamma - E_\infty,$$

$$\tan E_0 = \cot\Theta - \operatorname{cosec}\Theta\,(1 + h\, R_s^{-1})^{-1},$$

$$E = E' - (E' - E_0)\, \bar\varrho,$$

where R_s is the distance of the station from the centre of the earth, $h = r - R_s$ where r is the distance of the satellite from the centre of the earth, and $\bar\varrho$ is the density factor.

The radial distance r of the satellite is computed from the estimated orbit parameters.

In the case of parallactic refraction only, where E_∞ is observed,

$$8.4 \tan 2\eta = \cot E_\infty,$$

where
$$E' = E_\infty + a \tan\eta + b \tan^3\eta + c \tan^5\eta + d \tan^7\eta,$$

$$a = 956.''426, \quad b = 564.''095, \quad c = 175.''239, \quad d = 8.''216,$$

the remaining formulae being as above.

3.2.4 Conversion of elevation and azimuth to right ascension and declination. Topocentric right ascension α and declination δ are obtained from elevation E and azimuth A by the transformation

$$
\begin{pmatrix} \cos\alpha \, \cos\delta \\ \sin\alpha \, \cos\delta \\ \sin\delta \end{pmatrix}
=
\begin{pmatrix}
-\sin\beta \, \cos\lambda, & -\sin\lambda, & \cos\beta \, \cos\lambda \\
-\sin\beta \, \sin\lambda, & \cos\lambda, & \cos\beta \, \sin\lambda \\
\cos\beta, & 0, & \sin\beta
\end{pmatrix}
\begin{pmatrix} \cos A \, \cos E \\ \sin A \, \cos E \\ \sin E \end{pmatrix},
$$

where β is the latitude of the station, and $\lambda = \lambda_0 + t_s$, where λ_0 is the longitude of the station.

3.2.5 Transformation of right ascension and declination. When right ascension and declination are quoted relative to the mean equator and equinox at one of the Epochs 1855.0, 1875.0, 1900.0 or 1950.0 (the Epochs of the main Star Catalogues), a transformation is required to relate them to the true equator and equinox of date.

If α_0, δ_0 are the quoted values, and α, δ the values after the trans-formation, then [7]

$$\begin{pmatrix} \cos\alpha & \cos\delta \\ \sin\alpha & \cos\delta \\ & \sin\delta \end{pmatrix} = A_1 A_2 A_3 \begin{pmatrix} \cos(\alpha_0 + \xi_0) & \cos\delta_0 \\ \sin(\alpha_0 + \xi_0) & \cos\delta_0 \\ & \sin\delta_0 \end{pmatrix},$$

where A_1, A_2, A_3 are the matrices

$$A_1 \equiv \begin{pmatrix} 1, & -\Delta\psi\cos\varepsilon, & -\Delta\psi\sin\varepsilon \\ \Delta\psi\cos\varepsilon, & 1, & -\Delta\varepsilon \\ \Delta\psi\sin\varepsilon, & \Delta\varepsilon, & 1 \end{pmatrix},$$

and

$$A_2 \equiv \begin{pmatrix} \cos z, & -\sin z, & 0 \\ \sin z, & \cos z, & 0 \\ 0, & 0, & 1 \end{pmatrix}, \quad A_3 \equiv \begin{pmatrix} \cos\Theta, & 0, & -\sin\Theta \\ 0, & 1, & 0 \\ \sin\Theta, & 0, & \cos\Theta \end{pmatrix},$$

$$\left.\begin{aligned} \xi_0 &= a\,d + 2.2641 \times 10^{-10}\ d^2, \\ z &= \xi_0 + 5.9292 \times 10^{-10}\ d^2, \\ \Theta &= e\,d - 3.1933 \times 10^{-10}\ d^2, \end{aligned}\right\}$$

where d is the time in days from the quoted Epoch to date, and a, e are given by the following table:

Epoch	1855.0	1875.0	1900.0	1950.0
a	0.06307107	0.06307872	0.06308828	0.06310739
e	0.05489688	0.05489221	0.05488637	0.05487469

In the matrix A_1, $\Delta\psi$ and $\Delta\varepsilon$ are the nutations in longitude and obliquity as before, but expressed in radians.

3.2.6 The aberration/light-time correction. In the case of the most accurate observations errors due to the finite speed of light are not quite negligible. At time t let the satellite be at Q and the ground station at P. Then the light leaving Q reaches the ground station at time $t + \Delta t$, when it is at P', say. The observed time is $t + \Delta t$, and the observed station position is P', but due to the relative motion of station and sat-ellite, the apparent direction of the satellite is not along $P'Q$ but is parallel to PQ. We can therefore reproduce the condition of viewing Q from P at time t instantaneously by subtracting Δt from the observed time and rotating the station coordinates from P' to P.

The distance PQ is given by

$$PQ = -B + (B^2 + C)^{1/2},$$

where

$$B = X\cos\delta\cos\alpha + Y\cos\delta\sin\alpha + Z\sin\delta$$

and
$$C = r^2 - (X^2 + Y^2 + Z^2).$$

Then
$$\Delta t = P\,Q/c,$$

where
$$c = 299978 \text{ km/sec. is the speed of light.}$$

If w is the angular velocity of the earth, the corrected station coordinates are

$$X' = X + w \cdot \Delta t\, Y,$$

$$Y' = Y - w \cdot \Delta t\, X.$$

3.2.7 The weights of the observations. The weights of the observations are here defined as the multipliers of the residuals in the function to be minimised in the correction procedure (section 4.1 below), i. e. they are the square roots of the quantities usually defined as weights. In the normal procedure, the weights of the time and declination observations are the reciprocals of a priori estimates of the standard deviations of the observations, whilst the weight of a right ascension observation is equal to the weight of the corresponding declination observation multiplied by the cosine of the declination. In the computer the weights are determined from the code index word via a table look-up procedure.

3.2.8 The processed information block. The information stored on the selected data tape relating to each observation consists of eight computer words. Of these eight items, three are the corrected station coordinates, three are the corrected t, α and δ, one is zero and one is a special compound word.

The zero is an initial estimate of the residual in the time observation (introduced in section 4 below).

The first digit of the special compound word is 1 if the observation is the first of a burst, and 0 if not; the second is 0, but becomes 1 if the observation is rejected during orbit improvement; the third is 1 if the observed angles were azimuth and elevation, and 0 if they were right ascension and declination; the rest of the word contains the three weights.

4. The orbit improvement programme

4.1 The differential correction procedure

Observations are assumed to occur in bursts such that in each burst the time errors are equal. Let α_{ij}, δ_{ij}, t_{ij} denote the right ascension, declination and time of the jth observation in the ith burst. If there are l undetermined orbital parameters E_k $(k = 1, 2, \ldots, l)$ in the dynamic model and n observations, and if there were no errors in the observations or the model, then the observations would be related to the parameters by $2n$ equations of the form

$$\alpha_{ij} = \alpha(E_1, E_2, \ldots E_l, t_{ij}),$$

$$\delta_{ij} = \delta(E_1, E_2, \ldots, E_l, t_{ij}),$$

the functions α, δ depending on the dynamic model, to be described in section 4.3. When $2n > l$, the orbital parameters are over-determined by these equations and a least squares procedure is used to determine best values.

Let α_{ijc}, δ_{ijc}, t_{ijc} be values (called computed values) close to the observed values, such that

$$\alpha_{ijc} = \alpha(E_1, E_2, \ldots, E_l, t_{ijc}),$$

$$\delta_{ijc} = \delta(E_1, E_2, \ldots, E_l, t_{ijc}),$$

where $t_{ijc} - t_{ij}$ is independent of j.

Then the procedure adopted is to determine the values of the E_k and the t_{ijc} so as to minimise the quantity

$$H = \sum_{i,j} w_{\alpha ij}^2 (\alpha_{ijc} - \alpha_{ij})^2 + \sum_{i,j} w_{\delta i}^2 (\delta_{ijc} - \delta_{ij})^2 + \sum_{i} w_{ti}^2 (t_{ijc} - t_{ij})^2,$$

where $w_{\alpha ij}$, $w_{\delta i}$ and w_{ti} are the weights of the observations. It will be noted that $w_{\delta i}$, w_{ti} are independent of j, but $w_{\alpha ij} = w_{\delta i} \cos \delta_{ij}$ is not.

Suppose that $E_1, E_2, \ldots, E_l$ are current estimates of the orbital parameters and t_{ijc} current estimates of the times and let

$$\alpha_{ijc} = \alpha(E_1, E_2, \ldots, t_{ijc}),$$

$$\delta_{ijc} = \delta(E_1, E_2, \ldots, t_{ijc}),$$

$$R_{ij} = \alpha_{ijc} - \alpha_{ij},$$

$$S_{ij} = \delta_{ijc} - \delta_{ij},$$

$$T_i = t_{ijc} - t_{ij},$$

so that

$$H = \sum_{i,j} w_{\alpha ij}^2 R_{ij}^2 + \sum_{i,j} w_{\delta i}^2 S_{ij}^2 + \sum_{i} w_{ii}^2 T_i^2.$$

Consider now small changes δE_k and δt_i in E_k and t_{ijc} and let

$$\alpha'_{ijc} = \alpha(E'_1, E'_2, \ldots, t'_{ijc}),$$

$$\delta'_{ijc} = \delta(E'_1, E'_2, \ldots, t'_{ijc}),$$

where

$$t'_{ijc} = t_{ijc} + \delta t_i,$$

$$E'_k = E_k + \delta E_k.$$

Then, taking the TAYLOR expansions about the current point,

$$\alpha'_{ijc} = \alpha_{ijc} + \sum_k p_{kij}\,\delta E_k + f_{ij}\,\delta t_i + o(\delta E_k^2,\,\delta t_i^2),$$

$$\delta'_{ijc} = \delta_{ijc} + \sum_k q_{kij}\,\delta E_k + g_{ij}\,\delta t_i + o(\delta E_k^2,\,\delta t_i^2),$$

where

$$p_{kij} = \frac{\partial \alpha(E_1,\dots,t_{ijc})}{\partial E_k}, \qquad q_{kij} = \frac{\partial \delta(E_1,\dots,t_{ijc})}{\partial E_k},$$

$$f_{ij} = \frac{\partial \alpha(E_1,\dots,t_{ijc})}{\partial t_{ijc}}, \qquad g_{ij} = \frac{\partial \delta(E_1,\dots,t_{ijc})}{\partial t_{ijc}}.$$

The correction formulae are based on the assumption that $(\delta E_k)^2$, $(\delta t_i)^2$ are negligible. In practice this is not generally true, and the minimisation of H is obtained by repeated application of the procedure.

Let

$$R'_{ij} = \alpha'_{ijc} - \alpha_{ij},$$

$$S'_{ij} = \delta'_{ijc} - \delta_{ij},$$

$$T'_i = t'_{ijc} - t_{ij},$$

so that, neglecting $(\delta E_k)^2$, $(\delta t_i)^2$,

$$R'_{ij} = R_{ij} + \sum_k p_{kij}\,\delta E_k + f_{ij}\,\delta t_i, \tag{1}$$

$$S'_{ij} = s_{ij} + \sum_k q_{kij}\,\delta E_k + g_{ij}\,\delta t_i, \tag{2}$$

$$T'_i = T_i + \delta t_i, \tag{3}$$

and

$$H' = \sum_{i,j}(w_{\alpha ij}^2 R_{ij}'^2 + w_{\delta i}^2 s_{ij}'^2) + \sum_i w_{ti}^2 T_i'^2. \tag{4}$$

It is required to determine δE_k, δt_i so as to minimise H'. For this purpose, matrix notation is more convenient.

Let

$$
\begin{vmatrix}
w_{\alpha i1} & R_{i1} \\
w_{\alpha i2} & R_{i2} \\
\vdots & \\
w_{\delta i} & S_{i1} \\
w_{\delta i} & S_{i2} \\
\vdots &
\end{vmatrix} = Y_i,
\qquad
\begin{vmatrix}
w_{\alpha i1} & p_{1i1}, & w_{\alpha i1} & p_{2i1}, & \cdots \\
w_{\alpha i2} & p_{1i2}, & w_{\alpha i2} & p_{2i2}, & \cdots \\
\cdots & \cdots & \cdots & \cdots & \cdots \\
w_{\delta i} & q_{1i1}, & w_{\delta i} & q_{2i1}, & \cdots \\
w_{\delta i} & q_{1i2}, & w_{\delta i} & q_{2i2}, & \cdots \\
\cdots & \cdots & \cdots & \cdots & \cdots
\end{vmatrix} = M_i,
$$

$$
\begin{pmatrix}
\delta E_1 \\
\delta E_2 \\
\vdots
\end{pmatrix} = Z,
\qquad
\begin{vmatrix}
w_{\alpha i1} & f_{i1} \\
w_{\alpha i2} & f_{i2} \\
\vdots & \\
w_{\delta i} & g_{i1} \\
w_{\delta i} & g_{i2} \\
\vdots &
\end{vmatrix} = w_{ti}\,U_i, \qquad w_{ti}\,T_i = k_i,
$$

$$w_{ti}\,\delta t_i = \tau_i;$$

then Eqs. (1) to (4) become

$$Y'_i = Y_i + M_i Z + \tau_i U_i, \tag{5}$$

$$k'_i = k_i + \tau_i, \tag{6}$$

and

$$H' = \sum_i (Y'^T_i Y'_i + k'^2_i), \tag{7}$$

the index T standing for transposition.

The minimisation conditions are $\partial H'/\partial(\delta E_k) = 0$, $k = 1, 2, \ldots, l$

and $\partial H'/\partial \tau_i = 0$, $i = 1, 2, \ldots, m$,

where m is the number of bursts, l the number of orbital parameters, and these lead to the relations

$$\sum_i M^T_i Y'_i = 0 \tag{8}$$

$$U^T_i Y'_i + k'_i = 0. \tag{9}$$

Eqs. (5) to (9) are to be solved for Z and τ_i.

Let

$M^T_i M_i = A_i$ (an $l \times l$ symmetric matrix)

$M^T_i U_i = B_i$ (an $l \times 1$ matrix)

$M^T_i Y_i = C_i$ (an $l \times 1$ matrix)

$d_i = k_i + U^T_i Y_i$ (a scalar)

$e_i = 1 + U^T_i U_i$ (a scalar)

$f_i = e^{-1}_i d_i$ (a scalar)

$G_i = e^{-1}_i B_i$ (an $l \times 1$ matrix);

then, pre-multiplying (5) by M^T_i and summing over i gives, with (8),

$$\sum_i (C_i + A_i Z + \tau_i B_i) = 0. \tag{10}$$

Substitution from (5) and (6) in (9) gives

$$d_i + B^T_i Z + \tau_i e_i = 0, \tag{11}$$

so that

$$\tau_i = -f_i - G^T_i Z, \tag{12}$$

and substitution for τ_i in (10) gives

$$Z = L^{-1} J, \tag{13}$$

where

$$J = \sum_i (f_i B_i - C_i),$$

$$L = \sum_i (A_i - B_i G^T_i).$$

(13) gives the increments to be added to the orbital parameters and (12) the increments to be added to the computed times of observation.

The improvement procedure is then repeated, using the new orbital parameters and computed times as the starting values.

In translating the sequence of matrix operations into a convenient computer procedure, the observations are introduced one at a time and the matrices are built up step-by-step. The largest matrix held in the computer at any time is a matrix of l rows and $(l + 2)$ columns which is the combined matrix $(C_i \, B_i \, A_i)$.

4.2 Estimation of accuracy

In order to estimate the accuracy of the parameters determined by the differential correction procedure it is necessary to introduce the true values of the parameters, $\overline{E}_k$, say, and the true values of the observed quantities $\overline{\alpha}_{ij}$, $\overline{\delta}_{ij}$, $\overline{t}_{ij}$. These true values are, of course, unknown and only estimates of them are obtained. The set of observed values α_{ij}, δ_{ij}, t_{ij} may be regarded as one sample from all the possible samples from a $(2n + m)$ — dimensional space. If E_k', α_{ij}', δ_{ij}', t_{ij}' is a fixed set of quantities close to the true values and such that

$$\alpha_{ij}' = \alpha(E_1', \ldots, E_l', t_{ij}'),$$
$$\delta_{ij}' = \delta(E_1', \ldots, E_l', t_{ij}'),$$

then for each set of possible samples of observed values and these as initial computed values, a single stage of the correction procedure could be applied to give estimates $E_k' + Z_k$ of the parameters, where Z_k is the kth element of Z as given by equation (13).

Now from Eq. (13) Z may be expressed in the form

$$Z = Q\,Y,$$

where

$$Y = \begin{pmatrix} Y_1 \\ Y_2 \\ \vdots \\ Y_m \\ k_1 \\ \vdots \\ k_m \end{pmatrix}$$

and Q is an $l \times (2n + m)$ matrix the elements of which may be regarded as the same for all samples. A typical element of Y_i here would be $w_{\alpha ij}(\alpha_{ij}' - \alpha_{ij})$.

Considering all possible samples, the mean of Y is $\overline{Y}$ where, for example, the typical element is $w_{\alpha ij}(\alpha'_{ij} - \overline{\alpha}_{ij})$, and the mean of Z is $\overline{Z} = Q\,\overline{Y}$. Then the covariance matrix of Z is

$$\text{cov}\,Z = \mathcal{E}\{(Z - \overline{Z})(Z^T - \overline{Z}^T)\} = Q\,\mathcal{E}\{(Y - \overline{Y})(Y^T - \overline{Y}^T)\}Q^T,$$

where $\mathcal{E}$ indicates the expectation.

Since the typical element of $Y - \overline{Y}$ is $w_{\alpha ij}(\overline{\alpha}_{ij} - \alpha_{ij})$, it follows that $\mathcal{E}\{(Y - \overline{Y})(Y^T - \overline{Y}^T)\}$ is the covariance matrix of the weighted observations. In order to determine this certain assumptions have to be made. It is assumed that (1) the dynamical model, with parameters $\overline{E}_k$, exactly fits the true observations, (2) there are no systematic errors, for example in station coordinates or observing instruments, (3) the random errors are normal and uncorrelated (except for the time errors in a burst) and (4) the weighted observations have a common variance ε^2, that is to say, the standard deviations of the observations are equal to their a priori estimates multiplied by the factor ε.

With these assumptions,

$$\mathcal{E}\{(Y - \overline{Y})(Y^T - \overline{Y}^T)\} = \varepsilon^2\,I,$$

where I is the $2n + m$ unit matrix.

Thus,
$$\text{cov}\,Z = \varepsilon^2\,Q\,Q^T,$$

or, since $Q\,Q^T = L^{-1}$, as proved in Appendix 1,

$$\text{cov}\,Z = \varepsilon^2\,L^{-1}.$$

Since Z differs from the vector of estimated parameters by a constant, this is also the covariance matrix of the estimated parameters.

It remains to estimate ε from the given sample of observations. An unbiassed estimate of ε is obtained directly from H, the weighted sum of squares of the residuals. H is the sum of $(2n + m)$ terms and there are $(l + m)$ quantities determined (the parameters E_k and the t_{ijc}) so that the system has $(2n - l)$ degrees of freedom. It follows that $\mathcal{E}(H) = (2n - l)\,\varepsilon^2$, and an unbiassed estimate of ε^2 is given by $H/(2n - l)$.

Thus finally, we take, as the estimated covariance matrix of parameters,
$$\text{cov}\,Z = L^{-1}\,H/(2n - l).$$

If the a priori estimates of accuracy of the observations are correct, and the assumptions (1) to (4) are satisfied, $\varepsilon = 1$ and its estimated value should be close to 1.

4.3 The dynamical model of the satellite orbit

The inner core of the orbit improvement programme is a subroutine which determines the satellite's position at a given time, for given values of the parameters of the orbit. If the satellite were moving in a simple 'square-law' field, its orbit would be completely specified by six parameters, the 'elements' of the orbit, three of which specify the spatial position of the plane of the orbit, two give the shape and size of the orbit and the sixth relates the orbital motion to actual time. In practice, the satellite is subject to small perturbations and its motion may be specified by giving the six elements as functions of time. Certain of the variations of the elements can be specified as known functions of the elements themselves, but others can only be expressed in terms of parameters which must be determined from the observations.

As soon as perturbations are included in a dynamical system it becomes necessary to define the basic parameters with some care. In many astronomical applications, the well-known system of osculating elements is satisfactory, but in the case of near earth-satellites the osculating elements are subject to large short-period variations due to the oblateness of the earth. In the present programme, the oblateness effects are treated by the method given by MERSON [5], in which the elements are defined in such a way as to minimise the amplitudes of the short-period variations. Other secular perturbations of the elements are represented, if required, by polynomials in time, the coefficients of which are additional parameters to be determined.

In a preliminary model which was tried, polynomials in both semi-major axis and mean anomaly were used. It was found, however, that observations over a few days could not provide satisfactory values of even the first-order coefficients (the simultaneous equations solved in the correction procedure being highly ill-conditioned). It was therefore decided to dispense with the polynomial in semi-major axis, and determine the latter from KEPLER's equation $n^2 a^3 = \mu$, with $n = dM/dt$ (a = semi-major axis, n = mean motion, M = mean anomaly, μ = constant).

In a similar way, it was found that, although a variation in eccentricity is required for an adequate model representing a satellite's motion over a few days, this variation cannot be determined from observations over as short a period as this, at any rate in the case of low-perigee satellites. In order to avoid the introduction of such an indeterminate parameter, but at the same time include the variation in eccentricity, use was made of the fact that the change in perigee distance over a few days is very small, i.e. $a(1 - e)$ = constant. Thus if $a(1 - e)$

$$= a_0(1 - e_0),$$

$$e = e_0 + \left(1 - \frac{a_0}{a}\right)(1 - e_0).$$

In addition, a polynomial in time may be added to the variation in e, if required.

The complete model is represented by the following formulae, which relate the polar coordinates r, $\bar{\alpha}$, $\bar{\delta}$ (radius vector, geocentric right ascension and declination) to the orbital parameters and time.

$$e = e_0 + \left(1 - \frac{a_0}{a}\right)(1 - e_0) + \sum_{j=1}^{3} e_j(t - t_0)^j,$$

$$i = i_0 + i_s + \sum_{j=1}^{3} i_j(t - t_0)^j,$$

$$\Omega = \Omega_0 + \Omega_s + K_\Omega u + \sum_{j=1}^{3} \Omega_j(t - t_0)^j,$$

$$\omega = \omega_0 + \omega_s + K_\omega u + \sum_{j=1}^{3} \omega_j(t - t_0)^j,$$

$$n = n_0 + \sum_{j=1}^{3} n_j(t - t_0)^j,$$

where
- e (eccentricity),
- i (inclination,
- Ω (right ascension of node),
- ω (argument of perigee),

and
- n (mean motion),

are orbital elements at time t, the sixth element being

t_0 (time at a particular node);

$a = \mu^{1/3} n^{-2/3}$ (semi-major axis);

i_s, Ω_s and ω_s are short-period perturbations due to the earth's oblateness,

$$i_s = \frac{1}{2} J_2(R/p_0)^2 \sin 2 i_0 \{e_0 \cos\omega_0 - e_0 \cos(u - \omega_0)\},$$

$$\Omega_s = -2 J_2(R/p_0)^2 \cos i_0 \{e_0 \sin\omega_0 + e_0 \sin(u - \omega_0)\},$$

$$\omega_s = J_2(R/p_0)^2 \left[\frac{1}{8} \sin^2 i_0 \sin 2u + \frac{1}{2} e_0 \sin^2 i_0 \sin(u + \omega_0) - \right.$$

$$- \frac{1}{2} e_0 \sin^2 i_0 \sin\omega_0 + 2 e_0 \cos^2 i_0 \{\sin\omega_0 + \sin(u - \omega_0)\} +$$

$$\left. + \left\{\frac{3}{8} \sin^2 i_0 - \frac{1}{4}\right\} e_0^2 \{\sin 2\omega_0 + \sin 2(u - \omega_0)\}\right],$$

$K_\Omega u$, $K_\omega u$ are the main secular terms due to oblateness,

$$K_\Omega = -\frac{3}{2} J_2 (R/p_0)^2 \cos i_0,$$

$$K_\omega = 3 J_2 (R/p_0)^2 \left(1 - \frac{5}{4} \sin^2 i_0\right),$$

where J_2, μ, R are earth constants

(currently used values being $J_2 = 1082.8 \times 10^{-6}$, after KING-HELE[11]

and $\mu = 398602 \text{ km}^3/\text{sec}^2$,

$R = 6378.163 \text{ km}$, after KAULA [10].

u is the argument of latitude, and $p_0 = a_0 (1 - e_0^2)$ is the initial semi-latus rectum.

The suffix $_0$ is used throughout to indicate initial values, which are always at a node.

The parameters of the orbit (the E_k of section 4.1) are a_0, e_0, i_0, Ω_0, ω_0, t_0 and such of the e_j, i_j, Ω_j, ω_j, n_j as are required.

Then the polar coordinates r, $\bar{\alpha}$, δ are given by

$$\cos\bar\delta \cos(\bar\alpha - \Omega) = \cos u,$$

$$\cos\bar\delta \sin(\bar\alpha - \Omega) = \cos i \sin u,$$

$$\sin\bar\delta \qquad\qquad = \sin i \sin u,$$

$$\cos E \qquad\qquad = (e + \cos v)/(1 + e \cos v),$$

$$\sin E \qquad\qquad = (1 - e^2)^{1/2} \sin v/(1 + e \cos v),$$

$$M = E - e \sin E = M_0 + \int_{t_0}^{t} n\, dt,$$

$$p/r = 1 + e \cos v + J_2 z,$$

$$v = u - \omega,$$

$$p = a(1 - e^2),$$

where E (eccentric anomaly), M (mean anomaly) and v (true anomaly) are intermediate variables, and $J_2 z$ represents the perturbation in radial distance due to oblateness,

$$J_2 z = J_2 (R/r)^2 \left[-\frac{1}{4} \sin^2 i_0 \cos 2u + \left(\frac{1}{2} - \frac{3}{4} \sin^2 i_0\right)(1 - e \cos v)\right].$$

4.4 Topocentric coordinates

The coordinates of the satellite relative to the observing station, namely ϱ, α and δ, the slant range, right ascension and declination, are derived from the polar coordinates of the satellite and the cartesian station coordinates through the formulae

$$\varrho \cos\delta \cos\alpha = r \cos\bar{\delta} \cos\bar{\alpha} - X,$$

$$\varrho \cos\delta \sin\alpha = r \cos\bar{\delta} \sin\bar{\alpha} - Y,$$

$$\varrho \sin\delta \quad = r \sin\bar{\delta} \quad - Z.$$

4.5 The geocentric partial derivatives

In the differential correction procedure the partial derivatives of α and δ with respect to the orbital parameters and observation times are required. There are two ways in which the partial derivatives may be determined. In the first place, the subroutines for determining α, δ in terms of the E_k and t_c may be entered repeatedly, changing one of the parameters by a small amount each time and using a differencing technique. Alternatively, the algebraic formulae for the partial derivatives can be evaluated and programmed. The former method was discarded on two grounds. It wastes computer time and it would involve considerable study to determine the most suitable increments to use. It has, however, been used to check a programme based on the algebraic formulae.

The evaluation of the partial derivatives is simplified by considerations of the accuracy required. The partial derivatives enter in the form of the matrix Q in the correction equation $Z = Q Y$ where, it will be remembered, Y is the column vector of weighted residuals. Let us suppose first that Q is computed correctly and that the final iteration has been reached in which there is no further change in the parameters, i.e. $Z = Q Y = 0$. Now let the final values of the parameters E_k and times t_c be taken as the starting point for a further iteration using an incorrect Q, say Q'. Then there will be a small change in the parameters $Z = Q' Y = (Q' - Q) Y$. It is seen that these changes are equivalent to small biasses in Y and hence in the original observations of an order less than the standard deviations in the observations. It is therefore reasonable to omit the short-period perturbations in the calculation of partial derivatives and aim at an accuracy of the order of 1%.

The formulae for the geocentric partial derivatives are given in Appendix II.

4.6 The topocentric partial derivatives

The expressions for the topocentric partial derivatives are obtained by differentiation of the three equations of section 4.4. Using dashes to denote partial differentiation with respect to an arbitrary variable, this gives, after solving for α' and δ',

$$\varrho \cos\delta\, \alpha' = r' \cos\bar\delta \sin(\bar\alpha - \alpha) + r \cos\bar\delta\, \bar\alpha' \cos(\bar\alpha - \alpha) -$$
$$- r \sin\bar\delta\, \bar\delta' \sin(\bar\alpha - \alpha) + X' \sin\alpha - Y' \cos\alpha$$

$$\varrho\, \delta' = r'\{\sin\bar\delta \cos\delta - \cos\bar\delta \sin\delta \cos(\bar\alpha - \alpha)\} + r \cos\bar\delta\, \bar\alpha' \sin\delta \sin(\bar\alpha - \alpha) +$$
$$+ r\, \bar\delta'\{\cos\bar\delta \cos\delta + \sin\bar\delta \sin\delta \cos(\bar\alpha - \alpha)\} +$$
$$+ \sin\delta\,(X' \cos\alpha + Y' \sin\alpha) - Z' \cos\delta.$$

Z' is zero in all the required derivatives, and X', Y' are zero in all derivatives except those with respect to t. For these,

$$\frac{\partial X}{\partial t} = -w\,Y$$

and

$$\frac{\partial Y}{\partial t} = w\,X,$$

where w is the angular velocity of the earth.

5. The orbit improvement programme in operation

In order to illustrate a number of points which have not been brought out in previous sections, an actual series of iterations will now be given. These form a preliminary improvement of a predicted set of elements at the first node of the orbit of Sputnik 2 on November 8, 1957, and are based on a set of 24 observations.

During the operation of the *Select/Process* programme, the following output is obtained (the actual printed output has here been suitably annotated).

a_0	7302.558 km.	ω_0	58.064 deg.
e_0	0.097245	t_0	1 h. 32 m. 06 · 03 s. (on 1957 Nov. 8)
i_0	65.314 deg.	M_0	−48.9703 deg.
Ω_0	102.844 deg.	n_0 500833.7°/100 d. n_1 24000°/(100 d.)²

Serial No.	Obs. No.	Time (days)	
1	80114	−1.165	−18 REV +0.902
2	80067	−1.152	−17 REV +0.108
3	80080	−1.151	−17 REV +0.122
4	80112	−1.236	−19 REV +0.923
5	80016	−0.575	− 9 REV +0.132
6	80101	−0.575	− 9 REV +0.127
7	80119	−0.231	− 5 REV +0.886
8	80122	−0.159	− 4 REV +0.891
9	80124	−0.575	− 9 REV +0.126
10	80156	−0.303	− 6 REV +0.891
11	80166	−0.717	−11 REV +0.171

7*

Serial No.	Obs. No.	Time (days)	
12	80187	−0.575	− 9 REV +0.135
13	80508	−0.303	− 6 REV +0.892
14	80136	+0.627	+ 7 REV +0.832
15	80139	+0.704	+ 8 REV +0.893
16	80145	+0.776	+ 9 REV +0.902
17	80148	+0.705	+ 8 REV +0.914
18	80173	+0.702	+ 8 REV +0.878
19	80175	+0.774	+ 9 REV +0.870
20	80149	+1.708	+22 REV +0.871
21	80158	+1.709	+22 REV +0.894
22	80159	+1.638	+21 REV +0.896
23	80164	+1.297	+17 REV +0.183
24	78018	+0.059	− 1 REV +0.920

The first section above contains the values of the initial orbital parameters a_0, e_0, i_0, Ω_0, ω_0, t_0 and n_1. These values have been obtained by prediction from a previous determination on 1957 Nov. 6. In addition, the derived quantities M_0, n_0 are given.

The second section contains information about the observations used. The first column gives the serial number for the present run, the second gives the catalogue number of the observation, the third gives the time of the observation in days (measured from the midnight previous to the chosen nodal crossing) and the remaining information is the value of the argument of latitude u expressed in revolutions and fractions of a revolution .This information enables the operator to see at a glance the orbit coverage given by the observations.

Proceeding to the next stage, which includes the computations of the first iteration and the printing of residuals based on the initial orbital parameters, two decisions have to be made. Although the weights to be allotted to the observations in a final analysis have been specified by the code indices of the observations, it has been found to be unwise to use these until very good orbit parameters have been determined, especially when the observations vary considerably in accuracy. The reason for this is that rejection of observations is based on weighted residuals, and the orbit parameters need only be slightly wrong for an accurate observation to have a large residual and hence be rejected. A facility has therefore been introduced for running with equally weighted observations, based on nominal standard deviations of $0.^{s}2$ and 0.2 sec. in angle and time respectively. In the present case this facility was used.

The second decision that has to be made concerns the rejection level. If any of the weighted residuals of an observation exceed the rejection level the observation is immediately discarded and does not enter into further calculations. The rejection level is set as a 3-bit number on the computer handswitches. This allows 8 different levels, and the present list of levels is 8, 16, 32, 64, 128, 256, 512, 1024.

In the present instance the level selected was 32, and the following was the print-out during the first iteration and after its completion.

Serial No.	Weighted residuals			Actual residuals		
	(t)	(α)	(δ)	$t_{calc}-t_{obs}$	$\alpha_{calc}-\alpha_{obs}$	$\delta_{calc}-\delta_{obs}$
1	+0	+ 2	+ 1	+0.000	+ 0.410	+ 0.178
2	+0	+ 4	+ 16	+0.000	+ 0.667	+ 2.878
3	+0	− 10	+ 4	+0.000	− 1.752	+ 0.768
4	+0	+ 2	+ 0	+0.000	+ 0.306	− 0.028
5	+0	+ 5	+ 0	+0.000	+ 0.809	+ 0.082
6	Reject Angle	+ 33	+ 19		+ 5.765	+ 3.411
7	+0	+ 4	+ 4	+0.000	+ 0.617	+ 0.642
8	+0	+ 2	+ 1	+0.000	+ 0.438	+ 0.153
9	Reject Angle	− 2	+ 32		− 0.284	+ 5.539
10	+0	+ 1	+ 2	+0.000	+ 0.167	+ 0.398
11	Reject Angle	+ 37	+ 28		+ 6.475	+ 4.904
12	Reject Angle	+111	+ 45		+ 19.462	+ 7.837
13	+0	+ 29	+ 7	+0.000	+ 5.083	+ 1.216
14	Reject Angle	+844	+156		+148.401	+27.490
15	+0	+ 4	+ 4	+0.000	+ 0.655	+ 0.735
16	+0	− 4	+ 4	+0.000	− 0.652	+ 0.676
17	Reject Angle	+242	+291		+ 42.609	+51.089
18	+0	+ 1	+ 1	+0.000	+ 0.171	+ 0.220
19	+0	+ 0	− 2	+0.000	− 0.028	− 0.269
20	+0	+ 3	+ 1	+0.000	+ 0.502	+ 0.142
21	+0	+ 8	+ 5	+0.000	+ 1.354	+ 0.922
22	Reject Angle	+ 5	−219		+ 0.896	−38.427
23	+0	+ 9	− 10	+0.000	+ 1.563	− 1.767
24	+0	+ 4	+ 2	+0.000	+ 0.755	+ 0.318

$$H = 1716 \qquad \varepsilon = 7.974$$

$$a_0 = 7302.488 \qquad\qquad t_0 = 1\,\text{h. } 32\,\text{m. } 06.770\,\text{s.}$$
$$e_0 = + 0.097668 \qquad\qquad M_0 = -48.2872$$
$$i_0 = + 65.2134 \qquad\qquad n_0 = +500840.9$$
$$\Omega_0 = +102.7350 \qquad\qquad n_1 = +23734.7$$
$$\omega_0 = + 57.3395$$

In this print-out the first section gives the actual and the weighted residuals corresponding to the original sets of orbital parameters. The actual residuals are given in seconds ($t_{calc} - t_{obs}$) and degrees ($\alpha_{calc} - \alpha_{obs}$, $\delta_{calc} - \delta_{obs}$). It will be noted that the time residuals are zero. This is because the 'corrected' times are, of necessity, the 'observed' times for this first step. Seven observations, with weighted angular residuals greater than or equal to 32, were rejected. It is not proposed to argue here whether all seven should, in fact, be rejected. As the purpose of the present run was to obtain an intermediate improvement only, it should not matter.

Underneath the residuals are given the values of H, the weighted sum of squares of residuals, and $\varepsilon = [H/(2n - l]^{1/2}$. Here, $n = 24 - 7 = 17$ and l, the number of parameters, is 7.

The final section of print-out gives the corrected orbital parameters.

A second iteration followed, with the rejection level set at 8. Here is the result.

Serial No.	Weighted residuals			Actual residuals		
	(t)	(α)	(δ)	$t_{calc} - t_{obs}$	$\alpha_{calc} - \alpha_{obs}$	$\delta_{calc} - \delta_{obs}$
1	$+1$	-4	-2	$+0.142$	-0.706	-0.278
2	$+0$	$+1$	$+0$	-0.012	$+0.105$	$+0.005$
3	$+0$	-3	-1	$+0.074$	-0.521	-0.200
4	$+0$	$+1$	-2	$+0.058$	$+0.166$	-0.439
5	-1	$+4$	-2	-0.165	$+0.711$	-0.436
7	$+1$	-2	-3	$+0.153$	-0.412	-0.493
8	$+0$	-1	$+1$	-0.002	-0.111	$+0.136$
10	$+1$	-2	-2	$+0.125$	-0.280	-0.437
13	Reject Angle	$+17$	$+2$		$+3.071$	$+0.278$
15	$+0$	-1	$+0$	$+0.008$	-0.149	-0.002
16	$+0$	-6	$+4$	$+0.013$	-1.091	$+0.744$
18	$+0$	$+1$	$+0$	$+0.012$	$+0.167$	-0.044
19	Reject Angle	-8	$+2$		-1.425	$+0.330$
20	$+0$	$+1$	$+3$	-0.047	$+0.225$	$+0.462$
21	-2	$+0$	$+6$	-0.313	$+0.017$	$+1.051$
23	$+2$	$+1$	-2	$+0.340$	$+0.135$	-0.419
24	$+0$	-3	$+1$	$+0.063$	-0.487	$+0.103$

$$H = 210 \qquad \varepsilon = 3.024$$

$a_0 =$	7302.493	$t_0 =$	$1\,\text{h. }32\,\text{m. }06.111\,\text{s.}$
$e_0 =$	0.097483	$M_0 =$	-48.4213
$i_0 =$	65.2843	$n_0 =$	500840.4
$\Omega_0 =$	102.8160	$n_1 =$	23180.7
$\omega_0 =$	57.4716		

It will be seen that two more observations are rejected (again, this is immaterial for the present run) and there are now non-zero time residuals.

A third iteration gave the following result.

Serial No.	Weighted residuals			Actual residuals		
	(t)	(α)	(δ)	$t_{calc} - t_{obs}$	$\alpha_{calc} - \alpha_{obs}$	$\delta_{calc} - \delta_{obs}$
1	$+1$	-3	-2	$+0.133$	-0.517	-0.362
2	$+0$	$+1$	$+0$	-0.055	$+0.179$	$+0.064$
3	$+0$	$+0$	-1	-0.030	$+0.050$	-0.237
4	$+0$	$+2$	-3	$+0.061$	$+0.315$	-0.464
5	-1	$+4$	-1	-0.204	$+0.786$	-0.111

Serial No.	Weighted residuals			actual residuals		
	(t)	(α)	(δ)	$t_{calc}-t_{obs}$	$\alpha_{calc}-\alpha_{obs}$	$\delta_{calc}-\delta_{obs}$
7	$+0$	$+2$	-1	$+0.036$	$+0.430$	-0.127
8	$+0$	$+0$	$+1$	-0.013	$+0.043$	$+0.107$
10	$+0$	$+1$	-1	$+0.042$	$+0.139$	-0.153
15	-1	$+3$	$+2$	-0.097	$+0.555$	$+0.279$
16	$+0$	-5	$+4$	$+0.002$	-0.945	$+0.712$
18	$+0$	$+2$	$+1$	-0.003	$+0.293$	$+0.096$
20	$+0$	$+1$	$+1$	-0.027	$+0.118$	$+0.224$
21	-2	$+2$	$+3$	-0.258	$+0.438$	$+0.560$
23	$+3$	$+2$	-3	$+0.453$	$+0.301$	-0.513
24	$+0$	$+0$	$+1$	-0.043	$+0.045$	$+0.154$

$$H = 159 \qquad \varepsilon = 2.634$$

$$
\begin{aligned}
a_0 &= 7302.498 & t_0 &= 1\,\text{h. } 32\,\text{m. } 06.269\,\text{s.} \\
e_0 &= 0.097720 & M_0 &= -48.5101 \\
i_0 &= 65.2659 & n_0 &= 500839.8 \\
\Omega_0 &= 102.8044 & n_1 &= 23254.4 \\
\omega_0 &= 57.5952 &&
\end{aligned}
$$

This is as far as the present case was taken. The final orbital parameters here were accepted as being sufficiently good to start a run with a fuller set of observations. In fact, a set of 67 observations was taken, covering the same period (Nov. 8.0 ± 2 days). During the course of three iterations, 17 observations were rejected, i. e. about 25% of the original number. The final set of parameters obtained, with standard deviations for the 7 basic parameters, was as follows.

$$
\begin{aligned}
a_0 &= 7302.57 \pm .02\,\text{km,} & t_0 &= 1\,\text{h. } 32\,\text{m. } 7.9 \pm .6\,\text{s.} \\
e_0 &= 0.09747 \pm .00017 & M_0 &= -49.°1225 \\
i_0 &= 65.°293 \pm .009 & n_0 &= 500832.°3/100\,\text{d.} \\
\Omega_0 &= 102.°778 \pm .009 & n_1 &= 23815 \pm 405°/(100\,\text{d.})^2 \\
\omega_0 &= 58.°26 \pm .14 &&
\end{aligned}
$$

In this run, of course, the a priori estimates of the accuracy of the observations were used in the weighting. The final value of ε obtained was 2.345, the supposed standard deviation of an observation of unit weight. This figure is rather large (one hopes for a value between 0.5 and 1.5) and its magnitude can be attributed to either or both of two causes. Either the observations are, on the average, nearly $2\frac{1}{2}$ times less accurate than was thought or the model of 7 parameters is not good enough to describe the actual satellite motion. An inspection of the residuals (not given here) and a comparison of the orbit parameters with those of November 6 and 10 indicates that both causes are at work here. One can then conclude that the standard deviations in the orbit parameters, as listed above, are perhaps slightly optimistic.

6. Final remarks

Besides the normal procedure adopted for checking the computer programme, an interesting check on the statistical part of the programme has been obtained by first producing a set of pseudo-observations and then, after the addition of random errors using the set to recover the orbital parameters. A set of 30 basic observations, covering four days, was generated using the model of section 4.3 with 12 parameters. From these 30 'correct' observations the 12 parameters were recovered exactly to the number of digits printed out. A family of 10 sets of observations was then generated by adding random errors to the 'correct' observations and each was analysed by the differential correction procedure. A statistical assessment of the 10 resulting sets of orbital parameters proved extremely satisfactory and left us in no doubt about the accuracy of the statistical procedure.

From the little practical experience gained in operating the orbit improvement programme it is not yet possible to say how useful it will be. The only major drawback appears to be the rather large amount of computer time required. As a rough indication, a timing assessment can be based on 30 seconds per observation per iteration. Thus, four iterations using 30 observations takes about one hour.

In the immediate future it is hoped to analyse observations of certain of the short-lived satellites, for which good orbits have not yet been published.

The satellite section at the Radio Research Station, Slough, who are responsible for satellite orbit predictions in the United Kingdom, are also proposing to use the programmes for orbit studies.

References

[1] MERSON, R. H.: Techniques of Analysing Terrestrial Radio and Optical Observations of Earth Satellites. Astr. Acta, 5, No. 1 (1959).

[2] CORNFORD, E. C.: A Comparison of Orbital Theory with Observations made in the United Kingdom on the Russian Satellites. Paper presented at 5th C.S.A.G.I. Assembly, Moscow (1958).

[3] TAYLER, R. J.: Orbital Elements of Sputnik 3 (1958, δ 2), Wisseled Note (September, 1960).

[4] KING-HELE, D. G., and R. J. TAYLER: Orbital Elements of Sputnik 3 Rocket (1958, δ 1) Unissued note (June 1959).

[5] MERSON, R. H.: The Motion of a Satellite in an Axi-symmetric Gravitational Field. Geophys. J. 4 (1961).

[6] VEIS, G.: Research in Space Science, Special Report No. 23, Smithsonian Institution Astrophysical Observatory (March 1959).

[7] H. M. Nautical Almanac Office: Explanatory Supplement to the Astronomical Ephemeris. London, H. M. S. O. 1961.

[8] GARFINKEL, B.: An Investigation in the Theory of Astronomical Refraction. Astronom. J. 50, No. 8, 169—179 (Feb. 1944).

[9] BROWN, D. C.: RCA Data Reduction Technical Report No. 39. A Treatment of Analytical Photogrammetry (with emphasis on ballistic camera applications). AFMTC-TR-57-22. ASTIA DOCUMENT No. 124144. (August 1957).

[10] KAULA, W. M.: A Geoid and World Geodetic System based on a Combination of Gravimetric, Astrogeodetic and Satellite Data. Techn. Note NASA TN D-702 (May 1961).

[11] KING-HELE, D. G.: The Earth's Gravitational Potential, deduced from the Orbits of Artificial Satellites. Geophys. J. 4 (1961).

Appendix I

Proof of the formula $Q\,Q^T = L^{-1}$

The relevant equations of section 4 are

$$Z = Q\,Y = L^{-1}\,J,$$

with

$$L = \sum_i (M_i^T\,M_i - e_i^{-1}\,M_i^T\,U_i\,U_i^T\,M_i),$$

$$J = \sum_i (d_i\,e_i^{-1}\,M_i^T\,U_i - M_i^T\,Y_i),$$

$$d_i = k_i + U_i^T\,Y_i,$$

$$e_i = 1 + U_i^T\,U_i,$$

and Y is the column vector with elements $Y_1, Y_2, \ldots, Y_m, k_1, k_2, \ldots, k_m$.

Since d_i is a scalar we have

$$J = \sum_i (e_i^{-1}\,M_i^T\,U_i\,k_i + e_i^{-1}\,M_i^T\,U_i\,U_i^T\,Y_i - M_i^T\,Y_i),$$

or, putting

$$P_i = e_i^{-1}\,M_i^T\,U_i\,U_i^T - M_i^T, \qquad G_i = e_i^{-1}\,M_i^T\,U_i,$$

$$J = \sum_i (P_i\,Y_i + G_i\,k_i) = (P_1 \vdots P_2 \vdots \ldots \vdots P_m \vdots G_1 \vdots \ldots \vdots G_m)\,Y.$$

Comparing this with the equation $Q\,Y = L^{-1}\,J$ then gives

$$L\,Q = (P_1 \vdots P_2 \vdots \ldots \vdots P_m \vdots G_1 \vdots \ldots \vdots G_m)$$

and

$$L\,Q\,Q^T\,L^T = \sum_i (P_i\,P_i^T + G_i\,G_i^T).$$

Now

$$
\begin{aligned}
P_i\,P_i^T &= (e_i^{-1}\,M_i^T\,U_i\,U_i^T - M_i^T)(e_i^{-1}\,U_i\,U_i^T\,M_i - M_i) \\
&= M_i^T\,M_i + e_i^{-2}\,M_i^T\,U_i(U_i^T\,U_i)\,U_i^T\,M_i - 2e_i^{-1}\,M_i^T\,U_i\,U_i^T M_i \\
&= M_i^T\,M_i + e_i^{-2}\,M_i^T\,U_i(e_i - 1)\,U_i^T\,M_i - 2e_i^{-1}\,M_i^T\,U_i\,U_i^T\,M_i \\
&= M_i^T\,M_i - e_i^{-1}\,M_i^T\,U_i\,U_i^T\,M_i - e_i^{-2}\,M_i^T\,U_i\,U_i^T\,M_i,
\end{aligned}
$$

and

$$G_i\,G_i^T = e_i^{-2}\,M_i^T\,U_i\,U_i^T\,M_i,$$

so that
$$LQ\,Q^T\,L^T = \sum_i (M_i^T\,M_i - e_i^{-1}\,M_i^T\,U_i\,U_i^T\,M_i) = L.$$

Since L is a symmetric matrix, it follows that $L^T = L$ and so $Q\,Q^T = L^{-1}$.

Appendix II

The partial derivatives

After omitting the terms i_s, ω_s, Ω_s and $J_2\,z$ the formulae of section 4.3. are differentiated partially with respect to an arbitrary variable. Using dashes to denote derivatives,

$$\bar{\alpha}' = \Omega' + \alpha_i\,i' + \alpha_u\,u',$$

where $\alpha_i = -\sin\bar{\delta}\,\cos u/\cos^2\bar{\delta}$, $\alpha_u = \cos i/\cos^2\bar{\delta}$;

$$\bar{\delta}' = \delta_i\,i' + \delta_u\,u',$$

where $\delta_i = \sin(\bar{\alpha} - \Omega)$, $\delta_u = \sin i\,\cos(\bar{\alpha} - \Omega)$;

$$r' = a^{-1}\,r\,a' + r_e\,e' + r_v\,v',$$

where $r_e = (-2a\,e\,r - r^2\,\cos v)/p$, $r_v = (r^2\,e\,\sin v)/p$;

$$M' = M_e\,e' + M_v\,v' = M'_0 + n(t' - t'_0) +$$
$$+ n'_0(t - t_0) + \sum_{j=1}^{3} n'_j\,(t - t_0)^{j+1}/(j+1),$$

where

$$M_e = -r\,\sin v(1 + r\,p^{-1})/(a\,p)^{1/2}, \qquad M_v = r^2/a(a\,p)^{1/2};$$

$$M'_0 = M_{e_0}\,e'_0 - M_{v_0}\,\omega'_0; \qquad a' = -\frac{2}{3}\,a\,n'/n;$$

$$a'_0 = -\frac{2}{3}\,a_0\,n'_0/n_0; \qquad u' = v' + \omega';$$

$$e' = \left(\frac{a_0}{a}\right)e'_0 - \left(\frac{1 - e_0}{a^2}\right)(a\,a'_0 - a_0\,a') + e'_p;$$

$$i' = i'_0 + i'_p;$$

$$\Omega' = \Omega'_0 + K'_\Omega\,u + K_\Omega\,u' + \Omega'_p;$$

$$\omega' = \omega'_0 + K'_\omega\,u + K_\omega\,u' + \omega'_p;$$

$$n' = n'_0 + n'_p;$$

where $e_p = \sum_{j=1}^{3} e_j(t - t_0)^3$, etc.

Certain preliminary rearrangements can be made before proceeding to particular cases.

Eliminating v' from the expressions for r' and M', and substituting for a' in terms of n'

$$r' = -\frac{2}{3}\frac{r}{n}n' + \left(r_e - \frac{M_e}{M_v}r_v\right)e' + \left(\frac{r_v}{M_v}\right)M'.$$

After some manipulation of the expressions for r_e, M_e, r_v, M_v it can be shown that

$$r_e - \frac{M_e}{M_v}r_v = -a\cos v,$$

so that

$$r' = -\frac{2}{3}\frac{r}{n}n' - a\cos v \cdot e' + \left(\frac{r_v}{M_v}\right)M'.$$

From $u' = v' + \omega'$ and the expression for ω', we have

$$u'(1 - K_\omega) = v' + (\omega'_0 + K'_\omega u + \omega'_p)$$
$$= \left(\frac{M' - M_e\, e'}{M_v}\right) + (\omega'_0 + K'_\omega u + \omega'_p).$$

Since $K_\omega < .004$ we approximate this to

$$u' = \left(\frac{M' - M_e\, e'}{M_v}\right) + (\omega'_0 + K'_\omega u + \omega'_p).$$

In rearranging $\bar{\alpha}'$ and $\bar{\delta}'$ we note that the quantities required (section 4.5) are $r\cos\bar{\delta}\cdot\bar{\alpha}$ and $r\,\bar{\delta}'$, so we go straight to this form.

$$r\cos\bar{\delta}\cdot\bar{\alpha}' = r\cos\bar{\delta}[\Omega'_0 + K'_\Omega u + K_\Omega u' + \Omega'_p] + r\cos\bar{\delta}\cdot\alpha_i\, i' +$$
$$+ r\cos\bar{\delta}\cdot\alpha_u\cdot u' = A\,u' + C\,i' + r\cos\bar{\delta}(K'_\Omega u + \Omega'_0 + \Omega'_p),$$

where

$$A = r\cos\bar{\delta}(K_\Omega + \alpha_u)$$

and

$$C = r\cos\bar{\delta}\cdot\alpha_i.$$

Similarly,

$$r\,\bar{\delta}' = B\,u' + D\,i',$$

where

$$B = r\,\delta_u \quad\text{and}\quad D = r\,\delta_i.$$

Derivatives with respect to a_0. When a_0 is taken as the independent variable,

$$n' = n'_0 = -\frac{3}{2}\frac{n_0}{a_0}, \qquad a' = a\,n_0/a_0\,n,$$

$$e' = (1 - e_0)(n_0 - n)/a\,n, \quad i' = 0,$$

$$M' = -\frac{3}{2}\frac{n_0}{a_0}(t - t_0), \qquad K'_\omega = -\frac{2}{a_0}K_\omega,$$

$$K'_\Omega = -\frac{2}{a_0}K_\Omega, \quad \Omega'_0 = \omega'_0 = \omega'_p = \Omega'_p = 0,$$

and so

$$r' = \frac{r}{a_0}\left(\frac{n_0}{n}\right) - (1 - e_0)\left(\frac{n_0}{n} - 1\right)\cos v - \frac{3}{2}\frac{r_v}{M_v}\cdot\frac{n_0}{a_0}(t - t_0),$$

$$r\cos\bar\delta\cdot\bar\alpha' = A\,u' - \frac{2r\cos\bar\delta}{a_0}K_\Omega\,u,$$

and $r\,\bar\delta' = B\,u'$,
where

$$u' = M_v^{-1}\left[-\frac{3}{2}\frac{n_0}{a_0}(t - t_0) - \frac{M_e}{a}(1 - e_0)\left(\frac{n_0}{n} - 1\right)\right] - \frac{2}{a_0}K_\omega\,u.$$

Derivatives with respect to e_0

$$a_0' = a' = n_0' = n' = e_p' = i' = \Omega_0' = \Omega_p' = \omega_0' = \omega_p' = 0,$$

$$e' = a_0/a, \qquad K_\omega' = \left(\frac{4e_0}{1 - e_0^2}\right)K_\omega, \qquad K_\Omega' = \left(\frac{4e_0}{1 - e_0^2}\right)K_\Omega,$$

$$M' = M_0' = M_{e_0},$$

so that

$$r' = \frac{r_v}{M_v}\cdot M_{e_0} - a_0\cos v,$$

$$r\cos\bar\delta\cdot\bar\alpha' = A\,u' + \left(\frac{4e_0}{1 - e_0^2}\right)K_\Omega\,u,$$

and $r\,\bar\delta' = B\,u'$,
where

$$u' = \left(\frac{a\,M_{e_0} - a_0\,M_e}{a\,M_v}\right) + \left(\frac{4e_0\,r\cos\bar\delta}{1 - e_0^2}\right)K_\omega\,u.$$

Derivatives with respect to i_0

$$i' = 1, \qquad K_\Omega' = \frac{3}{2}J_2(R/p_0)^2\sin i_0,$$

$$K_\omega' = 5\sin i_0\cdot K_\Omega, \qquad n' = e' = M' = 0,$$

so that $r' = 0$

$$r\cos\bar\delta\cdot\bar\alpha' = A\,u' + C + r\cos\bar\delta\cdot K_\Omega'\,u,$$

$$r\,\delta' = B\,u' + D,$$

where

$$u' = 5\sin i_0\cdot K_\Omega\cdot u.$$

Derivatives with respect to Ω_0. These can be written down immediately.

$$r' = 0,$$

$$r\cos\bar\delta\cdot\bar\alpha' = r\cos\bar\delta,$$

$$r\,\bar\delta' = 0.$$

Derivatives with respect to ω_0

$$M' = M'_0 = -M_{v_0}, \qquad e' = 0, \qquad K'_\omega = K'_\Omega = 0,$$

so that

$$r' = -r_v M_{v_0}/M_v,$$

$$r \cos \bar{\delta} \cdot \bar{\alpha}' = A u',$$

$$r \bar{\delta}' = B u',$$

where

$$u' = 1 - \frac{M_{v_0}}{M_v}.$$

Derivatives with respect to t

$$M' = n, \qquad n' = n'_p, \qquad e' = -\frac{2}{3}(1 - e_0)\left(\frac{a_0}{a}\right)\frac{n'_p}{n} + e'_p,$$

$$\omega'_0 = \Omega'_0 = K'_\omega = K'_\Omega = 0, \qquad i' = i'_p,$$

so that

$$r' = -\frac{2}{3}\frac{r}{n} n'_p + \frac{2}{3}(1 - e_0) a_0 \cos v \frac{n'_p}{n} - a \cos v \, e'_p + \left(\frac{r_v}{M_v}\right) n,$$

$$r \cos \bar{\delta} \cdot \bar{\alpha} = A u' + C i'_p + r \cos \bar{\delta} \cdot \Omega'_p,$$

and

$$r \cdot \bar{\delta}' = B u' + D i'_p,$$

where

$$u' = \left(\frac{n - M_e e'}{M_v}\right) + \omega'_p.$$

A further study has shown that all the p-suffixed derivatives may be neglected, and we use the approximations

$$r' = r_v n/M_v,$$

$$r \cos \bar{\delta} \cdot \bar{\alpha}' = A n/M_v,$$

$$r \bar{\delta}' = B n/M_v,$$

The derivatives with respect to t_0 have the same magnitude as, but are opposite in sign to, the t derivatives.

Derivatives with respect to n_j

$$n' = (t - t_0)^j, \qquad e' = -\frac{2}{3}(1 - e_0)\frac{a_0}{a\,n}(t - t_0)^j,$$

$$M' = (t - t_0)^{j+1}/(j + 1),$$

so that

$$r' = \frac{(t - t_0)^j}{n\,M_v}\left[\frac{n\,r_v(t - t_0)}{j + 1} + \frac{2}{3} M_v\{a_0 \cos v(1 - e_0) - r\}\right]$$

$$r \cos \bar{\delta} \cdot \bar{\alpha}' = A u',$$

$$r \bar{\delta}' = B u',$$

where
$$u' = \frac{(t-t_0)^j}{n\,M_v}\left[\frac{n\,(t-t_0)}{j+1} - \frac{2}{3}\,M_e(1-e_0)\right].$$

Derivatives with respect to e_j

$$n' = M' = 0, \qquad e' = (t-t_0)^j, \qquad u' = -M_e(t-t_0)^j/M_v,$$

so that
$$r' = -a\cos v\,(t-t_0)^j,$$

$$r\cos\bar\delta\cdot\bar\alpha' = -A\,M_e(t-t_0)^j/M_v,$$

$$r\cdot\bar\delta' = -B\,M_e(t-t_0)^j/M_v.$$

Derivatives with respect to i_j

$$i' = (t-t_0)^j, \qquad n' = e' = M' = 0, \qquad u' = 0,$$

so that
$$r' = 0,$$

$$r\cos\bar\delta\cdot\bar\alpha' = C(t-t_0)^j,$$

$$r\cdot\bar\delta' = D(t-t_0)^j.$$

Derivatives with respect to Ω_j

$$\Omega' = (t-t_0)^j, \qquad n' = e' = M' = 0, \qquad u' = i' = 0,$$

so that
$$r' = 0,$$

$$r\cos\bar\delta\cdot\bar\alpha' = r\cos\bar\delta\,(t-t_0)^j,$$

$$r\cdot\bar\delta' = 0.$$

Derivatives with respect to ω_j

$$u' = \omega' = (t-t_0)^j, \qquad i' = n' = e' = M' = 0,$$

so that
$$r' = 0,$$

$$r\cos\bar\delta\cdot\bar\alpha' = A\,(t-t_0)^j,$$

$$r\cdot\bar\delta' = B(t-t_0)^j.$$

The determination of the small changes in the orbital elements of an earth satellite due to air drag

By

M. J. Davies

Applied Mathematics Department, University College of Wales,
Aberystwyth, England

Abstract. The increments of the orbital elements a, e, i, Ω, ω, χ of an earth satellite in one orbit due to air drag are computed and systematically tabulated for a rotating oblate atmosphere which possesses a diurnal "bulge". The expressions are computed to orders e^4, where e is the eccentricity of the orbit; δ^2, where δ is the rotation parameter; λ^4, where λ is an oblateness parameter; and to the first order in σ, where σ is a parameter characteristing the extent of the diurnal bulge.

1. Introduction

The purpose of this paper is to compute and tabulate in a systematic manner the small changes which occur in the orbital elements of an earth satellite under the influence of air during the extent of a single revolution. The method employed is to take as starting point the LAGRANGE formulation for the rates of change of the elements which may be found in any book or treatise on Celestial Mechanics. When this formulation is written in terms of an othogonal triad of forces at the satellite position, these forces can be identified with those produced by air drag on the satellite and may be expressed as functions of the orbital elements and time. Under the assumption that the orbital elements remain constant for the extent of one revolution, the equations for the rates of change become simple integrals with respect to time, or more conveniently, the eccentric anomaly E. The integrands are now expanded in terms of the small quantity, e, the eccentricity. Thus the general expression for the increment in any element s say, will be of the form:

$$\Delta s = \text{const} \int_0^{2\pi} \varrho \sum_n s_n(E)\, e^n\, dE, \tag{1.1}$$

which may be rearranged as

$$\Delta s = \text{const} \int_0^{2\pi} \varrho \sum_n S_n(e) \cos n\, E\, dE. \tag{1.2}$$

After some simplification in the term ϱ, the density, an exponential term, $\exp(x \cos E)$, is introduced; use is now made the standard integral

$$\int_0^{2\pi} \exp(x \cos E) \cos n \, E \, dE = 2\pi \, I_n(x), \qquad (1.3)$$

where I_n is the BESSEL function of the first kind and of imaginary argument; and certain other derived integrals.

The rotation of the atmosphere introduces a parameter δ into the expressions for the triad forces, while the oblateness and the diurnal bulge introduce parameters λ and σ respectively via the expression for the variation of atmospheric density with height away from some standard level. The orders retained of these various parameters conform with an overall plan for the retention of a given order of magnitude.

2. Atmospheric rotation

The orthogonal triad at the satellite position is defined by directions radially, in the plane of the orbit, and orthogonal to the plane of the orbit in the right handed sense. In this triad the velocity of the satellite relative to the medium is

$$\underline{v} = \underline{V}_{\text{sat}} - \underline{V}_{\text{air}}, \qquad (2.1)$$

where $\underline{V}_{\text{sat}}$ is the absolute velocity of the satellite, and $\underline{V}_{\text{air}}$ is the velocity of the atmosphere produced by the rotation of the earth.

The air drag is then taken to be

$$\text{Drag} = -\frac{1}{2} \varrho \, A \, C_d \, v^2, \qquad (2.2)$$

where:

    ϱ air density
    A Mean cross-sectional area
    C_d drag coefficient
    v^2 $|\underline{v}|^2$

The components of the total drag force in the triad now involve a parameter δ, where

$$\delta = \beta \, T_d \cos i, \qquad (2.3)$$

where T_d is the ratio of the period of the satellite to the period of the earth, and β is a numerical factor to allow for a lag in the atmospheric rotation. The constants A and C_d are written

$$A \, C_d = \varkappa \qquad (2.4)$$

in later work.

When these forces are introduced into the LAGRANGE equations, the independant variable transformed to E, and the integrands expanded in terms of the small quantities e and δ, the simple integrals required are attained. Orders retained in the expansions are e^4, $e^2 \delta$ and $e \delta^2$.

3. The air density functions

The air density is assumed to vary exponentially according to

$$\varrho = \varrho_0 \exp\left\{-\frac{Z - Z_0}{H}\right\}, \qquad (3.1)$$

where Z is the altitude, Z_0 is some standard altitude, ϱ_0 is the density at the standard altitude, and H is the scale height assumed constant. It is seen that this expression involves the tacit assumption that the atmospheric level surfaces of density are geocentric spheres. However if the level surfaces are not spheres it is necessary to assume some figure for the level surfaces and to assume an exponential variation of density away from them. This is the procedure in the cases of atmospheric oblateness and the diurnal bulge.

4. Atmospheric oblateness

The atmospheric level surfaces are assumed to be oblate spheroids, with axes in the polar direction, of ellipticity ε. Then the figure may be given by

$$r = a_1(1 - \varepsilon \sin^2 \psi_c), \qquad (4.1)$$

where terms $O(\varepsilon^2)$ are neglected, a_1 is the semi-major axis of the associated ellipse, and ψ_c is the geocentric latitude.

If a satellite perigee now occurs at a radial distance r_0 and at geocentric latitude ψ_{c0}, then the atmospheric level surface appropriate to the density at perigee is given by

$$r_0 = a_1(1 - \varepsilon \sin^2 \psi_{c0}), \qquad (4.2)$$

and in any other position

$$r = r_0(1 - \varepsilon \sin^2 \psi_c + \varepsilon \sin^2 \psi_{c0}). \qquad (4.3)$$

Thus if the satellite height is given by

$$z = a(1 - e \cos E), \qquad (4.4)$$

then

$$r_0 = a(1 - e), \qquad (4.5)$$

$$Z_0 = r, \qquad (4.6)$$

and after some reduction,

$$\varrho = \varrho_0 \exp\left\{-\frac{a e}{H} - \lambda \cos 2\omega\right\} \exp\left\{\frac{a e \cos E}{H} + \lambda \cos 2 u\right\} \qquad (4.7)$$

where

$$\lambda = \frac{1}{2} a(1 - e)\varepsilon \sin^2 i/H \qquad (4.8)$$

and

$$u = \omega + f, \qquad (4.9)$$

where f is the true anomaly in the satellite orbit.

The last term in the second exponent may be expanded in terms of λ, and it is necessary to retain $O(\lambda^4)$.

5. The diurnal bulge

For the purpose of accounting for the diurnal bulge, one may assume the atmospheric level surfaces to be ellipsoids of revolution about the major axis, with major axes directed along the line earth centre to a point on the equator lagging behind the sun's position by $\pi/6$, this angle being suggested by some figures published by PRIESTER et al [1]. The angle between this point and the satellite, which will be called the bulge aspect angle, ξ, is given by

$$\cos\xi = \cos u \cos\theta + \sin u \cos i \sin\theta. \qquad (5.1)$$

where

$$\theta = \alpha_s + \pi/6 - \Omega \qquad (5.2)$$

and α_s is the right ascension of the sun.

The figure of the atmosphere may now be written

$$r = \alpha_2(1 + \nu \cos\xi) \qquad (5.3)$$

neglecting $O(\nu^2)$, α_2 and ν being the semi-major axis and eccentricity of the associated ellipse.

Then if a satellite perigee occurs, as in section 4, at r_0, ξ_0, then

$$r_0 = \alpha_2(1 + \nu \cos\xi_0). \qquad (5.4)$$

In any other position

$$r = r_0(1 + \nu \cos\xi - \nu \cos\xi_0) + O(\nu^2). \qquad (5.5)$$

One now assumes the density to vary exponentially with constant scale height away from this surface. This is obviously an approximation since the variation of ν with height demonstrated in Ref. [1] implies exactly a variation of H with bulge aspect angle.

Substitution of (5.4) in (3.1) will now yield

$$\varrho = \varrho_0 \exp\left\{-\frac{a e}{H} - \eta \cos\xi_0\right\} \exp\left\{\frac{a e \cos E}{H} + \eta \cos\xi\right\}, \qquad (5.6)$$

where

$$\eta = a(1 - e)\nu/H. \qquad (5.7)$$

The second exponential in (5.5) can be conveniently developed, neglecting $O(e^2\nu)$, as

$$\exp\left[\frac{a e'}{H}\cos(E - \emptyset) - \frac{a e l}{H} + \sigma \cos E \sin(E + \psi)\right], \qquad (5.8)$$

where

$$e' = [(l + e)^2 + m^2]^{1/2}, \tag{5.9}$$

$$l = (1 - e)\, \nu\, \{\cos\omega\, \cos\theta + \sin\omega\, \cos i\, \sin\theta\}, \tag{5.10}$$

$$m = (1 - e)\, \nu\, \{-\sin\omega\, \cos\theta + \cos\omega\, \cos i\, \sin\theta\}, \tag{5.11}$$

$$n^2 = l^2 + m^2, \tag{5.12}$$

$$\sigma = a\, e\, n/H, \tag{5.13}$$

$$\tan\emptyset = \frac{m}{l + e}, \tag{5.14}$$

$$\tan\psi = l/m. \tag{5.15}$$

The last term in (5.8) is expanded and terms $O(\sigma^2)$ are neglected.

Thus the air density model becomes

$$\varrho = \varrho_0 \exp\left[-\frac{a\,e}{H} - \frac{a\,e\,l}{H} - \eta\, \cos\xi_0\right] \exp\left[\frac{a\,e'}{H}\cos(E - \emptyset)\right] \times$$
$$\times\, [1 + \sigma\, \cos E\, \sin(E + \psi)]. \tag{5.16}$$

In this way the basic integral type arising becomes

$$\int_0^{2\pi} \exp\left[\frac{a\,e'}{H}\cos(E - \emptyset)\right] \cos n\, E\, dE = 2\pi \cos n\, \emptyset\, I_n\left[\frac{a\,e'}{H}\right] \tag{5.17}$$

the related integrals being changed appropriately: in particular an integral, previously zero, is

$$\int_0^{2\pi} \exp\left[\frac{a\,e'}{H}\cos(E - \emptyset)\right] \cos n\, E\, \sin E\, dE$$
$$= \pi[\sin(n + 1)\,\emptyset\, I_{n+1} - \sin(n - 1)\,\emptyset\, I_{n-1}]. \tag{5.18}$$

It is necessary to examine the applicability of these results. The application of (5.18) would introduce to the basic formulae the entire set of previously vanishing terms involving $\sin E$ in the integrands, and in addition all terms I_n will change, becoming $\cos n\,\emptyset\, I_n$. Thus a simplification is sought. Obviously a lot will depend on the magnitude of $\emptyset$ given by (5.14).

Now m and l are $O(\nu)$, and for the moment assume $\nu = O(10^{-3})$ and $e = O(1/4)$, thus $\emptyset$ will be small. If e is smaller, this normally implies a lower orbit and a corresponding smaller ν. However no such limitations exist in general, but two obvious cases present themselves.

(1) $e \gg \nu$. Here all quoted powers of e must be retained, but $\emptyset$ is small.

$$\cos\emptyset = 1 + O(\nu^2/e^2),$$
$$\sin\emptyset = O(\nu/e),$$

the error terms being $O(10^{-2})$ say. Thus the leading terms in $\sin\emptyset$ must be retained, but all powers of e, δ, and λ associated with $\sin\emptyset$ are to be neglected. Thus only the six leading terms in the main formulae are affected.

(2) $e \sim \nu$. In this case, e will be so small that no powers of e need be retained. Thus the leading terms in δ, δ^2, $\lambda \ldots \lambda^4$ are all affected, but no others. Further reflection will show that these cases exhaust the applicability of the method. For if $e \sim \nu$ and ν is large such that some powers of e are to be retained and altered, the neglect of $e^2 \nu$, ν^2 and σ^2 will be invalid and the method will in any case break down. The present writer feels that with such sparse data as currently exists concerning the figure of the bulge, the retention of these terms would be placing too much confidence in the simple elliptic model employed. In addition to any deficiencies in the figure, it is obvious that a seasonal effect involving the declination of the sun must be considered.

6. Orders of magnitude retained

Some satellites have eccentricity as high as $1/4\,(0.25)$ and the retention of terms $O(e^4)$ means that an overall tolerance of 0.004 is to be maintained. The rotation parameter δ has a value of about $1/16$ or

$$\delta = O(e^2).$$

Thus it is necessary to retain terms $O(e^2\,\delta)$.

λ is found to be of the same order as e, or less, assuming that the atmospheric oblateness is bounded by and approximately equal to the value for the figure of the earth. Thus λ must be retained to the some order as e.

A very rough analysis of some figures given by PRIESTER [1] suggests that ν rises from zero about 200 km/s, to 10^{-2} at 500 km/s. Thus the neglect of σ^2 and other related terms, implying

$$\sigma^2 < 0.004$$

enables a rough estimate to be made of the equatorial perigee heights for which the formulae are adequate. This leads to

$$r_0 < 300 \text{ kms. (nightside)}$$

$$r_0 < 350 \text{ kms. (dayside)}.$$

These heights limit the applicability of the formulae.

7. Results

In the main formulae following, the changes in the orbital elements during one revolution are given, computed under the assumption that the elements remain constant for the extent of one revolution. This

assumption imposes a fundamental limit upon the smallness of the
· terms to be retained.

The argument of the BESSEL functions is $a\,e'/H$ as shown in Eq. (5.8)
et. seq.

The semi major axis

$$\Delta a = -2\pi a^2\,\varrho_0\,\varkappa\exp\left[-\frac{a\,e}{H} - \frac{a\,e\,l}{H} - \lambda\cos2\omega - \eta\cos\xi_0\right]\times$$

$$\times\varkappa[a_0 + a_1\,\delta + a_2\,\delta^2 + \lambda(a_3 + a_4\,\delta + a_5\,\delta^2) +$$

$$+ a_6\,\lambda^2 + a_7\,\lambda^3 + a_8\,\lambda^4 + a_9\,\sigma],\qquad(7.1)$$

where

$$a_0 = I_0 + 2e\,I_1 + \frac{3}{4}e^2(I_0 + I_2) + \frac{1}{4}e^3(3I_1 + I_3) +$$

$$+ \frac{7}{64}e^4(3I_0 + 4I_2 + I_4)$$

$$a_1 = -2\left[I_0 - \frac{1}{4}e^2(3I_0 + I_2)\right]$$

$$a_2 = I_0 + \frac{1}{4}\tan^2 i[I_0 + \cos(2\omega + 2\varnothing)\,I_2] -$$

$$- 2e\,I_1\left[1 + \frac{1}{4}\tan^2 i(1 + \cos2\omega)\right]$$

$$a_3 = \cos(2\omega + 2\varnothing)\,I_2 +$$

$$+ \cos2\omega\left[2e\,I_3 - \frac{1}{8}e^2(3I_0 + 2I_2 - 17I_4) - \frac{1}{4}e^3(3I_1 - 7I_5)\right]$$

$$a_4 = -2[\cos(2\omega + 2\varnothing)\,I_2 - e\cos2\omega(I_1 - I_3)]$$

$$a_5 = \left(1 + \frac{1}{4}\tan^2 i\right)\cos(2\omega + 2\varnothing)\,I_2 + \frac{1}{8}\tan^2 i[I_0 + \cos(4\omega + 4\varnothing)\,I_4]$$

$$a_6 = \frac{1}{4}(1 - 2\delta)\left[I_0 + \cos(4\omega + 4\varnothing)\,I_4 + e\{2I_1 - \cos4\omega(I_3 - 3I_5)\} +$$

$$+ \frac{1}{4}e^2\left\{3(I_0 + I_2) - \frac{1}{2}\cos4\omega(I_2 + 26I_4 - 39I_6)\right\}\right]$$

$$a_7 = \frac{1}{24}[3\cos(2\omega + 2\varnothing)\,I_2 + \cos(6\omega + 6\varnothing)\,I_6 +$$

$$+ 2e\{3\cos2\omega\,I_3 - \cos6\omega(I_5 - 2I_7)\}]$$

$$a_8 = \frac{1}{192}[3I_0 + 4\cos(4\omega + 4\varnothing)\,I_4 + \cos(8\omega + 8\varnothing)\,I_8]$$

$$a_9 = \frac{1}{2}\sin\psi(I_0 + I_2).$$

The eccentricity

$$\Delta e = -2\pi a\, \varrho_0\, \varkappa \exp\left[-\frac{a\,e}{H} - \frac{a\,e\,l}{H} - \lambda \cos 2\omega - \eta \cos \xi_0\right] \times$$

$$\times \varkappa[e_0 + e_1\,\delta + e_2\,\delta^2 + \lambda(e_3 + e_4\,\delta + e_5\,\delta^2) +$$

$$+ e_6\,\lambda^2 + e_7\,\lambda^3 + e_8\,\lambda^4 + e_9\,\sigma], \qquad (7.2)$$

where

$$e_0 = I_1 \cos\emptyset + \frac{1}{2}\,e(I_0 + I_2) - \frac{1}{8}\,e^2(5I_1 - I_3) -$$

$$- \frac{1}{16}\,e^3(5I_0 + 4I_2 - I_4) - \frac{1}{128}\,e^4(18I_1 + I_3 - I_5)$$

$$e_1 = -\left[2I_1 \cos\emptyset - \frac{1}{4}\,e(5I_0 + 3I_2) - 2e^2 I_1\right]$$

$$e_2 = \left(1 + \frac{1}{4}\tan^2 i\right) I_1 \cos\emptyset + \frac{1}{8}\tan^2 i\,[\cos(2\omega + \emptyset)\,I_1 +$$

$$+ \cos(2\omega + 3\emptyset)\,I_3] - \frac{1}{4}\,e\left[7I_0 + 5I_2 +\right.$$

$$\left.+ \frac{1}{2}\tan^2 i\left\{3(I_0 + I_2) + \frac{1}{2}\cos 2\omega(5I_0 + 6I_2 + I_4)\right\}\right]$$

$$e_3 = \frac{1}{2}\cos 2\omega\,[\sec 2\omega\{\cos(2\omega + \emptyset)\,I_1 + \cos(2\omega + 3\emptyset)\,I_3\} -$$

$$- \frac{1}{2}\,e(I_0 - 2I_2 - 3I_4) - \frac{1}{8}\,e^2(16I_1 + 3I_3 - 11I_5) +$$

$$+ \frac{1}{16}\,e^3(4I_0 - 25I_2 - 12I_4 + 17I_6)]$$

$$e_4 = -\cos(2\omega + \emptyset)\,I_1 - \cos(2\omega + 3\emptyset)\,I_3 +$$

$$+ \frac{1}{8}\,e \cos 2\omega(11I_0 + 10I_2 - 5I_4)$$

$$e_5 = \frac{1}{2}\left[\left(1 + \frac{1}{4}\tan^2 i\right)\{\cos(2\omega + \emptyset)\,I_1 + \cos(2\omega + 3\emptyset)\,I_3\} +\right.$$

$$+ \frac{1}{8}\tan^2 i(2\cos\emptyset\,I_1 + \cos 3\emptyset\,I_3 + \cos 5\emptyset\,I_5) -$$

$$\left.- \frac{1}{4}\tan^2 i \sin 2\omega\{\sin(2\omega + 3\emptyset)\,I_3 + \sin(2\omega + 5\emptyset)\,I_5\}\right]$$

$$e_6 = \frac{1}{4}(1 - 2\delta)\left[I_1 \cos\emptyset + \frac{1}{2}\,e(I_0 + I_2) - \frac{1}{8}\,e^2(5I_1 - I_3) +\right.$$

$$\left.+ \frac{1}{2}\{\cos(4\omega + 3\emptyset)\,I_3 + \cos(4\omega + 5\emptyset)\,I_5\}\right] -$$

$$- \frac{1}{4}\,e \cos 4\omega\left\{3I_2 - 2I_4 - 5I_6 - \frac{1}{4}\,e(5I_1 - 33I_3 - 9I_5 + 29I_7)\right\}$$

$$e_7 = \frac{1}{48}[3\{\cos(2\omega + \emptyset)\,I_1 + \cos(2\omega + 3\emptyset)\,I_3\} + \cos(6\omega + 5\emptyset)\,I_5 +$$
$$+ \cos(6\omega + 7\emptyset)\,I_7 - \frac{1}{2}\,e\{3\cos 2\omega(I_0 - 2I_2 - 3I_4) +$$
$$+ \cos 6\omega(5I_4 - 2I_6 - 7I_8)\}]$$

$$e_8 = \frac{1}{96}\left[\frac{3}{2}\,I_1\cos\emptyset + \cos(4\omega + 3\emptyset)\,I_3 + \cos(4\omega + 5\emptyset)\,I_5 +\right.$$
$$\left.+ \frac{1}{4}\{\cos(8\omega + 7\emptyset)\,I_7 + \cos(8\omega + 9\emptyset)\,I_9\}\right]$$

$$e_9 = \frac{1}{4}\sin\psi(3I_1 + I_3).$$

The argument of perigee

$$\Delta\omega = \Delta^1\omega - \cos i\,\Delta\Omega$$

$$\Delta^1\omega = -2\pi\frac{a\,\varrho_0\varkappa}{e}\exp\left[-\frac{a\,e}{H} - \frac{a\,el}{H} - \lambda\cos 2\omega - \eta\cos\xi_0\right] \times$$
$$\times \varkappa[\omega_0 + \omega_1\delta + \omega_2\delta^2 + \lambda(\omega_3 + \omega_4\delta + \omega_5\delta^2) +$$
$$+ \omega_6\lambda^2 + \omega_7\lambda^3 + \omega_8\lambda^4 + \omega_9], \tag{7.3}$$

where

$$\omega_0 = \sin\emptyset\,I_1$$

$$\omega_1 = -2\sin\emptyset\,I_1$$

$$\omega_2 = \left(1 + \frac{1}{4}\tan^2 i\right)\sin\emptyset\,I_1 - \frac{1}{8}\tan^2 i\left[\sin(2\omega + \emptyset)\,I_1 -\right.$$
$$\left.- \sin(2\omega + 3\emptyset)\,I_3 - \frac{1}{2}\,e\sin 2\omega(5I_0 - 4I_2 - I_4)\right]$$

$$\omega_3 = -\frac{1}{2}(\sin(2\omega + \emptyset)\,I_1 - \sin(2\omega + 3\emptyset)\,I_3) +$$
$$+ \frac{1}{4}\,e\sin 2\omega\left[I_0 - 4I_2 + 3I_4 + \frac{1}{4}\,e(10I_1 - 21I_3 + 11I_5) +\right.$$
$$\left.+ \frac{1}{8}\,e^2(2I_0 + 15I_2 - 34I_4 + 17I_6)\right]$$

$$\omega_4 = \sin(2\omega + \emptyset)\,I_1 - \sin(2\omega + 3\emptyset)\,I_3 - \frac{1}{8}\,e\sin 2\omega(11I_0 - 16I_2 + 5I_4)$$

$$\omega_5 = -\frac{1}{2}\left[\left(1 + \frac{1}{4}\tan^2 i\right)(\sin(2\omega + \emptyset)\,I_1 - \sin(2\omega + 3\emptyset)I_3) +\right.$$
$$\left.+ \frac{1}{8}\tan^2 i\{-2\sin\emptyset\,I_1 + \sin(4\omega + 3\emptyset)\,I_3 - \sin(4\omega + 5\emptyset)\,I_5\}\right]$$

$$\omega_6 = -\frac{1}{8}(1 - 2\delta)\left[-2\sin\emptyset\,I_1 + \sin(4\omega + 3\emptyset)\,I_3 - \sin(4\omega + 5\emptyset)\,I_5 -\right.$$
$$\left.- \frac{1}{2}\,e\sin 4\omega\{3I_2 - 8I_4 + 5I_6 - \frac{1}{4}\,e(5I_1 - 39I_3 + 63I_5 - 29I_7)\}\right]$$

$$\omega_7 = -\frac{1}{48}[3\sin(2\omega + \emptyset)I_1 - 3\sin(2\omega + 3\emptyset)I_3 + \sin(6\omega + 5\emptyset)I_5 -$$
$$- \sin(6\omega + 7\emptyset)I_7 - \frac{1}{2}e\{3\sin 2\omega(I_0 - 4I_2 + 3I_4) +$$
$$+ \sin 6\omega(5I_4 - 12I_6 + 7I_8)\}]$$

$$\omega_8 = -\frac{1}{96}\left[-\frac{3}{2}\sin\emptyset\,I_1 + \sin(4\omega + 3\emptyset)I_3 - \sin(4\omega + 5\emptyset)I_5 +\right.$$
$$\left.+ \frac{1}{4}\sin(8\omega + 7\emptyset)I_7 - \frac{1}{4}\sin(8\omega + 9\emptyset)I_9\right]$$

$$\omega_9 = \frac{1}{4}\cos\psi(I_1 - I_3).$$

Anomaly at epoch

$$\Delta\chi = +2\pi\frac{a\,\varrho_0\,\varkappa}{e}\exp\left[-\frac{a\,e}{H} - \frac{a\,e\,l}{H} - \lambda\cos 2\omega - \eta\cos\xi_0\right] \times$$
$$\times \chi\,[\chi_0 + \chi_1\,\delta + \chi_2\,\delta^2 + \lambda(\chi_3 + \chi_4\,\delta + \chi_5\,\delta^2) +$$
$$+ \varkappa_6\,\lambda_2 + \chi_7\,\lambda^3 + \chi_8\,\lambda^4 + \chi_9\,\sigma],\qquad\qquad(7.4)$$

where:

$\chi_0 = \omega_0$

$\chi_1 = \omega_1$

$\chi_2 = \omega_2$

$$\chi_3 = -\frac{1}{2}\left(\sin(2\omega + \emptyset)I_1 - \sin(2\omega + 3\emptyset)I_3\right) +$$
$$+ \frac{1}{4}e\sin 2\omega\left[I_0 - 4I_2 + 3I_4 + \frac{1}{4}e(6I_1 - 17I_3 + 11I_5) +\right.$$
$$\left.+ \frac{1}{8}e^2(10I_0 + I_2 - 26I_4 + 15I_6)\right]$$

$\chi_4 = \omega_4$

$\chi_5 = \omega_5$

$$\chi_6 = -\frac{1}{8}(1 - 2\delta)\left[-2\sin\emptyset\,I_1 + \sin(4\omega + 3\emptyset)I_3 - \sin(4\omega + 5\emptyset)I_5 -\right.$$
$$\left.- \frac{1}{2}e\sin 4\omega\left\{3I_2 - 8I_4 + 5I_6 - \frac{1}{4}e(5I_1 - 35I_3 + 59I_5 - 29I_7)\right\}\right]$$

$\chi_7 = \omega_7$

$\chi_8 = \omega_8$

$\chi_9 = \omega_9$

Right ascension of the ascending node

$$\Delta\Omega = -\pi a\,\varrho_0\,\varkappa\exp\left[-\frac{a\,e}{H} - \frac{a\,e\,l}{H} - \lambda\cos 2\omega - \eta\cos\xi_0\right] \times$$
$$\times \varkappa[\Omega_0 + \Omega_1\delta + \Omega_2'\delta^2 + \Omega_3\,\lambda + \Omega_4\lambda^2 + \Omega_5\lambda^3 + \Omega_6\lambda^4 + \Omega_7\sigma],\quad(7.5)$$

where:

$$\Omega_0 = 0$$

$$\Omega_1 = \frac{1}{2} \sec i \times$$

$$\times \left[\sin(2\omega + 2\emptyset) I_2 - 2e \sin 2\omega I_1 + \frac{1}{8} e^2 \sin 2\omega (11 I_0 - 2 I_2 - I_4) \right]$$

$$\Omega_2 = -\frac{1}{2} \sec i \left[\sin(2\omega + 2\emptyset) I_2 - e \sin 2\omega (3 I_1 + I_3) \right]$$

$$\Omega_3 = \frac{1}{4} \delta (1 - \delta) \sec i \left[\sin(4\omega + 4\emptyset) I_4 - e \sin 4\omega (3 I_3 - I_5) \right]$$

$$\Omega_4 = \frac{1}{16} \delta \sec i \left[\sin(2\omega + 2\emptyset) I_2 + \sin(6\omega + 6\emptyset) I_6 \right]$$

$$\Omega_5 = 0$$

$$\Omega_6 = 0$$

$$\Omega_7 = 0 .$$

Inclination of the orbit

$$\Delta i = - \pi a \varrho_0 \varkappa \exp \left[-\frac{a e}{H} - \frac{a e l}{H} - \lambda \cos 2\omega - \eta \cos \xi_0 \right] \times$$

$$\times \varkappa [i_0 + i_1 \delta + i_2 \delta^2 + i_3 \lambda + i_4 \lambda^2 + i_5 \lambda^3 + i_6 \lambda^4 + i_7], \qquad (7.6)$$

where:

$$i_0 = 0$$

$$i_1 = \frac{1}{2} \tan i \left[I_0 + \cos(2\omega + 2\emptyset) I_2 - 2e(1 + \cos 2\omega) I_1 + \right.$$

$$\left. + \frac{1}{4} e^2 \left\{ 3 I_0 + I_2 + \frac{1}{2} \cos 2\omega (11 I_0 - 2 I_2 - I_4) \right\} \right]$$

$$i_2 = -\frac{1}{2} \tan i \left[I_0 + \cos(2\omega + 2\emptyset) I_2 - 4e \left\{ I_1 + \frac{1}{4} \cos 2\omega (3 I_1 + I_3) \right\} \right]$$

$$i_3 = \frac{1}{4} \delta (1 - \delta) \tan i \left[I_0 + 2 \cos(2\omega + 2\emptyset) I_2 + \cos(4\omega + 4\emptyset) I_4 - \right.$$

$$\left. - 2e \left\{ I_1 + 2 \cos 2\omega I_1 + \frac{1}{2} \cos 4\omega (3 I_3 - I_5) \right\} \right]$$

$$i_4 = \frac{1}{16} \delta \tan i [2 I_0 + 3 \cos(2\omega + 2\emptyset) I_2 +$$

$$+ 2 \cos(4\omega + 4\emptyset) I_4 + \cos(6\omega + 6\emptyset) I_6]$$

$$i_5 = 0$$

$$i_6 = 0$$

$$i_7 = 0 .$$

8. Discussion

Some of the terms in Eqs. (7.1) to (7.6) have been given previously by many authors, and it would be a Herculean and pointless task to itemise the references. Generally in the present paper, the coefficients of e^4, δ^2, λ^3, λ^4 and σ are new together with the alterations involved in the bulge effect.

It is of importance to note the highly significant changes produced by the inclusion of the parameters δ, λ and σ; particularly λ. Though the oblateness and 'bulginess' of the atmosphere appear to be such small perturbations at first sight, a closer analysis reveals that λ is as large as e for a highly elliptical orbit, and that for satellites with perigees higher than about 300 kms at the equator the given theory of the bulge is inadequate. The present author hopes to publish a more sophisticated analysis of the bulge effect when more precise data on the figure of the bulge is available.

The author wishes to record his gratitude to Dr. G. V. GROVES, of University College, London, for his continuing interest; and to Dr. D. G. KING-HELE, of R. A. E., Farnborough, for the stimulation of his many papers on this subject.

Reference

[1] PRIESTER, W., H. A. MARTIN, und K. KRAMP: Sternwarte der Universität, Bonn. Nature 188, 4746, 200.

Integration of equations of celestial mechanics by Cowell's method with variable intervals

By

D. K. Kulikov

Institute for Theoretical Astronomy, Leningrad, U.S.S.R.

1. Numerical methods for the solution of different problems of celestial mechanics have acquired at present wide application and development. First of all they have been favoured by the introduction of fast program-controlled computers on the one hand and by practical requirements connected with investigations of the motion of artificial satellites on the other hand.

As it is well known, all the problems of celestial mechanics are reduced in the end to the solution of some differential equations, an analytical solution of which presents enormous difficulties. Whereas numerical methods being realized on fast computers make it possible to solve such problems efficiently and meet fully the requirement of practice.

Numerical methods of integration differential equations may be divided into two groups based on two different principles.

Methods of the first group are based on developments into TAYLOR-series, the higher-order derivatives being determined by differentiation of the given equation either analytically or numerically. The methods of EULER, ADAMS, STÜRMER, COWELL, MILN a.o. belong to this group.

Methods of the second group are also based on a development into series, but without using high-order derivatives in the computation. Instead at this step the right side of the equation is computed for different values of the variables. These methods are called by the names of their chief authors — RUNGE and KUTTA.

The above mentioned peculiarities enable to realize without great difficulties the methods of RUNGE-KUTTA on computers for the integration of differential equations with automatic choice of the interval; the results of calculations with normal and half-intervals serving as criteria for alterations of the interval.

Other more rational criteria may also be pointed out for alterations of the interval of integration [1].

Besides it should be noted, that using methods of the second group requires more substitutions into the differential equations, which involves more computational work, especially if the right sides of the equations are of a complicated structure. In addition a higher accuracy of the methods is connected with considerable difficulties, the structure of the working formulae becoming unwieldy and the formulae — of little use in practice.

Therefore in astronomy, where a high grade of accuracy is required, methods of the first group, not suffering from these shortcomings are widely adopted. Especially Cowell's method is practised on a large scale. Cowell devised his general method for the integration equations of motion of celestial mechanics in connection with the discovery of Jupiter's eighth satellite in 1908, when astronomers came to deal with a dynamic problem of general nature, as owing to great perturbations from the Sun, the motion of the VIII satellite hasn't any even remote resemblance with a motion conforming to Kepler's laws.

After a brilliant application of this method by Cowell and Crommelin [2, 3] to the investigation of the motion of Jupiter's VIII satellite and to the prediction of the return of Halley's famous comet in 1910, Cowell's method was generally recognized and took a firm position in astronomical practice.

Investigations carried out at the Institute for Theoretical Astronomy of the Academy of Sciences U.S.S.R. by V. F. Mjachin [4] on the evaluation of errors in methods of numerical integration displayed a high accuracy of Cowell's method in comparison with other, including Runge-Kutta's method.

Cowell's method has been widely and successfully used at the Institute for Theoretical Astronomy, both in its first-hand aspect [5] and modified by B. V. Numerov [6, 7] (extrapolation-method). Since 1954 Cowell's method and the extrapolation-method have been applied to the integration of the equations of motion of minor planets on the electronic computer BESM at the Computing Centre of the U.S.S.R. Academy of Sciences, as well as to the solution of other problems [8].

At the U.S.A. Naval Obervatory Cowell's modificated method has been adopted as basis for a very extensive work on the combined integration of the equations of motion of the outer planets, on the electronic computer IBM with an integration step of 40 days, for a time interval of 400 years [9].

However, the applications of Cowell's method to machine computations stated above have been performed for the simplest case, when the interval of integration remains constant.

Nevertheless, many problems of celestial mechanics, connected with an investigation of the motions fo comets, satellites, space rockets and other natural and artificial bodies of the Solar system require variable intervals.

Though the realization of Cowell's method on computing machines gives rize to certain difficulties, nevertheless a comparatively simple method may be pointed out, which enables to change the interval of integration and requires but little extra work.

2. As it is well known, the differential equations of motion in celestial mechanics may be written in vectorial form as

$$\frac{d^2 \bar{r}}{dt^2} = - \frac{k^2(1+m)\bar{r}}{r^3} + k^2 \sum_i m_i \left(\frac{\bar{r}_i - \bar{r}}{\varDelta_i^3} - \frac{\bar{r}_i}{r_i^3} \right) = \bar{R}(\bar{r}, t), \qquad (1)$$

where

$\bar{r}(x, y, z)$ is the heliocentric radius-vector of the body, the motion of which is being studied;

m the mass of this body;

$\bar{r}(x_i \, y_i \, z_i)$ the heliocentric radius-vector of the i-th perturbing body;

m_i the mass of the i-th perturbing body;

k the gravitational constant;

$$r^2 = x^2 + y^2 + z^2; \quad r_i^2 = x_i^2 + y_i^2 + z_i^2;$$

$$\varDelta_i^2 = (x_i - x)^2 + (y_i - y)^2 + (z_i - z)^2.$$

The peculiarity of Eq. (1) consists in that the right side doesn't contain first derivatives, therefore two integrations may be avoided — one to obtain $\dot{\bar{r}}$, and the other to get $\bar{r}$.

The basic idea of Cowell's method is to the effect, that one may immediately pass from the second derivative $\ddot{\bar{r}}$ to the unknown function $\bar{r}$ by one formula of double integration, using Stearling's interpolation formula to express the derivatives through differences.

If we denote by h the intervall of integration and put

$$\bar{f} = h^2 \, \bar{R}(\bar{r}, t), \qquad (2)$$

then we may write the basic integration formula in the form[1]:

$$r_k = \bar{f}_k^{-2} + \frac{1}{12} \bar{f}_k - \frac{1}{240} \bar{f}_k^2 + \frac{31}{60\,480} \bar{f}_k^4 - \frac{289}{3\,628\,800} \bar{f}_k^6 +$$

$$+ \frac{317}{22\,809\,600} \bar{f}_k^8 - \frac{6\,803\,477}{2\,615\,348\,736\,000} \bar{f}_k^{10} + \cdots, \qquad (3)$$

---

[1] Here we use the usual system of notations for the differences and sums adopted in astronomy [10].

where $\bar{f}_k^i$ are differences of the i-th order for line k, and $\bar{f}_k^{-2}$ is defined by:

$$\begin{aligned}
\bar{f}_k^{-2} &= \bar{f}_0^{-2} + \sum_{i=0}^{k-1} \bar{f}_{i+1/2}^{-1}, \\
\bar{f}_{i+1/2}^{-1} &= \bar{f}_{-1/2}^{-1} + \sum_{j=0}^{i} \bar{f}_j.
\end{aligned} \right\} \qquad (4)$$

We dispose of the arbitrariness in the choice of the initial values of the sums $\bar{f}_0^{-2}$ and $\bar{f}_{-1/2}^{-1}$ to satisfy the equalities:

$$\begin{aligned}
\bar{f}_0^{-2} &= \bar{r}_0 - \frac{1}{12} \bar{f}_0 + \frac{1}{240} \bar{f}_0^2 - \frac{31}{60480} \bar{f}_0^4 + \frac{289}{3628800} \bar{f}_0^6 - \cdots, \\
\bar{f}_{-1/2}^{-1} &= h\dot{r}_0 - \frac{1}{2} \bar{f}_0 + \frac{1}{12} \bar{f}_0^1 - \frac{11}{720} \bar{f}_0^3 + \frac{191}{60480} \bar{f}_0^5 - \frac{2497}{3628800} \bar{f}_0^7 + \\
&\quad + \frac{14797}{95800320} \bar{f}_0^9 - \frac{92427157}{2615348736000} \bar{f}_0^{11} + \cdots.
\end{aligned} \right\} \qquad (5)$$

Eqs. (3) and (4) with account of expression (5) represent the method of quadratures corresponding to COWELL's method.

The computation of the initial values of sums $\bar{f}_0^{-2}$ and $\bar{f}_{-1/2}^{-1}$ from given initial data $\bar{r}_0$ and $\dot{\bar{r}}_0$ relations (5) being accomplished, the further process of integration by formulae (3) and (4) is carried aut comparatively easy.

Let us consider the question about changing the interval of integration, when the process of integration is already adjusted. Usually this problem is solved by the computation of the values $\bar{r}_k$ and $\dot{\bar{r}}_k$ for a given moment t_k from the integration scheme; and using these values one starts the computation of the integration scheme for the changed interval with formulae (5).

However we choose another way to solve the problem raised; we build up a new integration scheme for a changed interval, on the basis of the former integration scheme, but without computing coordinates and velocities. This method is not only convenient for machine computations, but may be also successfully used for desk calculations.

Let ω be the new interval of integration.

Putting

$$\bar{F} = \omega^2 \bar{R}(\bar{r}, t)$$

and taking into consideration (2), we have

$$\bar{F} = \lambda^2 \bar{f}, \qquad (6)$$

where

$$\lambda = \frac{\omega}{h}. \qquad (7)$$

Thus the right side value of $\overline{F}$ with the new interval of integration is very simply expressed through the right side value of $\overline{f}$ at the former interval.

To find the initial values of the sums $\overline{F}_0^{-2}$ and $\overline{F}_{-1/2}^{-1}$, we apply to relations (5), which we write down in the form

$$
\left.\begin{aligned}
\overline{f}_0^{-2} &= \overline{r}_0 - \frac{1}{12}\overline{f}_0 + \sum_{i=1}^{5}\beta_{2i}\overline{f}_0^{2i} - \cdots, \\
\overline{f}_{-1/2}^{-1} &= h\dot{\overline{r}}_0 - \frac{1}{2}\overline{f}_0 + \sum_{i=0}^{5}\beta_{2i+1}\overline{f}_0^{2i+1} + \cdots.
\end{aligned}\right\} \tag{8}
$$

The coefficients β_j do not depend on the interval of integration.

Applying these relations to the function $\overline{F}$ and then eliminating $\overline{r}_0$ and $\dot{\overline{r}}_0$ by means of (8); further taking into account (6) and (7), we obtain:

$$
\left.\begin{aligned}
\overline{F}_0^{-2} &= \overline{f}_0^{-2} + \frac{1}{12}(1-\lambda^2)\overline{f}_0 + \sum_{i=1}^{5}\beta_{2i}(\overline{F}_0^{2i} - \overline{f}_0^{2i}) - \cdots, \\
\overline{F}_{-1/2}^{-1} &= \lambda\,\overline{f}_{-1/2}^{-1} + \frac{1}{2}\lambda(1-\lambda)\overline{f}_0 + \sum_{i=0}^{5}\beta_{2i+1}(\overline{F}_0^{2i+1} - \lambda\overline{f}_0^{2i+1}) + \cdots.
\end{aligned}\right\} \tag{9}
$$

Now we express the differences of function $\overline{F}$ through differences of the known function $\overline{f}$. Such a dependence may be established by means of STEARLING's interpolation formula, which can be written in the form:

$$
\overline{f}(t_0 + z\,h) = \overline{f}_0 + \sum_{i=1}^{\infty} A_i\,\overline{f}_0^i, \tag{10}
$$

where the coefficients A_i are determined by the recurrent relations:

$$
\left.\begin{aligned}
A_{2j+1} &= \frac{z^2 - j^2}{2j(2j+1)}\,A_{2j-1}, \\
A_{2j+2} &= \frac{z}{2j+2}\,A_{2j+1},
\end{aligned}\right\} \tag{11}
$$

where

$$
A_1 = z.
$$

We find from formula (10)

$$
h^n\left(\frac{d^n\overline{f}}{dt^n}\right)_0 = \sum_{i=1}^{\infty}\left(\frac{d^n A_i}{dz^n}\right)_{z=0}\overline{f}_0^i, \tag{12}
$$

where the variables t and z are connected by the relation

$$
t = t_0 + z\,h.
$$

Differentiating expression (11) n times and putting $z = 0$, we easily find the recurrent relations for the computation of the derivatives $A_i^{(n)}$ in formula (12). We have

$$
\left.
\begin{aligned}
A_{2j+1}^{(n)} &= \frac{1}{2j(2j+1)} \left[n(n-1) A_{2j-1}^{(n-2)} - j^2 A_{2j-1}^{(n)} \right], \\
A_{2j+2}^{(n)} &= \frac{n}{2j+2} A_{2j+1}^{(n-1)},
\end{aligned}
\right\} \tag{13}
$$

where

$$
A_1^{(1)} = 1.
$$

Owing to (11) and (12) it is easy to establish the following properties of the coefficients:

$$
\left.
\begin{aligned}
A_i^{(0)} &= 0; \qquad A_i^{(n)} = 0 \quad \text{at} \quad n > i; \\
A_n^{(n)} &= 1; \quad A_{2j+1}^{(2\sigma)} = 0; \qquad A_{2j+2}^{(2\sigma+1)} = 0; \\
A_{2j+1}^{(1)} &= (-1)^j \frac{(j!)^2}{(2j+1)!}; \qquad A_{2j+2}^{(2)} = (-1)^j \frac{2!(j!)^2}{(2j+2)!}; \\
A_{2j+1}^{(3)} &= (-1)^{j+1} \frac{3!(j!)^2}{(2j+1)!} \sum_{k=1}^{j} \frac{1}{k^2}; \\
A_{2j+2}^{(4)} &= (-1)^{j+1} \frac{4!(j!)^2}{(2j+2)!} \sum_{k=1}^{j} \frac{1}{k^2}.
\end{aligned}
\right\} \tag{14}
$$

As to the higher order derivatives, they must be found directly from the recurrent formulae (13).

So, expression (12) may be written in the form:

$$
h^n \left(\frac{d^n \bar{f}}{dt^n} \right)_0 = \sum_{i=n}^{\infty} A_i^{(n)} \bar{f}_0^i.
$$

We have for function $\bar{F}$

$$
\omega^n \left(\frac{d^n \bar{F}}{dt^n} \right)_0 = \sum_{i=n}^{\infty} A_i^{(n)} \bar{F}_0^i.
$$

From these relations, and taking into account formulae (6) and (7) we obtain

$$
\sum_{i=n}^{\infty} A_i^{(n)} \bar{F}_0^i = \lambda^{n+2} \sum_{i=n}^{\infty} A_i^{(n)} \bar{f}_0^i. \tag{15}
$$

It may be easily shown, that the matrix of the system is triangular and its determinant Δ is different from zero ($\Delta = 1$). Therefore the system (15) can be solved comparatively easily. After having solved system (15) and substituted the determined values $\bar{F}_0^i$ into (9), we obtain

$$\begin{aligned}
\bar{F}_0^{-2} &= \bar{f}_0^{-2} + \frac{1-\lambda^2}{12}\bar{f}_0 - \frac{1-\lambda^2}{240}(1+\lambda^2)\bar{f}_0^2 + \frac{1-\lambda^2}{60480}(31+31\lambda^2+10\lambda^4) \times \\
&\quad \times \bar{f}_0^4 - \frac{1-\lambda^2}{3628800}(289+289\lambda^2+121\lambda^4+21\lambda^6)\bar{f}_0^6 + \\
&\quad + \frac{1-\lambda^2}{159667200}(2219+2219\lambda^2+1031\lambda^4+261\lambda^6+30\lambda^8)\bar{f}_0^8 - \\
&\quad - \frac{1-\lambda^2}{2615348736000}(6803477+6803477\lambda^2+3344021\lambda^4+ \\
&\quad + 998821\lambda^6+179002\lambda^8+15202\lambda^{10})\bar{f}_0^{10} + \cdots,
\end{aligned}$$

$$\begin{aligned}
\bar{F}_{-1/2}^{-1} &= \lambda\bar{f}_{-1/2}^{-1} + \frac{\lambda(1-\lambda)}{2}\bar{f}_0 - \frac{\lambda(1-\lambda^2)}{12}\bar{f}_0^1 + \frac{\lambda(1-\lambda^2)}{720}(11+\lambda^2)\bar{f}_0^3 - \\
&\quad - \frac{\lambda(1-\lambda^2)}{60480}(191+23\lambda^2+2\lambda^4)\bar{f}_0^5 + \\
&\quad + \frac{\lambda(1-\lambda^2)}{3628800}(2497+337\lambda^2+43\lambda^4+3\lambda^6)\bar{f}_0^7 - \\
&\quad - \frac{\lambda(1-\lambda^2)}{95800320}(14797+2125\lambda^2+321\lambda^4+35\lambda^6+2\lambda^8)\bar{f}_0^9 + \\
&\quad + \frac{\lambda(1-\lambda^2)}{2615348736000}(92427157+13803157\lambda^2+2295661\lambda^4+ \\
&\quad + 307961\lambda^6+28682\lambda^8+1382\lambda^{10})\bar{f}_0^{11} - \cdots,
\end{aligned}$$

$$\bar{F}_k = \lambda^2 \bar{f}_{\lambda k}$$

(16)

Relations (16) enable to transform the integration scheme for any interval $\omega = \lambda h$. But in practice it is convenient to change the interval twice in the way of increase, as well as in the way of decrease. We shall confine ourselves below to the discussion of these particular cases.

Denoting the right sides of the equations for the increased interval by $\bar{F}$, and for the reduced one by $\bar{H}$, we obtain the formulae presented below.

Doubling the interval

$$\begin{aligned}
\bar{F}_0^{-2} &= \bar{f}_0^{-2} - \frac{1}{2^2}\bar{f}_0 + \frac{1}{2^4}\bar{f}_0^2 - \frac{1}{2^6}\bar{f}_0^4 + \frac{1}{2^8}\bar{f}_0^6 - \\
&\quad - \frac{1}{2^{10}}\bar{f}_0^8 + \frac{1}{2^{12}}\bar{f}_0^{10} - \cdots,
\end{aligned}$$

$$\begin{aligned}
\bar{F}_{-1/2}^{-1} &= 2\bar{f}_{-1/2}^{-1} - \bar{f}_0 + \frac{1}{2}\bar{f}_0^1 - \frac{1}{2^3}\bar{f}_0^3 + \frac{1}{2^5}\bar{f}_0^5 - \\
&\quad - \frac{1}{2^7}\bar{f}_0^7 + \frac{1}{2^9}\bar{f}_0^9 - \frac{1}{2^{11}}\bar{f}_0^{11} + \cdots,
\end{aligned}$$

(17)

$$\bar{F}_k = 4\bar{f}_{2k}.$$

Halving the interval

$$\bar{H}_0^{-2} = \bar{f}_0^{-2} + \frac{1}{2^4}\bar{f}_0 - \frac{1}{2^8}\bar{f}_0^2 + \frac{1}{2^{11}}\bar{f}_0^4 - \frac{5}{2^{16}}\bar{f}_0^6 + \frac{7}{2^{19}}\bar{f}_0^8 - \frac{21}{2^{23}}\bar{f}_0^{10} + \cdots,$$

$$\bar{H}_{-1/2}^{-1} = \frac{1}{2}\bar{f}_{-1/2}^{-1} + \frac{1}{2^3}\bar{f}_0 - \frac{1}{2^5}\bar{f}_0^1 + \frac{3}{2^9}\bar{f}_0^3 - \frac{5}{2^{12}}\bar{f}_0^5 +$$

$$\qquad\quad + \frac{35}{2^{17}}\bar{f}_0^7 - \frac{63}{2^{20}}\bar{f}_0^9 + \frac{231}{2^{24}}\bar{f}_0^{11} - \cdots,$$

$$\bar{H}_k = \frac{1}{4}\bar{f}_{k/2}.$$

$$\text{(18)}$$

Formulae (17) and (18) not only enable to transform comparatively easily the integration scheme by doubling or halving the interval of integration, but allow also to establish appropriate criteria for changing the interval of integration, which is especially important when computations are carried out on fast electronic computers.

3. Discussing the criteria for the choise of an interval of integration and working formulae for machine computations we must confine ourselves to a definite degree of accuracy for the development obtained above.

For the sake of simplicity we shall limit ourselves to accounting forth differences. By reason of (3) the basic integration formulae in this case will be

$$\bar{r}_k = \bar{f}_k^{-2} + \frac{1}{12}\bar{f}_k - \frac{1}{240}\bar{f}_k^2 + \frac{31}{60480}\bar{f}_k^4, \qquad (19)$$

where $\bar{f}_k^{-2}$ is defined by relation (4).

This, to compute the value $\bar{r}_k$ we must know besides $\bar{f}_k^{-2}$, the value of the right side or Eq. (2) for five moments located symmetrically in respect to t_k. Therefore, assuming conventionally $k = 0$, we must keep the integration scheme in the machine storage in the form of

$$\boxed{\bar{f}_0^{-2}, \bar{f}_{-1/2}^{-1}, \bar{f}_{-2}, \bar{f}_{-1} \;\vert\; \bar{f}_0, \bar{f}_1, \bar{f}_2} \qquad (20)$$

The quantities in the continuous frame are accurate[1]; the quantity $\bar{f}_0$ will be precised by a process of iteration, and the values $\bar{f}_1, \bar{f}_2$ will be computed approximately, the first one to account for $\bar{f}_0^2$, and the second one — for $\bar{f}_0^4$.

Thus, when integrating by formula (19) we must "glance" two steps in advance.

---

[1] The term "accurate value" has been used here in the sense, that the approximation process is completed for this quantity and the latter is not to be further precised.

In machine computations it is convenient to use ordinate formulae [11]. As

$$\bar{f}_k^n = \sum_{j=0}^{n} (-1)^j C_n^j \bar{f}_{k + n/2 - j},$$ (21)

so by reason of (19) we have

$$
\left.
\begin{aligned}
\bar{r}_0 &= \bar{f}_0^{-2} + \frac{1}{60480}(31\bar{f}_{-2} - 376\bar{f}_{-1} + 5730\bar{f}_0 - 376\bar{f}_1 + 31\bar{f}_2), \\
\bar{r}_1 &= \bar{f}_0^{-2} + \bar{f}_{-1/2}^{-1} + \\
&\quad + \frac{1}{60480}(31\bar{f}_{-2} - 124\bar{f}_{-1} + 60414\bar{f}_0 - 5420\bar{f}_1 - 221\bar{f}_2), \\
\bar{r}_2 &= \bar{f}_0^{-2} + 2\bar{f}_{-1/2}^{-1} + \\
&\quad + \frac{1}{60480}(-221\bar{f}_{-2} + 1136\bar{f}_{-1} + 118626\bar{f}_0 + 62624\bar{f}_1 + 431\bar{f}_2).
\end{aligned}
\right\}
$$ (22)

In constructing these formulae it was assumed, that the forth differences were constant.

The integration process, using formulae (22) is carried out on machines in the following way. First $\bar{r}_2$ is determined, by reason of (20) and (22), and then it is substituted in the right side of Eq. (2). As a result we get a more accurate value of $\bar{f}_2$, which is to be inserted into the integration scheme (20), after formation of a difference between the new value of $\bar{f}_2$ and its precedent value. Then we redetermine $\bar{r}_2$ and substitute it again in the right side of the equation. This process is to be repeated until a given relative accuracy ε will be attained, i.e. the condition

$$|(\bar{f}_k)_i - (\bar{f}_k)_{i-1}| < \varepsilon |f_k|$$ (23)

is fulfilled.

The inequality (23) being satisfied, we pass to a more accurate determination of $\bar{f}_1$ in quite the same way, as was shown above for $\bar{f}_2$. Finally we repeat the same procedure for a more accurate determination of $\bar{f}_0$. When the total of approximations for all the three values $\bar{f}_0, \bar{f}_1, \bar{f}_2$ amounts to more than four, then the whole process is to be repeated from the beginning. When the process of approximations is concluded, the integration scheme (20) must be shifted by one line and the process described above repeated for the next step, and so on. The integration with a constant interval is accomplished in that way.[1]

When the integration is carried out with a variable interval, then before shifting the integration scheme by one line, we must decide

---

[1] The iteration process can be speeded up by taking advantage of the method of TITJEN-SUBBOTIN [12].

whether to continue the integration with the former interval, or to change it.

Criteria for changing the interval of integration can be established on the grounds of (17), (18) and (3).

For doubling the interval of integration it is necessary and sufficient to fulfill the following conditions, resulting directly from (17) and (3):

$$\frac{2^{-5}\left|\overline{f_0^5}\right|}{\left|\overline{F_{-1/2}^{-1}}\right|} < \varepsilon, \qquad \frac{2^{-8}\left|\overline{f_0^6}\right|}{\left|\overline{F_0^{-2}}\right|} < \varepsilon, \qquad \frac{289}{3\,628\,800}\frac{\left|\overline{F_0^6}\right|}{\left|\overline{F_0^{-2}}\right|} < \varepsilon,$$

where ε is the permissible relative error of computation. As fifth and sixth differences are not available in the integration scheme, so we replace them in the above inequalities by fourth differences. Thereby we reinforce the requirements as to accuracy, because it is assumed, that the differences are diminishing. By reason of (15) and (17) we have approximately
$$\overline{F_0^{-2}} \approx \overline{f_0^{-2}}, \qquad \overline{F_{-1/2}^{-1}} \approx 2\overline{f_{-1/2}^{-1}}, \qquad \overline{F_0^4} \approx 2^6\overline{f_0^4}$$

and consequently

$$\left|\overline{f_0^4}\right| < 2^6\left|\overline{f_{-1/2}^{-1}}\right|\varepsilon, \qquad \left|\overline{f_0^4}\right| < 2^8\left|\overline{f_0^{-2}}\right|\varepsilon, \qquad \left|\overline{f_0^4}\right| < 2^7\left|\overline{f_0^{-2}}\right|\varepsilon.$$

It is easy to see, that if the first inequality is satisfied, the next two will also be fulfilled; therefore, if

$$\left|\overline{f_0^4}\right| < 2^6\left|\overline{f_{-1/2}^{-1}}\right|\varepsilon \tag{24}$$

the interval of integration must be doubled.

Halving the interval is to be done, when one of the following inequalities, resulting from (18) and (3) will be upset:

$$\frac{5\left|\overline{f_0^5}\right|}{2^{12}\left|\overline{H_{-1/2}^{-1}}\right|} < \varepsilon, \qquad \frac{5\left|\overline{f_0^6}\right|}{2^{16}\left|\overline{H_0^{-2}}\right|} < \varepsilon, \qquad \frac{289}{3\,628\,800}\frac{\left|\overline{f_0^6}\right|}{\left|\overline{f_0^{-2}}\right|} < \varepsilon.$$

By analogy with the foregoing wo have

$$\left|\overline{f_0^4}\right| < 2^9\left|\overline{f_{-1/2}^{-1}}\right|\varepsilon, \qquad \left|\overline{f_0^4}\right| < 2^{14}\left|\overline{f_0^{-2}}\right|\varepsilon, \qquad \left|\overline{f_0^4}\right| < 2^{13}\left|\overline{f_0^2}\right|\varepsilon.$$

We may see from these relations, that the first inequality will be upset first of all. Therefore halving the interval of integration must be performed as soon, as
$$\left|\overline{f_0^4}\right| \geqq 2^9\left|\overline{f_{-1/2}^{-1}}\right|\varepsilon. \tag{25}$$

Now we must check the coordination of the gained criteria. Let us assume, that criterion (25) is met. Then, after halving the interval of integration, applying of criterion (24) gives:

$$\left|H_0^4\right| \approx 2^{-6}\left|\overline{f_0^4}\right| \approx 2^3\left|\overline{f_{-1/2}^{-1}}\right|\varepsilon \approx 2^4\left|\overline{H_{-1/2}^{-1}}\right|\varepsilon < 2^6\left|\overline{H_{-1/2}^{-1}}\right|\varepsilon,$$

and hence the very first check of criterion (24) after halving the interval will lead to its doubling. Therefore criteria (24) and (25) are not coordinated.

To bring the said criteria into the necessary correspondence, we inforce criterion (24) four times, and weaken twice criterion (25). Weakening criterion (25) is justified by that we have replaced the fifth differences by fourth ones. So we get finally the following criterion for changing the interval of integration:

$$2^4|\bar{f}_{-1/2}^{-1}|\,\varepsilon < |\bar{f}_0^4| < 2^{10}|\bar{f}_{-1/2}^{-1}|\,\varepsilon, \tag{26}$$

where
$$\bar{f}_0^4 = \bar{f}_{-2} - 4\bar{f}_{-1} + 6\bar{f}_0 - 4\bar{f}_1 + \bar{f}_2.$$

If the inequality (26) is upset to the left, the interval of integration must be doubled, and if it is upset to the right, the interval must be halved. If the modulus of vector $|\bar{f}_0^4|$ satisfies the inequality (26), the integration must be performed without changing the interval.

The recomputation of the integration scheme at changing the interval of integration is done by means of formulae (17) or (18); it is expedient to express them through ordinates. It must be emphasized, that after a recomputation and shifting by one line the integration scheme assumes the form (20).

4. Now it remains to discuss the computation of the initial integration scheme (20), according to the initial conditions of motion $\bar{r}_0$ and $\dot{\bar{r}}_0$ at the moment t_0.

In computing the initial values of the sums $\bar{f}_0^{-2}$ and $\bar{f}_{-1/2}^{-1}$ we must increase the requirements to accuracy. This is especially important in respect to $\bar{f}_{-1/2}^{-1}$, as an error of this quantity will increase in proportion to time.

Besides, taking into account forth differences in COWELL's integration formula secures an accuracy to the fifth differences through; hence the fifth differences must be taken into account in the determination of the initial sums by means of relation (5); therefore we compute the right sides of Eq. (2) for seven moments, located symmetrically in respect to t_0.

Assuming the sixth differences to be constant, and considering relations (3), (4) and (2), we obtain the following working formulae for the computation of coordinates at the indicated moments:

$$\begin{aligned}
\bar{r}_1 &= \bar{f}_0^{-2} + \bar{f}_{-1/2}^{-1} + \gamma(-289\bar{f}_{-3} + 1734\bar{f}_{-2} - 2475\bar{f}_{-1} + \\
&\quad + 3612020\bar{f}_0 + 339465\bar{f}_1 - 20826\bar{f}_2 + 1571\bar{f}_3), \\
\bar{r}_2 &= \bar{f}_0^{-2} + 2\bar{f}_{-1/2}^{-1} + \gamma(1571\bar{f}_{-3} - 11286\bar{f}_{-2} + 34725\bar{f}_{-1} + \\
&\quad + 7200140\bar{f}_0 + 3667005\bar{f}_1 + 306474\bar{f}_2 - 9829\bar{f}_3), \\
\bar{r}_3 &= \bar{f}_0^{-2} + 3\bar{f}_{-1/2}^{-1} + \gamma(-9829\bar{f}_{-3} + 70374\bar{f}_{-2} - 217695\bar{f}_{-1} + \\
&\quad + 11265140\bar{f}_0 + 6856125\bar{f}_1 + 3873414\bar{f}_2 + 237671\bar{f}_3),
\end{aligned} \tag{27}$$

where

$$\gamma = \frac{1}{3\,628\,800}.$$

As to $\bar{r}$ with negative indexes, they are easily received from corresponding expressions for $\bar{r}$ with positive indexes, owing to the symmetry of the scheme in respect to the zero line.

If we write formula (27) in its general from

$$\bar{r}_k = \bar{f}_0^{-2} + k\,\bar{f}_{-1/2}^{-1} + \sum_{i=-n}^{+n} A_{ki}\,\bar{f}_i \tag{28}$$

the corresponding formula for $\bar{r}_{-k}$ will be

$$\bar{r}_{-k} = \bar{f}_0^{-2} - k(\bar{f}_{-1/2}^{-1} + \bar{f}_0) + \sum_{i=-n}^{+n} A_{ki}\,\bar{f}_{-i}. \tag{29}$$

Formulae (5) serve to compute the initial values of the sums; with account of sixth differences they may be written in the form:

$$\left.\begin{aligned}
\bar{f}_0^{-2} &= \bar{r}_0 + \gamma(289\bar{f}_{-3} - 3594\,\bar{f}_{-2} + 26\,895\,\bar{f}_{-1} - \\
&\quad - 349\,580\bar{f}_0 + 26\,895\bar{f}_1 - 3\,594\,\bar{f}_2 + 289\bar{f}_3), \\
\bar{f}_{-1/2}^{-1} &= h\,\dot{\bar{r}} + \gamma(-5730\bar{f}_{-3} + 50\,640\,\bar{f}_{-2} - 235\,290\,\bar{f}_{-1} - \\
&\quad - 1\,814\,400\bar{f}_0 + 235\,290\bar{f}_1 - 50\,640\bar{f}_2 + 5730\bar{f}_3).
\end{aligned}\right\} \tag{30}$$

The stated problem is solved by the iteration method. First, we compute $\bar{f}_0$ for the moment t_0 from a given value of $\bar{r}_0$. As a first approximation we assume, that the values of the right sides of equations are constant and equal to $\bar{f}_0$. Further we compute with these values of the right sides and the initial conditions $\bar{r}_0$ and $\dot{\bar{r}}_0$ the quantities $\bar{f}_0^{-2}$ and $\bar{f}_{-1/2}^{-1}$, using formula (30). Then we compute by means of formulae (27)–(29) and (2) the values $\bar{f}_k$ for all moments t_k (except t_0). After that we recompute again all the quantities $\bar{f}_0^{-2}$ and $\bar{f}_{-1/2}^{-1}$ and so on. This iteration process will be repeated until the differences between the following and the preceding values of $\bar{f}_0^{-2}$, $\bar{f}_{-1/2}^{-1}$, $\bar{f}_k$ become less than a given relative accuracy ε.

As to the interval h, it has been chosen beforehand. But if there is no confidence, that the interval has been chosen correctly, one must make use of criterion (26).

As a result of the process described above, we obtain an initial scheme in such a form:

$\bar{f}_0^{-2},\ \bar{f}_{-1/2}^{-1}$	$\bar{f}_{-3},\ \bar{f}_{-2}$	$\bar{f}_{-1},\ \bar{f}_0,\ \bar{f}_1$	$\bar{f}_2,\ \bar{f}_3$

The quantities in the continuous frames are accurate. This scheme may be easily transformed into an initial integration scheme (20); hence we have solved the stated problem.

Thus we have discussed in detail the integration process with a variable interval by Cowell's quadrature method, taking into account forth differences.

In conclusion it may be noticed, that formulae for the integration with sixth and eighth differences are listed by the author in article [13].

References

[1] Мячин, В. Ф.: Об одном критерии смены шага при численном интегрировании уравнений небесной механики по обобщенному методу Рунге. Бюлл. Ин-та теор. астр. т. VIII, № 2 (1961).

[2] Cowell, P. H., and A. D. Crommelin: The orbit of Jupiter's Eighth Satellite. M. N. 68, 576 (1908).

[3] Cowell, P. J., and A. D. Crommelin: Essay on the return of Halley's comet. Astronom. Ges., 23 (1910).

[4] Мячин, В. Ф.: Оценка погрешности численных методов интегрирования уравнений небесной механики. Бюлл. Ин-та теор. астр. т. VIII, № 8 (1962).

[5] Куликов, Д. К.: Численные методы небесной механики в применении к изучению движения VIII спутника Юпитера. Бюлл. Ин-та теор. астр. т. IV, № 7 (1950).

[6] Numerov, B.: Méthode nouvelle de la détermination des orbites et le calcul des éphémérides en tenant compte des perturbations. Труды главн. Российск. астрофизич. обс. т. 2, 188 (1923).

[7] Самойлова-Яхонтова, Н. С., и С. Г. Маковер: Вычисление частных возмущений малых планет на счётно-аналитических машинах. Бюлл. Ин-та теор. астр. т. V, № 3 (1952).

[8] Извеков, В. А.: Вычисление эфемерид малых планет на электронной машине БЭСМ. Бюлл. Ин-та теор. астр. т. VII, № 9 (1960).

[9] Eckert, W. J., D. Brouwer, and G. M. Clemence: Coordinates of the five outer planets 1653—2060. Astr. Papers XII (1951).

[10] Субботин, М. Ф.: Курс небесной механики, т. 2, ОНТИ. НКТП. М-Л. (1937).

[11] Токмалаева, С. С.: Ординатные формулы численного интегрирования обыкновенных дифференциальных уравнений 1-го порядка. Вычислительная математика № 5 (1959).

[12] Subbotin, M.: Sur le calcul des coordonnées héliocentriques des planètes et des comètes au moyen des quadratures. Poulk. Observ. Circ. No. 9 (1933).

[13] Куликов, Д. К.: Интегрирование уравнений движения небесной механики на электронных вычислительных машинах по квадратурному методу Коуэлла с автоматическим выбором шага. Бюлл. Ин-та теор. астр. т. VII, № 10 (1960).

The determination of atmospheric drag on artificial satellites

By

Luigi G. Jacchia

Smithsonian Institution Astrophysical Observatory
Cambridge, Mass., U.S.A.

Abstract. While all the forces acting on a satellite in orbital motion produce short-periodic perturbations in the major axis, only solar-radiation pressure and atmospheric drag are capable of producing long-periodic and secular variations. In the absence of direct drag-pressure measurements, the atmospheric drag on a satellite can be determined from the secular variation of its period of revolution (the orbital acceleration) when the effect of solar-radiation pressure is removed.

Orbital accelerations computed as by-products of orbit determinations are bound to be inaccurate for a number of reasons. Much better results are obtained by a method involving the analysis of the mean anomaly over long time-intervals. Analytical time functions are fitted to all the final elements; no systematic residuals are permitted, except in the mean anomaly, where residuals ΔM up to 10^{-3} revolutions are tolerated. The individual values of ΔM are plotted and a smooth curve is drawn through the points. The period variation is determined from the reference equation for the mean anomaly and from the second time-derivative of the ΔM curve, with allowance for the acceleration of the perigee.

1. The purpose of this paper is to describe a practical method for determining the atmospheric drag on artificial satellites. The most practical method, of course, consists in measuring the aerodynamic pressure on the satellite by means of a gauge that transmits signals to ground station. One such gauge flown on Sputnik III seems to have yielded satisfactory results.

Most satellites, however, do not have such a gauge aboard. Even if they did, it is doubtful whether the telemetered material would be accurate and continuous enough to account for the orbital variations caused by the drag. Furthermore, in view of the limited sensitivity of pressure gauges, the results would be reliable only at relatively low heights. Thus, the gauge on Sputnik III recorded pressures only up to a height of 500 km, even though we know that orbital changes caused by atmospheric drag can be measured quite accurately when the perigee of a satellite is 1000 km and more above the surface of the earth.

The pressures directly measured on Sputnik III have been useful because they have confirmed the atmospheric densities previously obtained from orbital data; we have to recognize, however, that nearly all we know today about atmospheric densities and their variations at heights above 200 km has been derived from analysis of the secular decrease in period of a number of artificial satellites. It is with this method that I shall be concerned in this paper.

2. Atmospheric drag is only one of several perturbing forces acting on a satellite in motion. By "perturbing force" I mean any force that has the effect of deflecting the satellite from the Keplerian orbit that a body would describe in the idealized two-body problem in which the two bodies are point masses. Perturbing forces arise from the irregular shape of the earth, from the pressure of solar radiation, and from the attraction of the nearest celestial bodies. All these forces may cause both short-periodic and long-periodic perturbations, as well as secular perturbations.

In view of the different ways in which the perturbing forces act on a satellite, their effects on a specific orbital element bears very little relation to their relative intensities. This, as we shall see, is a very fortunate circumstance for the determination of air drag. The perturbing forces arising from the departure of the earth's figure from spherical symmetry are of the order of 1 dyne, while the atmospheric drag of the Vanguard I satellite is of the order of 10^{-5} dynes. Yet, the atmospheric drag on Vanguard I produces a very noticeable secular decrease in the semimajor axis of the orbit, while the gravitational anomalies leave the major axis secularly undisturbed. As a matter of fact, KOZAI (1959a, 1959b) and BROUWER (1959) have shown that gravitational perturbations, whether arising from the earth's figure or from the attraction of the sun and the moon, never produce any appreciable long-periodic perturbations upon the orbital semimajor axis (and, therefore, upon the anomalistic period) of a close earth satellite. Thus, provided that in the analysis of the period variations we do not go to time intervals smaller than one revolution of the satellite, we are left with only two contending effects: atmospheric drag and solar-radiation pressure.[1] The separation of these two effects is an easy matter, assuming that we have the right program to compute radiation-pressure perturbations.

3. Solar-radiation pressure affects the orbital period when the satellite spends part of the time in the earth's shadow, which is commonly

---

[1] Note added in proof. In this paper it was tacitly assumed that the observations used to determine the drag are field-reduced photographic observations or Minitrack observations with an accuracy of a few minutes of arc. When the precision of the observations is of a few *seconds* of arc, as in the case of photo-reduced BAKER-NUNN camera positions, twelve-hour oscillations, due to the ellipticity of the equator, emerge in the mean anomaly as well as in the other orbital elements.

the case. For a relatively close satellite with a moderately eccentric
orbit $(0.1 < e < 0.2)$, the variations in period dP/dt caused by solar-
radiation pressure are of the order of $\pm 1 \times 10^{-7} A/m$, when the area/
mass ratio A/m is expressed in cm²/g. For comparison, the atmospheric
drag at intermediate heights gives rise to values of dP/dt of the order
of $-1 \times 10^9 \varrho A/m$, where ϱ is the atmospheric density in g/cm³. Thus,
when ϱ is of the order of 10^{-16} g/cm³, the effect of solar-radiation pressure
may equal that of atmospheric drag. At times of sunspot maximum,
this will occur at a height of 900 km; at times of low solar activity,
however, when the atmosphere is appreciably contracted, it will occur
as low as 500 km above the earth. If we want to determine atmospheric
drag with a 10-percent accuracy or better, we must take into account
solar-radiation pressure whenever the perigee height of the satellite is
greater than 400 km.

The program for computing perturbations due to solar-radiation
pressure must be one in which the *observed* orbit is used as a basis all the
time. This avoids the tedious iterations and the limitations in the useful
time interval when an initial orbit is modified by continually adding
computed perturbations — a procedure which, because of the irregularities in the atmospheric drag, can never be very accurate. The program we use at the Smithsonian Astrophysical Observatory was prepared
by Dr. KOZAI and follows his theory of the effect (KOZAI, 1961).

Only for spherical satellites can the effect of solar-radiation pressure
be accounted for with some degree of precision. For any other shape,
the computation based on an average A/m ratio is bound to be somewhat
in error. Even supposing A/m were known for any given instant, the
computation of the radiation pressure would be exceedingly complicated.

4. Our problem is thus reduced to the determination of the change
in orbital period of the satellite. This is not a simple operation, and we
must start by examining some of the prerequisites for the determination
of the orbital period itself. If we write

$$M = M_0 + \dot{M}_0(t - t_0) + \frac{\ddot{M}_0}{2}(t - t_0)^2 + \cdots, \tag{1}$$

where M is the mean anomaly, and the subscript 0 refers to the time
$t = t_0$, we clearly have
$$\dot{M}_0 = n_0 = 1/P_0,$$
        (n mean motion, in revolutions; P anomalistic period)
and
$$\dot{P}_0 = -\ddot{M}_0/\dot{M}_0^2. \tag{2}$$

The secular acceleration $\dot{P} = dP/dt$, a non-dimensional quantity,
is always relatively small, of the order of 10^{-5} for low satellites and 10^{-7}

to 10^{-8} for higher ones; therefore, $\dot{M}$ changes very little compared to $\ddot{M}$, and over reasonable time intervals, $\dot{P}$ is proportional to $-\ddot{M}$.

Even for orbits of modest eccentricity, M does not vary smoothly in the course of one revolution because the atmospheric-drag perturbation is concentrated in the vicinity of the perigee. We can, however, treat this short-period variation of the drag in the same fashion as we do the gravitational short-periodic perturbations; a complete and practical theory of the effect has been given by IZSÁK (1960). The Differential Orbit Improvement (D.O.I.) program of the Smithsonian Astrophysical Observatory (VEIS and MOORE, 1960) computes mean elements in which the effect of short-periodic perturbations is removed, but not that of long-periodic perturbations. For gravitational effects, the known numerical values of the zonal harmonics of the earth's potential are used. For the short-periodic drag effects, we can by an iteration process determine the amplitude from the empirical coefficients in Eq. (1). If M is expressed in revolutions, the maximum error ΔM that arises from neglecting the periodic drag perturbations is smaller than $\dot{P}/8$; this means that it may exceed 1 second of arc only for values of $\dot{P}$ larger 10^{-5}. Once we have removed the short-periodic drag perturbations, or satisfied ourselves that they are negligible, we can treat ϱ as a slowly varying function of time and identify $\Delta P/\Delta t$ (the change of period during one revolution) with dP/dt.

5. The least laborious, but also the least accurate, way to determine $\dot{P}$ is to derive it directly, as a by-product of orbital computations, by treating as unknowns all the coefficients of the time expansion (1) of the mean anomaly M. Although for very low satellites strongly affected by atmospheric drag, this method may yield satisfactory accelerations, it generally has the following disadvantages:

1. Since at least 7 unknowns and often several more (the other orbital elements and their time variations) are involved in the solution, the error in $\ddot{M}_0$ is, in part, a compound of the errors in the other unknowns.

2. If the observations are not well distributed along the orbit, the orbital elements can become uncertain; this uncertainty is reflected in the determination of $\ddot{M}_0$.

3. To obtain reliable orbital elements, we may need observations taken within a relatively long interval, often as long as several days. During this time the drag may undergo lively fluctuations, in which case the computed acceleration is a smoothed version of the true acceleration. We should not increase the number of unknowns in the M expression to account for such a possibility, because if we do we may have,

instead of a smoothed acceleration, merely a larger error caused by the excessive number of unknowns.

By this method, even quite satisfactory observations are wasted if they are not sufficiently numerous or sufficiently well distributed to allow a good orbit determination. Since the orbital elements of a satellite vary rather slowly, it appears more logical to accept only the most reliable orbits and use interpolated elements for an analysis of the individual observations. This is the basis of the method I have followed since the early days of atmospheric-drag determinations; for a sketchy description, see Jacchia (1961) and Jacchia and Slowey (1962). A more detailed description of the method is given below.

6. As a preliminary, orbital elements are determined at convenient time intervals by an iterative process consisting of several steps: (a) Prediction orbits, which are usually computed from observations over two-week intervals, are first used as input in the D.O.I. program to compute "intermediate" orbits over shorter time intervals (2 to 6 days); the variations of the elements of these orbits are represented by analytical time functions whose coefficients are *unknowns* of the problem. (b) The "intermediate" elements are assembled and new analytical time functions are fitted to them by least squares over convenient time intervals. (c) These time functions, with the exception of the absolute terms (the elements themselves), are used in a third and final determination, a series of orbits in which the time variation of the elements except the mean anomaly are treated as *known* quantities. These final orbits are generally computed from observations over the same time intervals as the "intermediate" orbits.

The final elements are again assembled, and analytical time functions are fitted to each of them over intervals of 30 to 120 days. Great care should be exercised to avoid leaving systematic residuals in any of the elements except the mean anomaly, in which systematic residuals up to 10^{-3} revolutions can be tolerated. This condition is, in itself, a limiting criterion for the time interval to be used, if we bar excessively complicated time functions in the least-squares analysis.

Having thus obtained an analytical representation of all the elements over lengthy intervals of time, we feed the equations into a modified D.O.I. program that gives, for each observation, a residual ΔM in the mean anomaly from a set of given elements. The residuals are plotted against time and a smooth curve drawn through them. Ordinates on the curve are read off at equal time intervals and differenced; from the differences we compute $d^2(\Delta M)/dt^2$. Let $\bar{M}(t)$ be the function used to represent M, so that $M = \bar{M} + \Delta M$. We then have

$$\ddot{M} = \frac{d^2 M}{dt^2} = \frac{d^2 \bar{M}}{dt^2} + \frac{d^2}{dt^2} \Delta M. \tag{3}$$

If the argument of perigee is represented by a non-linear equation, we must correct for the acceleration of the perigee, which is the origin of M. The corrected value $\ddot{M}_c$ of the acceleration is then

$$\ddot{M}_c = \ddot{M} + \left(\frac{dM}{dv}\right)_0 \ddot{\omega}, \tag{4}$$

where $\left(\dfrac{dM}{dv}\right)_0$ is the derivative of M with respect to the true anomaly v at perigee ($M = v = 0$); this derivative is a function of the orbital eccentricity e and is given by

$$\left(\frac{dM}{dv}\right)_0 = 1 - 2e + \frac{3}{2}e^2 - \cdots. \tag{5}$$

The rate of change of the anomalistic period, or secular acceleration, is clearly given by

$$\dot{P} = \frac{dP}{dt} = -\frac{\ddot{M}_c}{\dot{M}^2}. \tag{6}$$

7. If $\overline{M}(t)$ represented the observations without any systematic deviations, we would have $\Delta M = \text{const} = 0$. In that case, the major axis of the orbit, which is derived from the linear term of $\overline{M}(t)$, or the mean motion n, would always be correctly computed. If, however, ΔM varies, an error is introduced into the major axis of the orbit, and through it into the range, which is used to compute the true anomaly and the mean anomaly from the observed positions of the satellite. The error in the major axis $\underline{a}$ is evidently given by

$$\frac{\Delta a}{a} = -\frac{2}{3}\frac{\Delta n}{n}. \tag{7}$$

By keeping ΔM within 10^{-3} revolutions, we can be practically certain that $d\Delta M/dt = \Delta n$ is smaller than 10^{-4} revolutions/day. Since n is of the order of 10 to 15 revolutions per day for a typical satellite, we see that the corresponding $\Delta a/a$ turns out to be smaller than 10^{-6}; i.e., we have an error of only a few meters in the major axis of the orbit.

Several machine programs are necessary for a successful application of this method; I shall list them below.

1. An orbit-computing program such as the Smithsonian D.O.I.

2. A program that computes ΔM when all orbital elements are given as time functions. This could be a modification of the orbit-computing program.

3. A program for the computation, by least squares, of the coefficients of empirical time functions to be fitted to the individual elements. This should include power polynomials and simple functions such as trigonometric functions and exponentials.

4. A program to compute, for given values of the independent variable, the second derivative of a simple function. This program is used to compute $\dfrac{d^2\overline{M}}{dt^2}$ and $\left(\dfrac{dM}{dv}\right)_0 \ddot{\omega}$.

5. A program to compute the effect of solar-radiation pressure on the mean anomaly.

In view of the large number of points to be plotted, an automatic plotter is definitely advisable. In spite of all this automation, there is one operation in which the human element seems to be irreplaceable. This is the drawing of the curves through the $\varDelta M$ residuals. If we had to deal only with slow atmospheric fluctuations, this operation could be replaced by the use of means, graduation formulas, and curve fitting. Since, however, we must count on unexpected, abrupt atmospheric disturbances, any degree of blind smoothing is apt to distort the picture of the true drag variations.

Concerning abrupt perturbations in M such as those caused by atmospheric heating during magnetic storms, we must keep in mind that their effect on all other orbital elements is smaller by a factor of 10^3 or more, and is therefore practically undetectable. Thus we should not expect any error to arise from the fact that, except for the mean anomaly, we treat all the orbital elements as smooth functions of time.

References

BROUWER, D. (1959): Solution of the problem of artificial satellite theory without drag. Astronom. J. **64**, 378—397.

IZSÁK, I. G. (1960): Periodic drag perturbations of artificial satellites. Astronom. J. **65**, 355—357.

JACCHIA, L. G. (1961): The atmospheric drag of artificial satellites during the October 1960 and November 1960 events. Smithsonian Astrophys. Obs., Special Report No. 62.

JACCHIA, L. G., and J. SLOWEY (1962): Preliminary analysis of the atmospheric drag of the twelve-foot balloon satellite (1961, δ 1). Smithsonian Astrophys. Obs., Special Report No. 84.

KOZAI, Y. (1959) a: The motion of a close earth satellite. Astronom. J. **64**, 367—377. (1959) b: On the effects of the sun and the moon upon the motion of a close earth satellite. Smithsonian Astrophys. Obs., Special Report No. **22**, 7—10. (1961): Effects of solar radiation pressure on the motion of an artificial satellite. Smithsonian Astrophys. Obs., Special Report **56**, 25—33.

VEIS, G., and C. E. MOORE (1960): The Smithsonian Astrophysical Observatory differential orbit improvement program. Seminar Proceedings, Tracking Programs, and Orbit Determination, Jet Propulsion Laboratory.

Discussion of atmospheric heat sources based on the analysis of satellite drag data[1]

By

Wolfgang Priester

Universitäts-Sternwarte Bonn, Deutschland,
and Institute for Space Studies, NASA Goddard Space Flight Center,
New York 27, N.Y., U.S.A.

Abstract. This review describes the main effects found to affect the physical properties of the upper atmosphere: the solar activity effect, the diurnal variation, the geomagnetic activity effect and the semiannual and annual variation. Furthermore a theoretical investigation of the energy balance of the upper atmosphere has been carried out by solving the time-dependent heat conduction equation combined with the hydrostatic law. Three reasons are given that the heating of the thermosphere due to absorption of solar extreme ultraviolet radiation alone cannot explain the observed diurnal variation of density and temperature. Thus it is necessary to have another heat source in addition. There is evidence that this source derives its energy ultimately from the solar corpuscular radiation and the solar wind. If an additional heat source is used, which has a maximum at about 9^h local time and a flux of 1 erg cm^{-2} sec^{-1}, a time-dependent model of the upper atmosphere is obtained that is in good agreement with the observed densities.

I. The physical behavior of the upper atmosphere deduced from satellite data

Since 1958 four main effects have been found to affect the physical properties of the upper atmosphere. This information was obtained from analysis of the fluctuations in the orbital periods of artificial satellites. They are:

a) the solar activity effect,
b) the diurnal variation,
c) the geomagnetic activity effect, and
d) the semiannual and annual variation

---

[1] The theoretical part of this review is based on investigations of upper atmosphere physics carried out under the auspices of Dr. ROBERT JASTROW, chief of the Institute for Space Studies and the Theoretical Division of NASA Goddard Space Flight Center in cooperation with Dr. ISADORE HARRIS. The results are published (June 1962) in detail in the NASA Technical Notes D 1443 and D 1444 by ISADORE HARRIS and WOLFGANG PRIESTER. These notes contain also complete tables of the properties of the upper atmosphere as function of height, local time and solar activity.

a) **The solar activity effect** (27-day variation) is defined as the correlation found between density fluctuations with a mean period of 27 days in the altitude range from 200 to 1600 km and the solar flux in the decimeter wavelength range (3 to 30 cm) [W. PRIESTER (1959); L. G. JACCHIA (1959a, b); H. K. PAETZOLD (1959); W. PRIESTER and H. A. MARTIN (1960); M. ROEMER (1961a, b); P. E. ZADUNAISKY, I. I. SHAPIRO and H. M. JONES (1961); R. BRYANT (1961)]. This correlation has again brilliantly been confirmed by L. G. JACCHIA and J. SLOWEY (1962) in their analysis of the orbit of the twelve-foot ballon satellite (1961 delta 1).

The solar decimeter radiation cannot be the physical cause of the fluctuations but is merely an index of it. This radiation in the 3 to 30 cm wavelength range is the so-called "slowly varying component" which is produced, according to M. WALDMEIER and H. MUELLER (1950), by thermal emission from condensations in the solar corona. This flux is proportional to $A \int n_e n_i \, ds$, integrated along the ray path through the condensation. The symbols n_e and n_i represent the number densities of electrons and ions respectively and A is the projected area of the condensation.

It is believed that the solar activity effect in the upper atmosphere is caused to a large extent by the heating due to absorption (photoionization) of solar emission lines and continuous radiation in the extreme ultraviolett (EUV). H. E. HINTEREGGER (1961) has shown from rocket observations that the absorption of this radiation takes place in the altitude range between 150 and 300 km.

The total intensities of the emission lines can be expected to be proportional to the above mentioned integral. It is therefore reasonable to expect a close correlation between the decimeter flux and the strongest lines in the extreme ultraviolet range of the solar spectrum (He II 304 Å, He I 584 Å and numerous lines of highly ionized atoms), since these lines should also orginate mostly in the coronal condensations.

b) **The diurnal density variation** has an amplitude which increases with altitude. At 210 km it is only a few percent of the mean value. This result was obtained from the orbital analysis of 1958 delta 1 and 2 (Sputnik III rocket and satellite) by W. PRIESTER, H. A. MARTIN and K. KRAMP (1960). At an altitude of 650 km, however, the amplitude reaches a factor of almost ten, as found by W. PRIESTER and H. A. MARTIN (1960), L. H. JACCHIA (1960), H. K. PAETZOLD and H. ZSCHOERNER (1960), D. G. KING-HELE and D. M. C. WALKER (1960). The data showing this effect were obtained mostly from satellites 1958 beta 2 (Vanguard I) and 1959 alpha 1 (Vanguard II). These data also revealed that the diurnal variation in density reaches its peak density at approximately 14^h local time followed by a decline; then at about sunrise the density begins to increase rapidly to its peak value again. This behavior results from the combined action of a time-dependent heat source and thermal conduction as pointed out by M. NICOLET (1960). Figure 1

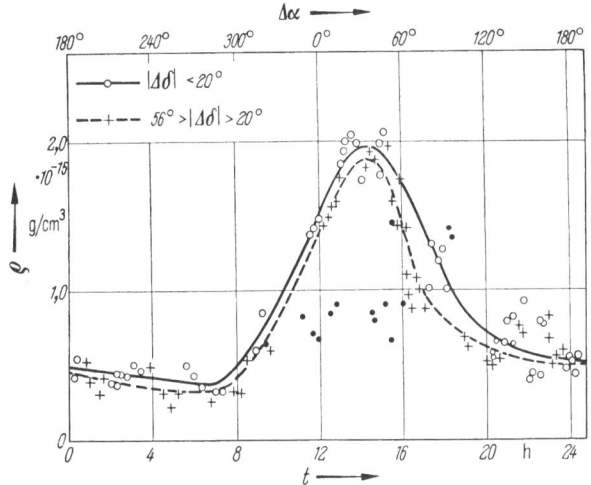

Fig. 1. The diurnal variation of the density ϱ at an altitude of 562 km above the Earth's ellipsoid, derived from satellite 1959 alpha 1 (Vanguard 2).

The lower x-scale gives the local time of the perigee for which the density was calculated, the upper x-scale the difference in right ascension between the perigee and the Sun. The small seasonal effect (dependence on $\Delta \delta$) is indicated by the circles and crosses. During June to August 1960 the diurnal variation is disturbed by the semiannual effect. The densities for this time interval are plotted as dots. Only values with $|\Delta \delta| < 20°$ have been used in this case.

shows the diurnal variation of the density at an altitude of 562 km, derived from satellite 1959 alpha 1 (Vanguard II) [see H. A. MARTIN, et al. (1961)].

c) **The geomagnetic activity effect** in the upper air densities was noted by L. G. JACCHIA (1959) as a correlation between short-lived density fluctuations and the geomagnetic activity represented by the K_p or A_p indices. It was confirmed by density data obtained from seven satellites during the so-called "November 1960 events" [L. G. JACCHIA (1961), G. V. GROVES (1961)]. It was also confirmed by R. JASTROW and R. BRYANT (1961), who used the data from Echo I and was further verified by H. K. PAETZOLD (1960) using the data from Sputnik II (1958 delta 2). The closest correlation of this kind has been obtained with data from the twelve-foot balloon satellite (1961 delta 1) by L. G. JACCHIA and J. SLOWEY (1961).

Figure 2 gives these data. They show clearly the geomagnetic activity effect and the solar activity effect by comparison with the A_p numbers and with the 10.7 cm and 20 cm solar radiation fluxes.

This effect strongly suggests the existence of another heat source in addition to the absorption of solar EUV-radiation. It seems plausible

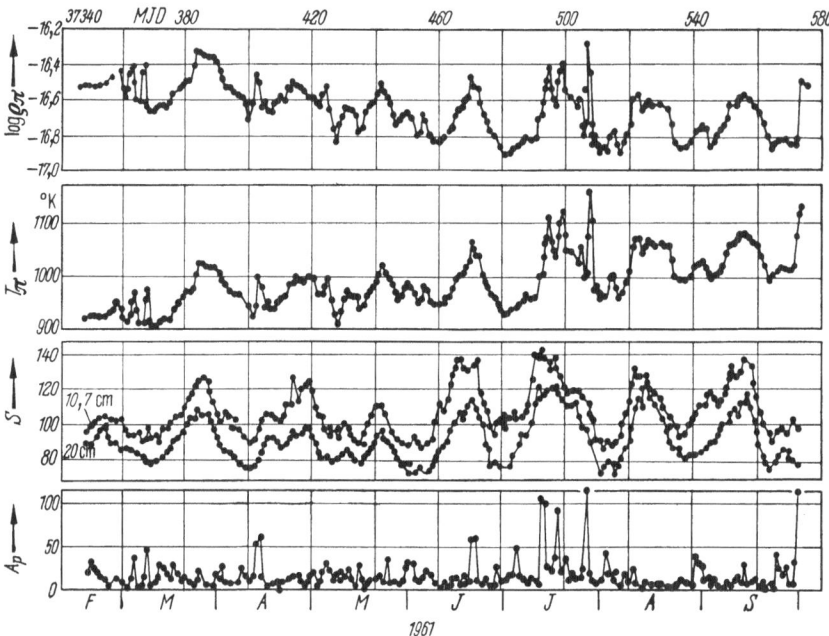

Fig. 2. Density ϱ and temperature T derived from the orbital analysis of the satellite 1961 delta 1 by L. G. JACCHIA and J. SLOWEY (1961) for the time interval February through September 1961. For comparison the 10.7 cm and 20 cm solar radiation flux and the geomagnetic A_p numbers are given.

to attribute this additional heat source to energy that is ultimately derived from the solar corpuscular radiation or its "steady" component, the solar wind.

d) The existence of such an additional heat source also is suggested by the fourth effect, a *semiannual variation* in atmospheric density found by H. K. PAETZOLD and H. ZSCHOERNER (1960). They observed variations in which the densities have maxima in March and September and minima in June and July and also December and January. This behavior is quite similar to the semiannual variation of geomagnetic activity, found by A. L. CORTIE (1912) and discussed in great detail by J. BARTELS (1932).

A fifth *effect with an annual period* has been found also by H. K. PAETZOLD (1962). The statistical significance of this effect is still somewhat poor. Recently, however, M. ROEMERs careful analysis of the drag acting on the Echo I satellite (M. ROEMER 1962) not only confirms the semiannual variation with minima in January and June-July, but shows also the annual variation with a minimum in spring for the altitude range from 1000 to 1500 km (see Figure 3). Figure 3 gives ROEMERS results. Plotted are the remaining deviations of the density after the solar activity effect (27-day variation) has been corrected for and after the dependence of height, local time and long-term solar cycle variation has been removed by reduction using the theoretical models of I. HARRIS and W. PRIESTER (1962b).

H. K. PAETZOLD and H. ZSCHOERNER (1961) estimate that a decrease of 0.2 erg cm^{-2} sec^{-1} in the over-all flux used for heating during the periods of minima in June-July is required to explain the observed decrease in density. This suggests that the second heat source, which

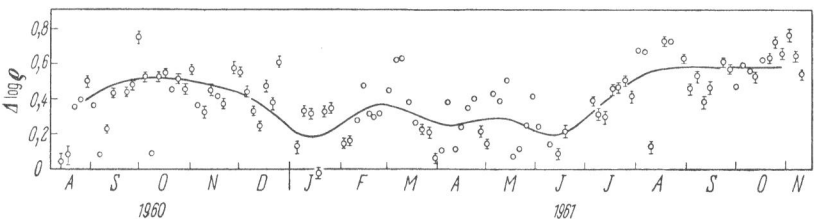

Fig. 3. Density variations due to semiannual and annual variations derived from the Echo I satellite by M. ROEMER after corrections have been made for the other effects. The zero level in the ordinate scale is arbitrary. The bars indicate the uncertainty of the reduction for the solar activity effect.

we shall call a "corpuscular" heat source, normally provides an energy flux which is a few times larger than the above mentioned value of 0.2 erg cm^{-2} sec^{-1}. A crude estimate may also be obtained from the absolute value of and the variation of the geomagnetic u-measure which was defined by J. BARTELS (1932) as the "inter-diurnal variability of the horizontal component at the geomagnetic equator". In a recent paper by W. PRIESTER and D. CATTANI (1962) the semiannual variation of the u-measure was related to a model of the solar corpuscular radiation dependent on heliographic latitude.

In view of this relationship and with roughly 20 percent semiannual variation in the amplitude of the u-measure, we might expect a total flux in the order of 1 erg cm^{-2} sec^{-1} for the "corpuscular" heat source. As the measurements from Explorer X by H. S. BRIDGE et al. (1961) revealed a flux of about 5 erg cm^{-2} sec^{-1} for the solar wind outside the earth's magnetosphere, our estimate of 1 erg cm^{-2} sec^{-1} for the "corpuscular" heat source seems to be plausible. Since this heat source is likely to have a diurnal variation, the estimated flux given refers to the diurnal peak value.

These conclusions are supported by the results obtained from our calculations of the energy balance of the upper atmosphere in which the time-dependent heat conduction equation was used with the condition of quasi hydrostatic equilibrium. These results showed that a theoretical explanation of the observed atmospheric densities can only be obtained by taking into a account a second heat source, which contributes roughly the same amount of heat to the upper atmosphere as the EUV heat source.

II. Theoretical interpretation

The energy balance of the upper atmosphere is determined mainly by absorption of solar energy and the heat transfer by conduction. These processes are described by the time-dependet equation for heat conduction and by the heat source functions. If these equations are combined with the equation for hydrostatic balance they yield the physical properties of the upper atmosphere as function of time and altitude.

We have studied the solutions of these equations in detail by a process of numerical integration, and compared the results with time-dependent models derived from satellite density data, in order to obtain some information as to the nature of the heat sources of the upper atmosphere.

The basic equation is

$$\frac{\partial}{\partial z}\left(K(T)\frac{\partial T}{\partial z}\right) - \varrho\, C_p\frac{\partial T}{\partial z}\, T\int_{z_0}^{z}\frac{1}{T^2}\frac{\partial T}{\partial z'}\,dz' + Q_{\mathrm{EUV}} + Q_{\mathrm{ox}} + Q' = \varrho\, C_p\frac{\partial T}{\partial t}. \tag{1}$$

The derivation of this equation is given in NASA TND 1443 [I. Harris and W. Priester (1962a)]. The first term is the heat transport due to thermal conduction, the second term is the energy transport due to the diurnal expansion-contraction of the upper atmosphere. The term on the right hand side is the time-dependence of the energy content.

In Eq. (1) T is the temperature, ϱ is the density, z is the altitude, z_0 is the lower boundary, C_p is the specific heat at constant pressure and K is the thermal conductivity given by

$$K(T) = \frac{\sum^i A_i\, n_i(z)}{\sum^i n_i(z)}\, T_{(z)}^{1/2}. \tag{2}$$

A_i is a constant depending upon the constituent i, and n_i is the number density of atoms or molecules of the ith constituent.

We have from S. Chapman and T. G. Cowling (1960) [see also M. Nicolet (1961)] the following values in erg $\mathrm{cm}^{-1}\ \mathrm{sec}^{-1}\ (^{\circ}K)^{-3/2}$:

$$A\,(\mathrm{H}) = 2.1 \times 10^3,$$
$$A\,(\mathrm{He}) = 9.0 \times 10^2,$$
$$A\,(\mathrm{O}) = 3.6 \times 10^2,$$

and
$$A\,(O_2,\,N_2) = 1.8 \times 10^2.$$

The quantities Q are the heat sources.

Q_{EUV} is the heat source due to the absorption of the solar EUV-radiation given by

$$Q_{EUV} = \sum^i \varepsilon_i\, n_i(z) \int_0^\infty F_\lambda\, c_i(\lambda)\, e^{-\tau(\lambda,\,z,\,t)}\, d\lambda\,. \tag{3}$$

where

$$\tau(\lambda,\,z,\,t) = \sum^i \int_z^\infty \sigma_i(\lambda)\, \frac{n_i(z)}{\cos\theta}\, dz\,. \tag{4}$$

$\sigma_i(\lambda)$ is the cross-section for absorption by the ith constituent of radiation of wavelength λ in the region $d\lambda$, F_λ is the incident flux of wavelength λ in the region $d\lambda$ at the top of the atmosphere and ε_i is an efficiency factor for the conversion into thermospheric heat of energy in the extreme ultraviolet absorbed by the ith constituent. θ is the zenith angle of the sun.

Q_{ox} is the heat loss due to the cooling by atomic oxygen radiating in the infrared and is given according to D. R. BATES (1951) by

$$Q_{ox} = -\, n_0\, f(T)\,, \tag{5}$$

where

$$f(T) = E_1 A_{12} \left[\frac{W_1 \exp\left[-\dfrac{E_1}{k\,T}\right]}{W_2 + W_1 \exp\left[-\dfrac{E_1}{k\,T}\right] + W_0 \exp\left[-\dfrac{E_0}{k\,T}\right]} \right]. \tag{6}$$

n_0 is the number density of atomic oxygen, E_1 is the difference in energy between the $3P_1$ and $3P_2$ levels of atomic oxygen, E_0 is the difference in energy between the $3P_0$ and $3P_2$ levels of atomic oxygen and W the statistical weights of the various levels and A_{12} the EINSTEIN coefficient for the transition $3P_1 - 3P_2$.

The observations of the geomagnetic acticity effect and of the semiannual effect strongly indicate the existence of another heat source the energy of which is very likely provided finally by the solar corpuscular radiation and /or its "steady" component, the solar wind, for which we reserve a quantity Q' in our basic formula. Detailed consideration about this source are given later.

We need further the expression for the total heat capacity at constant pressure, which we take as

$$\varrho\, C_p = \sum^i n_i(z)\, k\, B_i\,, \tag{7}$$

where k is the Boltzmann constant and B_i is a constant depending upon the constituent. We have

$$B_i = 3.5 \text{ for diatomic molecules}$$
$$B_i = 2.5 \text{ for monatomic molecules.}$$

We shall integrate Eq. (1) numerically, linearizing by evaluating the non-linear factors from the values of the temperatures and densities at the previous time step in the integration procedure. All quantities in Eq. (1) depend on altitude z and time t. Once the temperature profile is determined, the number densities n_i are calculated by means of the barometric relationship under the assumption of diffusive equilibrium

$$n_i(z,\ t) = n_i(z_0)\,\frac{T(z_0)}{T(z,t)}\,\exp\left[-\int_{z_0}^{z}\frac{m_i\,g(z')}{k\,T(z',t)}\,dz'\right] \qquad (8)$$

The boundary conditions are a given initial temperature distribution, the temperature and densities of the constituents at the lower boundary (120 km) held constant in time, and a zero gradient of the temperature at the upper boundary. The numerical integration was performed on an IBM 7090 computer.

In our first attempts to obtain a theoretical time-dependent model of the upper atmosphere properties by solving the heat conduction equation for the atmosphere in quasi hydrostatic equilibrium [Eqs. (1)–(8)] — a model which should represent the observed densities satisfactorily — we used as the only heat source the absorption of solar EUV-radiation combined, however, with the heat loss due to the infrared reradiation by oxygen atoms. The loss due to reradiation has only a small influence on the temperature distribution, as previously noted by D. C. HUNT and T. E. VAN ZANDT (1961). The source Q' in Eq. (1) was taken equal to zero in these first attempts. The calculations yield a diurnal maximum of temperature and density at 17^h local time, not at 14^h as observed from satellite drag data. Losses due to conduction and infrared radiation by oxygen atoms are not sufficient to balance the heat input due to the EUV-absorption in order to yield a maximum of temperature at 14^h local time. Furthermore, the ratio of daytime maximum temperature or density to night-time minimum temperature or density respectively is much larger than the observed ratio. Our calculations using the EUV absorption as the only heat source yield a value of 2.6 for the temperature ratio, while the observed value is about 1.5. These results are qualitatively independent of boundary conditions. For testing we used two different sets of number densities which differ considerably in the important ratio of O to O_2 at the lower boundary. Diffusive equilibrium was assumed to be valid for all altitudes above 120 km. The numerical values for the boundary conditions at 120 km are given in the detailed publication NASA TND 1443.

The diurnal variation of the temperature calculated with the EUV heat source only and using the first set of boundary values is given in Figure 4 for an altitude of 600 km by the dotted line, together with the temperature distribution (solid line), which well represents the observed data.

Thus it can be concluded that no agreement can be obtained between the observations and a theory which is based on an EUV heat source alone. Furthermore, an extremely high efficiency (70 to 90 per cent) would be required for the conversion of the EUV flux into thermospheric heat (about 2 erg cm^{-2} sec^{-1}) if we are to compare with HINTEREGGER's

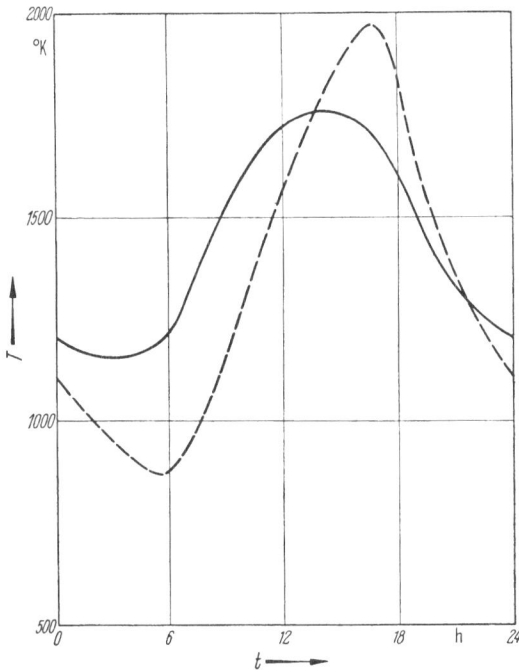

Fig. 4. Diurnal variation of the exospheric temperature calculated with an EUV-heat source alone (dotted curve) and combined with our additional ,,corpuscular'' heat source (solid curve) as function of local time.

rocket measurements (1961) which yield a total flux of 2.5 erg cm^{-2} sec^{-1} in the range from 44 Å to 1,000 Å.

Therefore the existence of a second heat source with the following properties is strongly suggested: a maximum in the mid-morning and a minimum in the early or mid-afternoon, a small amount of heating during the night, and an average magnitude comparable to the provided by the EUV-flux.

A heat source with these properties is required to represent the density observed in those times where a strong diurnal bulge is found. For the comparison we used the model of Bonn Observatory 1961 [H. A. MARTIN, et al. (1961)]. This model is in very good agreement with the density data obtained by other groups. This model is reduced to years of high solar activity, represented by a 10.7 cm solar radiation flux of

$$S = 200 \times 10^{-22} \ \mathrm{W/m^2 \ c/s}$$

or by the corresponding flux

$$S = 170 \times 10^{-22} \ \mathrm{W/m^2 \ c/s}$$

of the 20 cm solar radiation.

The time variation of this new heat source, which we shall call the "corpuscular" heat source, might be correlated with some other geomagnetic phenomena, which should be an indicator of corpuscular activity. The intensities of the micropulsations observed by W. H. CAMP-BELL (1959) in California show the main maximum at about 9^h local time, and thus have the principal property required. A physical connection of these pulsations with the heat source we found necessary is plausible, but direct observations of micropulsations in the ionospheric F-layer would be desirable. In addition the geomagnetic field strength at the equator also has approximately the time-varying properties required.

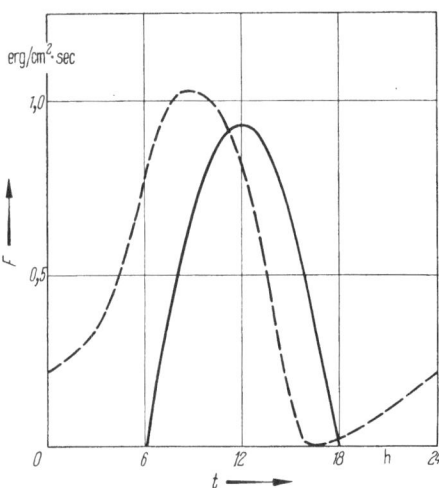

Fig. 5. Diurnal variation of the fluxes of the EUV-heat source (solid curve) and of the "corpuscular" heat source (dotted curve).

The diurnal variation of the flux for the corpuscular heat source which we have chosen is given in Figure 5, together with the fraction of the solar EUV flux converted into heat in the thermosphere having a peak value of 0.93 ergcm^{-2} sec^{-1}. This implies an efficiency factor of 37 percent.

The peak value of the flux for the corpuscular heat source that yields a good agreement with the observed densities is 1.03 erg cm^{-2} sec^{-1}. This value is in good agreement with the estimates obtained from the semiannual variation. It agrees also with the magnitude of the heat source proposed by J. DESSLER (1959) — a source which is due to the dissipation of hydromagnetic waves generated by the solar corpuscular radiation. The energy dissipation takes place in the F region of the ionosphere. For the altitude dependence of our corpuscular heat source Q' we used an analytic approximation to the dissipation curve given by DESSLER. This form is also similar to the shape of the heat source due to absorption of solar EUV radiation. Our expression for Q' is:

$$Q' = \frac{F}{s_1 - s_2} \left[\exp\left(-\frac{z - z_0}{s_1}\right) - \exp\left(-\frac{z - z_0}{s_2}\right) \right] f(t), \qquad (9)$$

where F is the flux of this heat source

$$F = \int_{z_0}^{\infty} Q'(z, t_{max}) \, dz \qquad (10)$$

at the time of its peak value with

$$s_1 = 60 \text{ km},$$
$$s_2 = 40 \text{ km},$$
$$z_0 = 120 \text{ km}.$$

$f(t)$ is the diurnal variation as given in Figure 5, but normalized to a peak value equal to unity. The Q' function has a maximum at 170 km.

Using this heat source in addition to the EUV heat source we obtained a good agreement between the observed and calculated densities. The solid line in Figure 4 shows the calculated diurnal variation of the exospheric temperature. In Figure 6 the derived diurnal variations of densities are compared with the densities of the model of Bonn Observatory

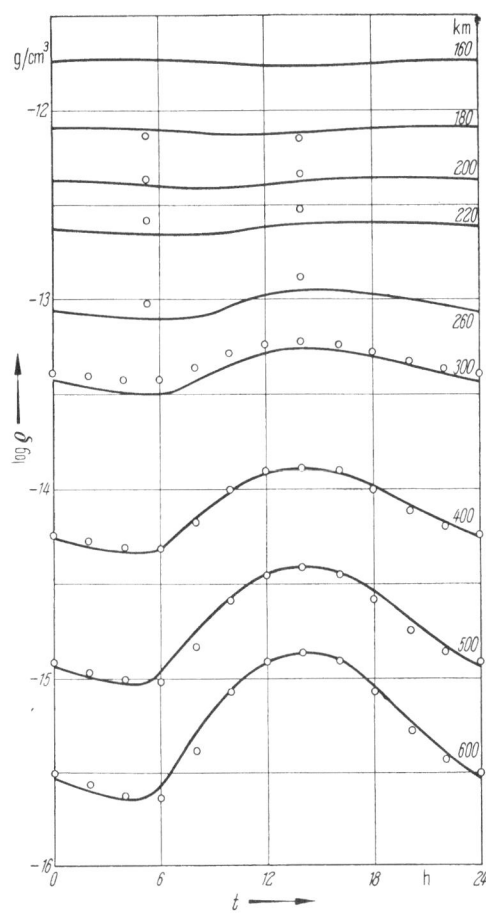

Fig. 6. Diurnal variation of density as function of local time for selected altitudes from 160 to 600 km. The solid curves give our calculated values. The circles are densities taken from the observational model of Bonn University 1961 [H. A. MARTIN, et al. (1961)].

1961. In the altitude range from 400 to 600 km, where the Bonn model is based on the most reliable density determinations, we have an almost perfect agreement. For altitudes below 400 km the agreement is also satisfactory.

The temperature and the mean molecular weight are given as function of altitude for four selected local times of the day in Figures 7 and 8. It may be noticed that the 10^h — and the 22^h — temperature curves in Figure 7 cross at about 170 km. This is related to the fact that the daytime-densities in the altitude range from 130 to 190 km are lower than the nighttime-densities. This is in agreement with the

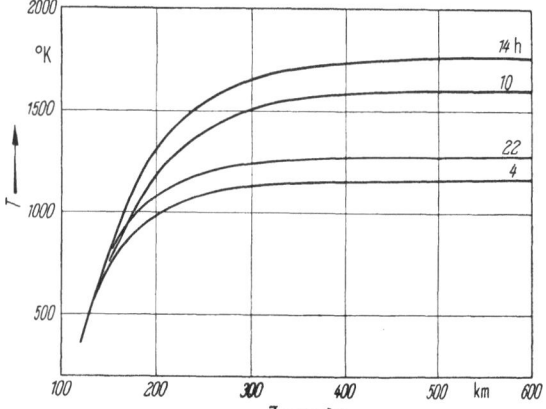

Fig. 7. Temperature (°K) as function of altitude from 120 km to 600 km for four selected local times.

observational model by H. A. MARTIN, et al. (1961). The mass difference between nighttime and daytime densities below 190 km is sufficient to

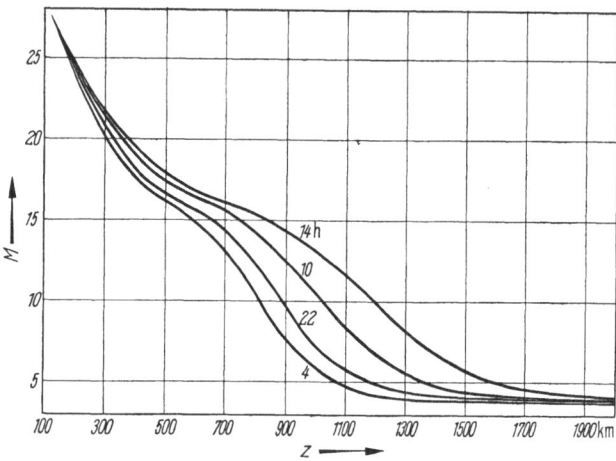

Fig. 8. The mean molecular weight as function of altitude from 120 km to 2,000 km for four selected local times.

provide the mass for the diurnal bulge above that height, since our model conserves the total mass.

Our model pertains to the equatorial and temperate zones of the earth and is valid for those years when the solar activity can be represented by an average flux of 200×10^{-22} W/m² c/s for the 10.7 cm radiation (or a flux of 170×10^{-22} W/m² c/s for the 20 cm radiation).

We wish to emphasize again that to obtain a good aggreement with the observational model a heat source in addition to the solar EUV flux is required. It is probable that this heat source derives its energy ultimately from the solar corpuscular radiation.

In our calculations we obtained good aggreement with the observational models with a flux having a diurnal average of 0.44 erg cm^{-2} sec^{-1} and a peak value of 1.03 erg cm^{-2} sec^{-1} at 9^h local time, combined with an EUV heat source having an average flux of 0.30 erg cm^{-2} sec^{-1} and a peak value of 0.93 at 12^h local time.

Both the EUV- and the corpuscular heat sources are expected to vary considerably during the solar cycle. In order to obtain information as to how these heat sources vary in comparison with the indices of solar activity, we calculated theoretical models of the physical properties of the upper atmosphere for five different sets of flux-values for the EUV heat source and the corpuscular heat source. These models then must be compared with the density data deduced from satellite orbital analysis for a complete cycle of solar activity. A detailed description of these theoretical models is given in NASA TND 1444 by I. HARRIS and W. PRIESTER (1962 b).

I am very much obliged to Dr. ROBERT JASTROW for his permanent stimulation and to Dr. ISADORE HARRIS for his fine cooperation. I am further grateful to the National Academy of Sciences — National Research Council whose Senior Research Associateship provided for me the opportunity to stay with the NASA Institute for Space Studies for research in upper atmosphere physics.

I also have to thank Mr. MAX ROEMER for the permission to publish some of his results of the Echo I satellite prior to publication.

References

BARTELS, J. (1932): Terrestrial-magnetic activity and its relations to solar phenomena. Terr. Magn. **37**, 1—52.

BATES, D. R. (1951): The temperature of the upper atmosphere. Proc. Roy. Soc. London, Sect. B **64**, 805—821.

BRIDGE, H. S., C. DILWORTH, A. J. LAZARUS, C. F. LYON, B. ROSSI, and F. SCHERB (1961): Direct observations of the interplanetary plasma. Lab. Nucl. Sci. and Dep. Phys. M. I. T. Preliminary Rep., 13 pp.

BRYANT, R. (1961): A comparison of theory and observation of the Echo I satellite. J. Geophys. Res. **66**, 3066—3069.

CAMPBELL, W. H. (1959): Studies of magnetic field micropulsations with periods of 5 to 30 seconds. J. Geophys. Res. **64**, 1819—1826.

CHAPMAN, S., and T. G. COWLING (1960): The mathematical theory of non-uniform gases. Cambridge, 431 pp.

CORTIE, A. L. (1912): Sunspots and terrestrial magnetic phenomena, 1898—1911. Monthly Notices **73**, 52—60.

DESSLER, A. J. (1959): Ionospheric heating by hydromagnetic waves. J. Geophys. Res. **64**, 397—401.

GROVES, G. V. (1961): Correlation of upper atmosphere air density with geomagnetic activity, November 1960. Space Research II (H. C. VAN DE HULST, C. DE JAGER, and A. F. MOORE, ed.) Amsterdam, 751—753.

HARRIS, I., and W. PRIESTER (1962)a: Time-dependent structure of the upper atmosphere. NASA Technical Note D 1443 (June 1962). 71 pp.

HARRIS, I., and W. PRIESTER (1962)b: Theoretical models for the solar-cycle variation of the upper atmosphere. NASA Technical Note D 1444 (June 1962). 261 pp.

HINTEREGGER, H. E. (1961): Preliminary data on solar extreme ultraviolet radiation in the upper atmosphere. J. Geophys. Res. **66**, 2367—2380.

HUNT, D. C., and T. E. VAN ZANDT (1961): Photoionization heating in the F-region of the atmosphere. J. Geophys. Res. **66**, 1673—1682.

JACCHIA, L. G. (1959)a: Atmospheric fluctuations of solar origin revealed by satellites. Harvard Observatory Announcement Card No. 1423.

JACCHIA, L. G. (1959)b: Solar effects on the acceleration of artificial satellites. Smithsonian Astrophysic. Obs. Spec. Rep. No. 29, 15 pp.

JACCHIA, L. G. (1959)c: Corpuscular radiation and the acceleration of the artificial satellites. Nature **183**, 1662—1663.

JACCHIA, L. G. (1960): A variable atmospheric-density model from satellite accelerations. Smithsonian Astrophysic. Obs. Spec. Rep. No. 39, 15 pp.

JACCHIA, L. G. (1961): Satellite drag during the events of November 1960. Space Research II (H. C. VAN DE HULST, C. DE JAGER, and A. F. MOORE, ed.) Amsterdam, 747—750.

JACCHIA, L. G., and J. SLOWEY (1962): Preliminary analysis of the atmospheric drag of the twelve-foot balloon satellite (1961 delta 1). Smithsonian Astrophysic. Obs. Spec. Rep. No. 84, 18 pp.

JASTROW, R., and R. BRYANT (1961): Effects of a severe solar storm on the orbit of Echo I. IGY Bulletin No. 44, 6—7.

JASTROW, R., and L. KYLE (1961): The earth atmosphere. Handbook of astronautical engineering (H. H. KOELLE, ed.) New York, Toronto, London, Sect. 2, pp. 2—13.

KING-HELE, D. G., and D. M. C. WALKER (1961): Upper-atmosphere density during the years 1957—1961 determined from satellite orbits. Space Research II (H. C. VAN DE HULST, C. DE JAGER, and A. F. MOORE, ed.) Amsterdam, 918—957.

MARTIN, H. A., W. NEVELING, W. PRIESTER, and M. ROEMER (1961): Model of the upper atmosphere from 130 through 1600 km derived from satellite orbits. Mitt. Sternwarte Bonn No. 35, 16 pp.; and Space Research II (H. C. VAN DE HULST, C. DE JAGER, and A. F. MOORE, ed.) Amsterdam, 902—917.

NICOLET, M. (1960): Les variations de la densité et du transport de chaleur par conduction dans l'atmosphere superieure. Space Research I (H. KALLMANN-BIJL, ed.) 46—89.

NICOLET, M. (1961): Structure of the thermosphere. Planetary Space Sci. **5**, 1—32.

PAETZOLD, H. K. (1959): Observations of the Russian satellites and the structure of the outer terrestrial atmosphere. Planetary Space Sci. **1**, 115—124.

PAETZOLD, H. K. (1960): Proceedings XI. Internat. Astronautical Congress, Stockholm. Springer Verlag, Wien.

PAETZOLD, H. K., and H. ZSCHOERNER (1960): Bearings of Sputnik III and the variable acceleration of satellites. Space Research I (H. KALLMANN-BIJL, ed.) Amsterdam, 24—36.

PAETZOLD, H. K., and H. ZSCHOERNER (1961): The structure of the upper atmosphere and its variations after satellite observations. Space Research II (H. C. VAN DE HULST, C. DE JAGER, and A. F. MOORE, ed.) Amsterdam, 958—973.

PAETZOLD, H. K. (1962): Solar activity effects on the upper atmosphere deduced from satellite observations. Space Research III (W. PRIESTER, ed.) Amsterdam (to be published).

PRIESTER, W. (1959): Sonnenaktivität und Abbremsung der Erdsatelliten. Mitt. Univ.-Sternwarte Bonn, No. 24, 4 pp; and Naturwiss., 46, 197—198.

PRIESTER, W., and H. A. MARTIN (1960): Solare und tageszeitliche Effekte in der Hochatmosphäre aus Beobachtungen künstlicher Erdsatelliten. Mitt. Univ.-Sternwarte Bonn No. 29, 53 pp. Engl. translation: Royal Aircraft Establishment Farnborough library translation No. 901, 20 pp.

PRIESTER, W., H. A. MARTIN, and K. KRAMP (1960): Diurnal and seasonal density variations in the upper atmosphere. Nature 188, 202—204.

PRIESTER, W., and D. CATTANI (1962): On the semiannual variation of geomagnetic activity and its relation to the solar corpuscular radiation. J. Atmospheric Sci. 19, 121—126.

ROEMER, M. (1961)a: Modell der Exosphäre im Höhenbereich 1000—1700 km berechnet aus den Bahnänderungen des Satelliten Echo I. Mitt. Univ.-Sternwarte Bonn No. 37, 34 pp. Engl. translation: Royal Aircraft Establishment Farnborough library translation No. 954, 32 pp.

ROEMER, M. (1961)b: Terrestrial exosphere at heights of 1000—1700 km. Nature 191, 238—240.

ROEMER, M. (1962): (paper in preparation).

WALDMEIER, M., and H. MÜLLER (1950): Die Sonnenstrahlung im Gebiet von $\lambda = 10$ cm, Z. Astrophysik 27, 58—66.

ZADUNAISKY, P. E., I. I. SHAPIRO, and H. M. JONES (1961): Experimental and theoretical results on the orbit of Echo I. Smithsonian Astrophysic. Obs. Spec. Rep. No. 61, 22 pp.

Some problems of motion of artificial satellites about the centre of mass

By

V. V. Beletsky

The U.S.S.R. Academy of Sciences, Moscow, U.S.S.R.

A number of scientific tasks connected with the creation of artificial Earth satellites requires knowledge of the satellite motion about the centre of mass. For instance, investigation of solar radiations is possible only when the satellite-borne instruments are illuminated by the Sun, while conditions of illumination depend on motion of satellites about the centre of mass. Indications of different instruments designed for exploring the composition and structure of the upper atmosphere depend on the position of the satellite with respect to the incident stream. The satellite postition with respect to the Earth's magnetic field influences the indications of magnetometers. The motion about the centre of mass also influences the average resistance coefficient and hence the retardation of the satellite in its onward motion along the orbit. There are other problems requiring knowledge of the satellite orientation in space. Investigation of the satellite motion about the centre of mass is also of independent interest.

The satellite will be considered as an absolutely solid body.

The satellite motion with respect to the centre of mass can be divided into two basic types. If the kinetic energy of the satellite is small with respect to the action of outer forces, the motion of the libration type is possible, i.e. oscillations of the satellite about the position of relative equilibrium in the coordinate system connected with the radius-vector of the orbit. Such motion is caused by the stabilizing action of the moments of gravitational forces. Motion of the Moon about the Earth represents precisely this type of motion.

If the kinetic energy of the satellite rotation is large as compared with the action of outer forces, motion will be of another character. Due to insignificant action of outer forces motion at a small time interval will be close to undisturbed one, i.e. to the EULER motion of a free solid body. Moments of outer forces will introduce small disturbances in the motion which, however, will be of secular character, i.e. accumulate with time. For instance, the Earth's rotation axis under the influence of the lunar and solar attraction slowly precesses (with a precession period of about 26,000 years).

In classic celestial mechanics the theory of celestial bodies motions about the centre of mass is developed relating to the concrete bodies (the Moon, the Earth) [5] which enables us to make a number of simplifications absent in the general case. For instance, in the theory of the lunar-solar precession of the Earth's axis the small value of the difference of the equatorial and polar moments of the Earth's inertia [6] and the small value of the nutation angle [7] are used, i.e. the fact is utilized that the undisturbed motion of the Earth is a stationary rotation about the dynamic symmetry axis fixed in an absolute space. In a general case this supposition is not justified for artificial satellite, and the undisturbed motion of the satellite is the EULER motion of a free solid body.

Besides gravitational moments a number of disturbing factors influence an artifical satellite rotation. These disturbing factors are aerodynamics moments, the influence of the evolution of the satellite orbit, dissipative effects connected with the satellite-atmosphere friction and with the interaction of metallic envelope and the satellite's own magnetic field with the Earth's magnetic field, etc.

Therefore in the general case the artificial satellites rotations problem is a very complicated one. However, the basic effects and general laws of motion of a satellite about the centre of mass are determined by the action of the main factors; by gravitational and aerodynamic moments, by the influence of the orbit evolution. Therefore it is reasonable to conduct investigations of motion of a satellite about the centre of mass when only the main disturbing factors are taken into account. Such investigation was carried out in the papers [1] to [3], [12]. Some of the results of this investigation are given below.

As mentioned above, at low kinetic energy of a satellite rotation about the centre of mass a libration type motion is possible, i.e. oscillations about the position of relative equilibrium. Stable relative equilibrium and libration are caused by the interaction of the Newtonian attraction forces of the particles of the body to the gravitating centre, and by the action of the centrifugal forces produced by motion of the mass centre of the body along the orbit. Since the central Newtonian force acts on each point of the body, the force function found by the integration with respect to the whole volume will depend, firstly, on the distance of its centre of mass to the attracting centre and, secondly, on the position of the body with respect to the radius-vector of the centre of mass. Equations of motion of the body mass centre and equations of motion about the centre of mass generally speaking are not divided into two independent systems. The onward and rotational motions are correlated.

The equations of motion have the first integrals: the energy integral and three integrals of the momentum.

It turns out that equations of motion under definite simple conditions imposed on the force function permit the following partial solution: the centre of mass of the body moves with constant angular velocity on a circular orbit of random radius. The main central axes of the body inertia are directed during motion along the radius-vector, the tangent and the binormal to the orbit. Such motion of the body represents relative equilibrium, i.e. equilibrium in the coordinate system connected with the radius-vector of the orbit. This motion is considered as undisturbed one. The LYAPUNOV function positively determined under certain conditions is compiled by means of the above integrals for disturbed motion. According to LYAPUNOV's stability theory these conditions will be sufficient ones for stability of undisturbed motion. It is shown that in real cases these sufficient conditions of stability are as follows: in undisturbed motion the major axis of the body inertia ellipsoid should be directed along the radius-vector of the orbit, the minor axis is directed along the normal to the plane of the orbit (and, therefore, the middle axis is directed along the tangent to the orbit). Then undisturbed motion will be stable with respect to small disturbances of motion of the centre of mass and motion about the centre of mass.

It is convenient to simplify the formulation of the problem indicated above for the description of the satellite motion about the centre of mass in the gravitational field. Due to the small size of the satellite as compared to the distance to the centre of attraction, one can consider that motion of the centre of mass does not depend on rotation around the centre of mass. The centre of mass moves on the conventional KEPLER's orbit and the problem is reduced to finding the satellite rotation whose motion of the centre of mass is known. The smallness of the satellite enables us to limit ourselves to the main terms of expansion of thel force function. This limited formulation of the problem is conventiona in celestial mechanics. Such a formulation makes it possible to study the satellite oscillations about the position of relative equilibrium comparatively simply. The conditions of stability indicated earlier follow directly from the first integral of the equations of motion — the integral of the JACOBI type. (It is considered that in disturbed motion the orbit does not vary as compared to the orbit of undisturbed motion).

This integral takes place only for the case of a circular orbit and can be written in the following form

$$A \, \overline{p}^2 + B \, \overline{q}^2 + C \, \overline{r}^2 + 3\omega^2 \{(A - C) \, \gamma^2 + (B - C) \, \gamma'^2\} +$$

$$+ \omega^2 \{(B - A) \, \beta^2 + (B - C) \, \beta''^2\} = h. \qquad (1)$$

Here ω is angular velocity of motion of the mass centre of the satellite, A, B, C are the main central moments of the satellite inertia. In the
position of relative equilibrium the axis of the moment of inertia B
coincides with the normal to the plane of the orbit, and the axis of
the moment of inertia C coincides with the radius-vector of the orbit,
$\bar{p}$, $\bar{q}$, $\bar{r}$ are components of the satellite relative angular velocity along
its main central axes, γ, γ', β, β'' are direction cosines of the radius-
vector and the normal to the plane of the orbit with the satellite main
central axes. In a position of relative equilibrium $p = q = r = \gamma = \gamma'$
$= \beta = \beta'' = 0$.

This motion, as follows from (1), is stable, if

$$B > A > C, \tag{2}$$

which gives the stable distribution of the satellite main central axes.

Integral (1) enables us to obtain the estimates of the amplitude of
spatial nonlinear oscillations, fully consider plane nonlinear oscillations.
Spatial oscillations are considered in [2, 12] in a linear formulation.
The satellite oscillations take place slowly: the lowest possible periods
of oscillations has an order of the satellite period of revolution on the
orbit. It turns out that the libration motion is possible only at very
low initial angular velocities of the satellite rotation about the centre
of mass, namely on the order of angular velocity of the satellite motion
on the orbit (0.01–0.1°/sec).

Investigation of oscillations on an elliptic orbit have shown that the
main difference from the oscillations on a circular orbit is displayed in
this case in the appearance of forced oscillations in the plane of the
orbit with an amplitude proportional to the eccentricity of the orbit
(eccentricity oscillations).

If e is the orbit eccentricity, $n^2 = 3\dfrac{A-C}{B}$, ν is the true anomaly,
forced oscillations for angle ϑ_e between the satellite axis and the radius-
vector of the orbit are given with an accuracy up to e^3 by the formula

$$\vartheta_e = \frac{2e \sin \nu}{1 + e \cos \nu} \left\{ \frac{1}{n^2 - 1} + \frac{e}{n^2 - 4} \cos \nu \right\} \tag{3}$$

if only n^2 is not close to unity. At $n^2 \approx 1$ the resonance effect appears
and instead of formula (3) we should use a more precise formula obtained from the nonlinear equation of plane oscillations. The first
harmonic of forced oscillations is given in this case by formulas

$$\vartheta_e = \pm \alpha \sin \nu, \qquad n^2 = \frac{\alpha \pm 2e}{I_1(2\alpha)}, \tag{4}$$

where $I_1(a)$ is the BESSEL function of the first kind. The second formula
(4) makes it possible to determine the amplitude of oscillations α as a

function of n^2 and e. For instance, for $e = 0.01$ at the most unfavourable value $n^2 \approx 1.12$ the amplitude α reaches the maximum value $\alpha_{\max} = 31°$. The second essential effect of oscillations on an elliptic orbit is the possibility of the appearance of parametric resonance both in plane and in spatial oscillations. This resonance appears at the expense of periodicity of the coefficients of oscillation equations when parameters of the equations get into some narrow (on the order of e) region.

Let us consider now motion of the second type — the case when the action of outer forces is small as compared to the kinetic energy of the satellite rotation about the centre of mass. In this case outer moments of forces will introduce only low disturbances in the main undisturbed motion of the satellite. These disturbances can be accumulated for a long time interval leading to considerable deviations from undisturbed motion. Such secular disturbances are of the greatest interest.

The investigation of such motion can be conveniently carried out by the purturbation theory methods. The solution of equations of undisturbed motion contains integration constants. The solution of equations of disturbed motion is sought in the same kind, but the above mentioned constants are considered as time functions.

The choice of variable constants having clear mechanical sense is of considerable significance. For satellites with dynamic symmetry with respect to one axis (such are exactly or approximately all the space vehicles launched) parameters describing the value and the position (with respect to the orbit) of the vector of kinetic moment and the satellite rotation about this vector are taken as variable constants of motion. In undisturbed motion, i.e. in the absence of the moments of outer forces, these parameters are constant, and the satellite regularly precesses around the fixed kinetic moment vector. In disturbed motion these parameters are variable and are found by different methods from differential equations which connect them. Secular disturbances are found by averaging with respect to periods of revolutions which gives differential equations for secular disturbances. In a number of cases these differential equations are integrated precisely [1, 3].

The main disturbing action on the satellite rotation is realized by gravitational and aerodynamic moments of forces. In calculating aerodynamic moments to a first approximation effects can be ignored connected with thermal velocities of molecules and with low linear velocities of the satellite surface caused by its rotation in favour of effects connected with high velocity of the motion of the satellite mass centre. It turns out that gravitational and aerodynamic moments do not produce secular changes in the angle between the satellite axis and the kinetic moment vector (ϑ is the nutation angle), in angular velocity $\dot{\psi}$ of the precession of the satellite axis around the kinetic moment vector, in angular velocity r_0 of the satellite rotation around its symmetry axis, in value $|L|$ of the kinetic moment vector. The whole secular effect is felt in the secular precession-nutation motion of the kinetic moment vector. In other words, the satellite regularly precesses around the axis slowly changing its direction in space. It turns out that gravitational moments cause precession of the kinetic moment vector around the direction of the normal to the orbit plane with angular velocity $\dot{\lambda}_1$

$$\dot{\lambda}_1 = \frac{3}{2}\,\omega^2\,\frac{A-C}{C\,r_0}\,\cos\Theta_1\,\cos\vartheta\left(1-\frac{3}{2}\sin^2\vartheta\right), \tag{5}$$

where Θ_1 is an angle between the kinetic moment vector and the normal to the plane of the orbit. Small nutation oscillations are added on an elliptic orbit.

Aerodynamic moments cause chiefly precession around the direction of the velocity vector of the satellite mass centre in the orbit perigee. In the simplest case angular velocity $\dot{\lambda}$ of aerodynamic precession of the kinetic moment vector can be calculated from the formula

$$\left.\begin{aligned}\dot{\lambda} &= -\frac{N}{C\,r_0}\,I\cos^2\vartheta\,, \qquad N = \frac{1}{2}\,S\,C_R\,Z_0\,\varrho_\pi\,V_k^2\,, \\[2mm] I &= \frac{1}{2\pi}\int\limits_0^{2\pi}\frac{\varrho\,(\cos\nu)}{\varrho_\pi}\,\sqrt{1+e^2+2e\cos\nu}\,(e+\cos\nu)\,d\nu\,. \end{aligned}\right\} \tag{6}$$

Here ϱ is the atmosphere density; ϱ_π is the value of this density in the orbit perigee ($\nu = 0$); V_K^2 is the square of circular velocity at a distance of the orbit perigee; S is the characteristic area of the satellite; C_R is the aerodynamic drag coefficient; Z_0 is distance from the satellite centre of mass to centre of pressure. In more complicated cases the aerodynamic precession velocity λ depends on the angle Θ between the kinetic moment vector and the perigee velocity vector.

Aerodynamic moments cause also small nutation oscillations. Generally speaking precession also can take place coinciding in character with gravitational precession around the normal to the orbit plane. For a satellite of Sputnik III type the gravitational precession is less than $4°$ per one revolution of the satellite on the orbit, the aerodynamic precession is lower than $1°$ for the same period.

The combined action of gravitational and aerodynamic disturbances leads to various classes of motions of the kinetic moment vector. These motions are investigated and classified in [1] and [3] by means of the secular motion equations integral.

Since the value of gravitational and aerodynamic disturbances depend on the satellite position with respect to the orbit, and the orbit due to the influence of the Earth's oblateness changes its position in space, this orbit regression influences the satellite motion about the centre of mass. Investigations of the influence of the orbit regression have shown that this influence for actual satellites is small as compared to the influence of aerodynamic and gravitational moments. The kinetic moment vector motion in real cases is stable with respect to the disturbing action of the orbit regression. Motion with respect to the regressing orbit differs little from motion with respect to the fixed orbit and reduces chiefly to the small (by several per cent) shift of the centre and to the change of the amplitude of precession-nutation motion of the kinetic moment vector.

The developed theory of the satellite perturbed rotation constitutes the basis of the method for determining actual motion about the centre of mass of Sputnik III. This investigation was made together with Yu. V. ZONOV [4]. Parameters of Sputnik III rotation and orientation were determined from indications of the magnetometer installed on this satellite [8] and were corrected in accordance with indications of other instruments. The magnetometer indications gave the possibility to determine the satellite position with respect to the Earth's magnetic field. Comparison of actual and theoretical dependencies describing this position has made it possible to determine fully the satellite rotation and orientation parameters at the large time interval.

It has turned out that Sputnik III motion about the centre of mass represents regular precession with slowly changing parameters and that the revealed secular motion of the kinetic moment vector agrees well with theoretical one. It was also revealed that dissipative electromagnetic effects causing the drag of Sputnik III rotation played a significant role.

Sputnik III precession around the kinetic moment vector direction is close to the "tumbling" regime: the angle between the satellite axis and the kinetic moment vector is close to $90°$ deviating from this value at different circuits more than by $6°$. The precession period T_ψ (the "tumbling" period) slowly increases from 135–140 sec. at the first ($N = 1 - 5$) circuits to 195 sec. at the 283th circuit (Figure 1a). Besides, the satellites slowly rotates around its own symmetry axis. Angular velocity $\dot\varphi$ of this rotation diminished from $0.375°$/sec. at the first circuit to zero approximately at the 20th circuit after which the direction of rotation changed to the opposite one, and the values of angular velocity of the own rotation at individual circuits varied about the average value on the order of $0.1°$/sec deviating from this value not more than by $0.1°$/sec. (Figure 1b).

The kinetic moment vector slowly changed its direction in space with mean angular velocity on the order of $1°$ per one circuit. At the beginning of motion the angular velocity is larger, and at the end of the motion it is lower. Motion of the kinetic moment vector is such that for the considered interval of circuits it tends to the satellite mass centre velocity vector at the orbit perigee. By the 100–110th circuits the

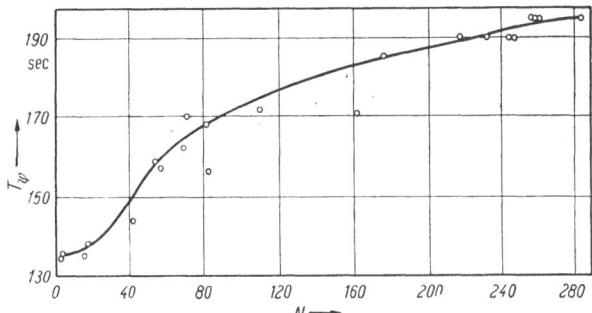

Fig. 1 a. The period of the satellite precession.

kinetic moment vector nearly coincides with the velocity vector at the perigee. The trajectory of the trail of the kinetic moment vector at the single sphere is shown in Figure 1 c. The pole of the piece of the sphere surface shown in the figure is the trail of the velocity vector at the orbit

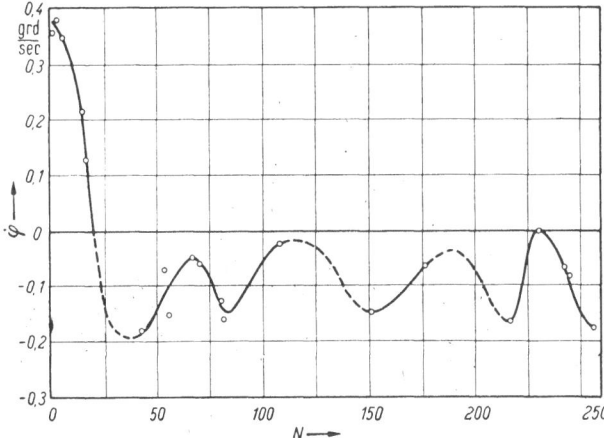

Fig. 1 b. The angular velocity of the satellite own rotation. The dashed line indicates portions of the curve obtained extrapolation.

perigee. Numbers along the trajectory represent numbers of circuits, i.e. complete satellite revolutions on the orbit.

Knowledge of Sputnik III rotation and orientation parameters has made it possible to calculate orientation of different instruments installed at the satellite. Orientation of instruments can be calculated with respect to any direction (for instance, orientation in absolute space or with respect to the Sun; orientation with respect to the incident stream; orientation with respect to the magnetic field, etc.). The data on the instruments orientation were used in turn for interpretation of the results of the experiments (see, for instance, [*9, 10, 11*]).

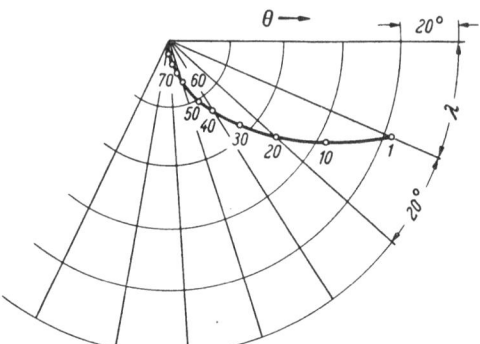

Fig. 1 c. The trajectory of the kinetic moment vector with respect to the orbit.

The dependence on the time $\cos\overline{\Theta}$ is illustrated in Figure 2 where Θ is the angle between the "axis" of the manometer and the satellite mass centre velocity vector. Indications of the manometer on a logarithmic scale are given for the sake of comparison.

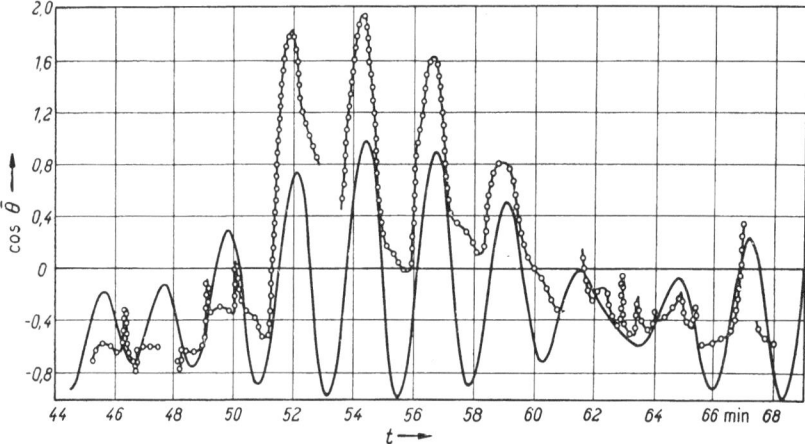

Fig. 2. Orientation of the axis of the ionization manometer with respect to the velocity vector on the 15 th circuit and comparison of this orientation with the record of the manometer. The solid line represents the charge of $\cos\overline{\Theta}$. The line circles represents indications of the manometer on the logarithmic scale.

References

[1] Beletsky, V. V.: Motion of an Artificial Earth Satellite About Its Centre of Mass, "Artificial Earth Satellites" published by the USSR Academy of Sciences Publishing House, vol. 1, 1958.

[2] Beletsky, V. V.: Onthe Libration of a Satellite, "Artificial Earth Satellites", vol. 3, 1959.

[3] BELETSKY, V. V.: Classification of Motion of an Artificial Earth Satellite About Its Mass Centre, "Artificial Earth Satellites", vol. 6, 1961.

[4] BELETSKY, V. V., and YU. V. ZONOV: Sputnik III Rotation and Orientation, "Artificial Earth Satellites", vol. 7, 1961.

[5] TISSERAND, F.: Traité de mecanique céleste, II, Paris, 1891.

[6] CHARLIER, C. V. L.: Eine neue Methode zur Behandlung des Rotations-problems. Meddelande fran Lunds Astr. Observatorium, 1908.

[7] POISSON, S. D.: Mémoire sur le mouvement de la Terre autour de son centre des gravite. Mémoires de l'Institut, 7 (1827); 9 (1830).

[8] DOLGINOV, S. SH., L. N. ZHUZGOV, and N. V. PUSHKOV: Preliminary Report on Geomagnetic Measurements Carried out from Sputnik III.

[9] KRASSOVSKY, V. I., I. S. SHKLOVSKY, YU. I. GALPERIN, E. M. SVETLITSKY, YU. M. KUSHNIR, and G. A. BORDOVSKY: Detection of Electrons with Energies of about 10 keV in the Upper Atmosphere, "Artificial Earth Satellites", vol. 6, 1961.

[10] ISTOMIN, V. G.: Variation of Concentration of Positive Ions with Height from Data of Mass Spectrometer Measurements on Sputnik III, "Artificial Earth Satellites", vol. 6, 1961.

[11] MIKHNEVICH, V. V., B. S. DANILIN, A. I. REPNEV, and V. A. SOKOLOV: Some Results of the Determination of the Structural Parameters of the Atmosphere Using Sputnik III.

[12] BELETSKY, V. V.: The Libration of a Satellite on an Elliptic Orbit (in print).

On the approximated analysis of the orbit evolution of artificial satellites

By

M. L. Lidov

The U.S.S.R. Academy of Sciences, Moscow, U.S.S.R.

Methods of classic celestial mechanics for an approximated study of the limited three body problem by averaging the disturbing accelerations [1] acquire great significance for the missions connected with artificial satellites. Both the qualitative picture of the evolution, which can be obtained from averaged equations, and approximated quantitative estimates based on simplified formulas are very useful for such purposes.

At present the evolution of individual orbits of artificial satellites at a considerable time interval can be calculated on computers by numerical integration of the precise set of differential equations. Comparison of these solutions with the results obtained from approximated formulas is a sufficiently reliable practical estimate of the accuracy of approximated equations. Such comparison (alongside the estimates of the orders of magnitude of ignored values) permits to conclude about the applicability of the obtained approximated equations for considering the evolution of some classes of the orbits.

For many practical problems connected with artificial satellites element variation estimates are required at a limited time interval with accuracies relatively not high. This enables us to use approximated equations for a broad class of artificial satellites orbits. The latter in its turn stimulates a detailed study of approximated averaged equations.

In the present paper we shall not consider the more precise averaging scheme with respect to one or several revolutions which allowed us to obtain formulas for calculating the element variation depending on the satellite successive revolution and formulas determining the orbital elements change during the time of one revolution of the disturbing body. This problem was analysed in the author's paper [2]. If element variations during the time of one revolution of the disturbing body turn to be sufficiently small, a scheme is applicable (in any case for qualitative estimates) of independent twofold averaging of disturbing accelerations with respect to the mean motion of the satellite and the mean motion of the disturbing body. In the theory of a first approximation the satellite orbital elements variation at an averaging interval is not taken into account when equations of evolution are obtained.

1. The formulation of the problem

The problem of the satellite orbit evolution is considered in the following formulation:

(a) The satellite rotates around the central body which has the axisymmetric gravitational potential V determined by the formula (3)

$$V = \frac{f M}{r} - \frac{f M \delta}{3 r^3} (3 \operatorname{Sin}^2 \psi - 1),$$

$$\delta = R^2 \left(\alpha - \frac{m}{2} \right); \qquad m = \frac{\Omega^2 R}{G_R}; \qquad (1)$$

f is the gravitational constant.

M is the mass of the central body.

ψ is the latitude of the point calculated from the plane of gravitational symmetry. We shall call this plane the equatorial plane.

R is the equatorial radius of the central body.

α is the oblateness of the central body.

Ω is the angular velocity of the body own rotation.

G_R is the gravity acceleration at the equator.

(b) The satellite expperiences the disturbances of the outer gravitating point with mass M_b which rotates along an elliptical orbit with small eccentricity e_b and period T_b.

(c) The maximum distance of the satellite from the central body $r_{\max}$ is much lower than the value of semiaxis a_b of the disturbing body orbit. Disturbing accelerations are considered to a first approximation with respect to the value $\dfrac{r_{\max}}{a_b} \ll 1$.

This formulation of the problem after twofold averaging leads to the following system of evolution equations:

$$\frac{d a}{d n} = 0,$$

$$\frac{d \varepsilon}{d n} = -(1 - \varepsilon) \varepsilon^{1/2} \operatorname{Sin}^2 i \operatorname{Sin} 2\omega,$$

$$\frac{d i}{d n} = -\frac{1}{2} \frac{(1 - \varepsilon)}{\varepsilon^{1/2}} \operatorname{Sin} i \operatorname{Cos} i \operatorname{Sin} 2\omega + \beta \frac{\operatorname{Sin} I \operatorname{Sin} \Omega}{\varepsilon^2} \operatorname{Cos} i_{\mathrm{eq}},$$

$$\frac{d \Omega}{d n} = -\frac{\operatorname{Cos} i}{\varepsilon^{1/2}} \left[(1 - \varepsilon) \operatorname{Sin}^2 \omega + \frac{\varepsilon}{5} \right] +$$

$$+ \beta \frac{\operatorname{Cos} i \operatorname{Sin} I \operatorname{Cos} \Omega - \operatorname{Sin} i \operatorname{Cos} I}{\operatorname{Sin} i \, \varepsilon^2} \operatorname{Cos} i_{\mathrm{eq}}, \qquad (2)$$

$$\frac{d \omega}{d n} = \frac{(\operatorname{Cos}^2 i - \varepsilon) \operatorname{Sin}^2 \omega + \dfrac{2}{5} \varepsilon}{\varepsilon^{1/2}} +$$

$$+ \frac{1}{2} \frac{\beta}{\varepsilon^2} \left(5 \operatorname{Cos}^2 i_{\mathrm{eq}} - 1 - \frac{2 \operatorname{Sin} I \operatorname{Cos} \Omega}{\operatorname{Sin} i} \operatorname{Cos} i_{\mathrm{eq}} \right).$$

$$\text{Cos}\,i_{eq} = \text{Cos}\,i\,\text{Cos}\,I + \text{Sin}\,i\,\text{Sin}\,I\,\text{Cos}\,\Omega\,,$$

$$n = A\,N, \qquad A = \frac{15}{2}\,\pi\,\frac{M_b}{M}\left(\frac{a}{a_b}\right)^3 \varepsilon_b^{-\frac{3}{2}},$$

$$\beta = \frac{4}{15}\,\frac{\delta\,a_b^3\,\varepsilon_b^{3/2}}{a^5}\left(\frac{M}{M_b}\right), \qquad \varepsilon = 1 - e^2, \qquad \varepsilon_b = 1 - e_b^2. \tag{3}$$

Here M is mass of the central body, a, e, i, Ω, ω are generally accepted designations for artificial satellite orbital elements. Angular elements are calculated from the disturbing body orbit plane. The node position Ω is calculated from the descending node of the disturbing body orbit in the plane of the equator.

I is the inclination of the equatorial plane to the plane of the disturbing body orbit.

N is the order number of satellite revolutions.

Besides the trivial integral $a = $ const the considered system has the integral [1]

$$\overline{w} = \text{const},$$

where $\overline{w}$ is the twice averaged total potential of disturbing accelerations of the outer attracting point and of noncentricity of the main field.

In the considered approximation this integral takes the form:

$$(1 - \varepsilon)\left(\frac{2}{5} - \text{Sin}^2\omega\,\text{Sin}^2 i\right) + \frac{\varepsilon\,\text{Cos}^2 i}{5} + \frac{\beta}{\varepsilon^{3/2}}\left(\text{Cos}^2 i_{eq} - \frac{1}{3}\right) = \text{const}. \tag{4}$$

2. The disturbance of the gravitating point

In the case when $\beta = 0$ the set of equations has three integrals:

$$\left.\begin{array}{c} a = \text{const}, \\[4pt] \varepsilon\,\text{Cos}^2 i = c_1, \\[4pt] (1 - \varepsilon)\left(\dfrac{2}{5} - \text{Sin}^2\omega\,\text{Sin}^2 i\right) = c_2. \end{array}\right\} \tag{5}$$

The range of possible values of constants c_1 and c_2 is represented in Figure 1.

In the case $c_2 > 0$ the argument of the latitude of the pericentre (angular distance of the pericentre from the node) ω monotonously increases: $\delta\,\omega > 0$. ε and i oscillate.

Maximum $\varepsilon_{\max}$ and minimum $\varepsilon_{\min}$ are realized at $\text{Sin}\,2\omega = 0$ and determined by formulas:

$$\left.\begin{array}{l} \varepsilon_{\max} = 1 - \dfrac{5}{2}\,c_2, \\[10pt] \varepsilon_{\min} = \dfrac{1}{2}\left[1 + \dfrac{5}{3}(c_1 + c_2) - \sqrt{\left[1 + \dfrac{5}{3}(c_1 + c_2)\right]^2 - \dfrac{5}{3}\cdot 4c_1}\,\right]. \end{array}\right\} \tag{6}$$

The corresponding extremal values for the inclination are found from the second integral (5) in the form

$$(\text{Cos}^2 i)\,\text{min} = \frac{c_1}{\varepsilon_{\max}}\,; \qquad (\text{Cos}^2 i)\,\text{max} = \frac{c_1}{\varepsilon_{\min}}\,.$$

In the case $c_2 < 0$ the argument of the latitude of the perigee ω also varies within limited bounds. Extremal values ω_{extrem} are found from the relationships:

$$\text{Sin}^2\omega_{\text{extrem}} = \frac{\dfrac{2}{5}}{1 - \dfrac{c_1}{\varepsilon^{*\,2}}}, \qquad (7)$$

where ε^* is determined from the equation

$$-\left(c_2 + \frac{2}{5}c_1\right)(\varepsilon^*)^2 +$$

$$+ \frac{4}{5}c_1\,\varepsilon^* - c_1\left(\frac{2}{5} - c_2\right) = 0 \quad (8)$$

and the condition $0 \leq \varepsilon^* \leq 1$.

Two values $\omega_{\min}$ and $\omega_{\max}$ are obtained as two roots of Eq. (7) which

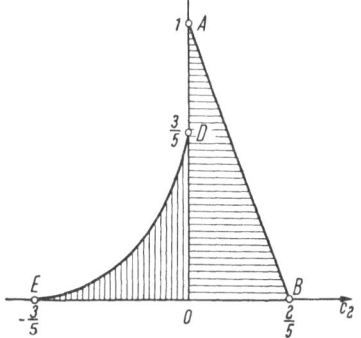

Fig. 1. The region of possible values of constants c_1 and c_2.

are symmetrical with respect to $\pi/2$ [if $\omega_0 \subset (\omega_1^*, \omega_3^*)$] or $\dfrac{3}{2}\pi$ [if initial $\omega_0 \subset (\omega_4^*, 2\pi + \omega_2^*)$]. In these formulas and subsequently the following designations for special values of ω are introduced:

$$\omega_1^* = \frac{1}{2}\,\text{arc Cos}\frac{1}{5}\,; \qquad \omega_2^* = -\omega_1^*;$$

$$\omega_3^* = \pi - \omega_1^*; \qquad \omega_4^* = \pi + \omega_1^*. \qquad \left.\right\} \qquad (9)$$

In the case $c_2 < 0$ extremal values $\varepsilon_{\text{extrem}}$ are two roots of a quadratic equation:

$$\varepsilon_{\text{extrem}}^2 - \left[1 + \frac{5}{3}(c_1 + c_2)\right]\varepsilon_{\text{extrem}} + \frac{5}{3}c_1 = 0. \qquad (10)$$

Dependences of extremal values ε and ω from constants c_1 and c_2 are represented in Figures 2, 3, 4.

If $\varepsilon_{\max} = \sqrt{1 - \varepsilon_{\min}} > 1 - \dfrac{R}{a}$ the fall of the satellite onto the central body takes place.

The limit of the range of permissible values of constants c_1 and c_2 corresponds to a number of limit cases:

1. Line AB $c_2 = \dfrac{2}{5}(1 - c_1)$ corresponds to the case when the initial inclination i_0 is equal to 0 or $180°$.

In this case $\varepsilon = \varepsilon_0$, the latitude of the pericentre φ which is calculated from the disturbing body orbit plane also does not vary and is
equal to zero.

The longitude of the pericentre α changes with velocity

$$\frac{d\alpha}{dn} = \pm \frac{1}{5} \varepsilon_0^{1/2} \tag{11}$$

(sign plus corresponds to $i_0 = 0$, sign minus corresponds to $-i_0 = 180°$).

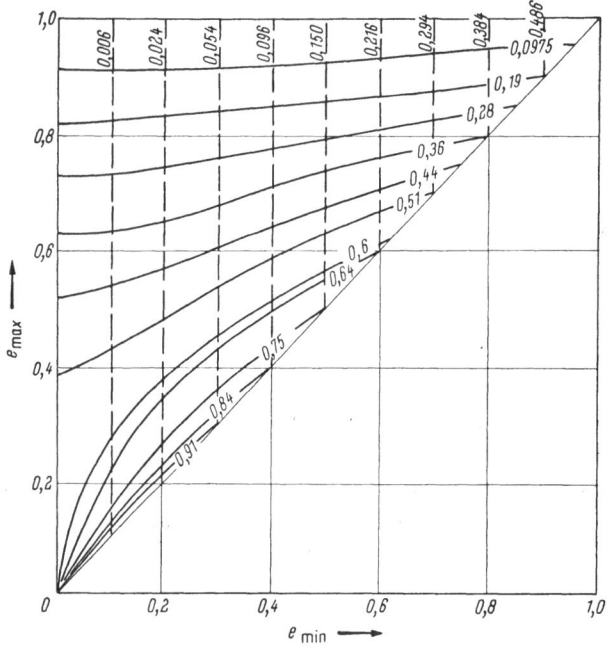

Fig. 2.   Extremal values of the eccentricity depending on c_1 (solid lines) and $c_2 > 0$
(dashed lines). Numbers near the lines denote constants c_1 and $1.5 c_2$.

2. Line ED $c_2 = -\dfrac{3}{5}\left(1 - \sqrt{\dfrac{5}{3} c_1}\right)^2$ corresponds to special initial
data

$$Sin^2 \omega_0 = 1, \qquad Cos^2 i_0 = \frac{3}{5} \varepsilon_0.$$

In this case $\varepsilon = \varepsilon_0$, $i = i_0$, $\varphi = \varphi_0$ and the evolution will represent
only the turn of the orbit about the normal to the plane of the orbit of
the disturbing body with a velocity

$$\frac{d\alpha}{dn} = \pm \frac{1}{5} \sqrt{\frac{5}{3}\left(\frac{12}{3} \varepsilon_0 - 3\right)} \tag{12}$$

("+" if $Cos\, i_0 > 0$,   "−" if $Cos\, i_0 < 0$).

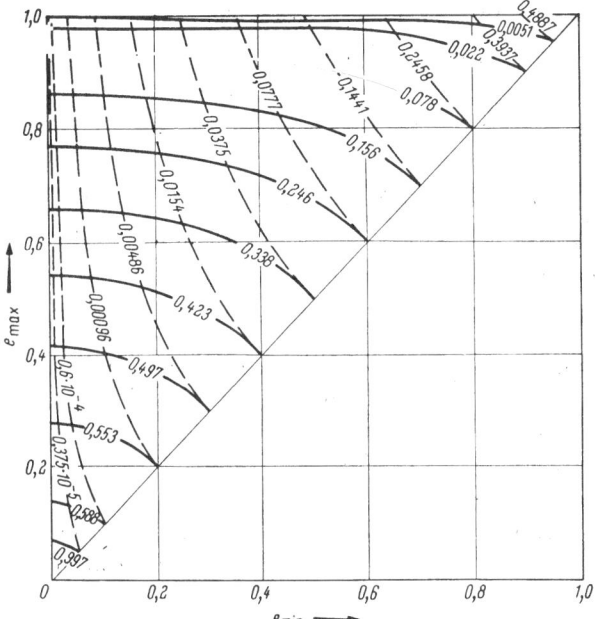

Fig. 3. Extremal values of the eccentricity depending on c_1 (solid lines) and $c_2 < 0$ (dashed lines). Numbers near the lines denote constants c_1 and c_2.

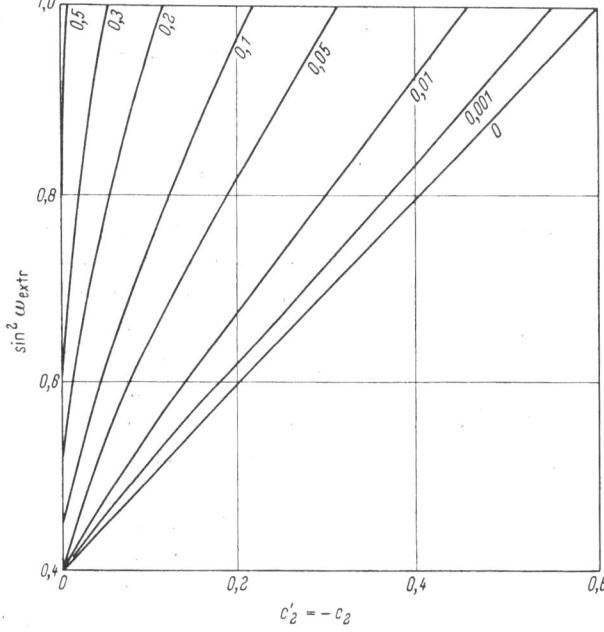

Fig. 4.
Extremal values of $\sin^2 \omega$ depending on $c_2 < 0$ at different values of c_1 (numbers near the lines).

3. Line OA corresponds to the case when

$$\mathrm{Sin}^2\,\varphi_0 = \mathrm{Sin}^2 i_0\,\mathrm{Sin}^2\omega_0 = \frac{2}{5}\,.$$

In this case $\omega \to \omega_3^*$ [if $\omega_0 \subset (\omega_3^*,\,\omega_1^*)$] or $\omega \to \omega_2$ [if $\omega_0 \subset (\omega_4^*,\,2\pi + \omega_2^*)$]. The orbit tends to a circular one, i.e. $e \to 0$.

If $\omega_0 < \dfrac{\pi}{2}$ (or correspondingly $\omega_0 < \dfrac{3}{2}\pi$) then in the process of the evolution the eccentricity local maximum is observed. It is determined by the formula

$$e_{\max} = \sqrt{1 - \frac{5}{3}\varepsilon_0\,\mathrm{Cos}^3 i_0}\,. \tag{13}$$

4. Line BE corresponds to the case when the initial inclination i_0 is equal to $90°$. In this case $i = 90° = \mathrm{const}$.

In the process of evolution ω tends to special values ω_1^* [in the case, if $\omega_0 \subset (\omega_2^*,\,\omega_3^*)$] or $\omega \to \omega_4^*$ [if $\omega_0 \subset (\omega_3^*,\,2\pi + \omega_2^*)$]. The eccentricity of the orbit tends to unity.

For a central body with radius R the fall to the surface will take place at $\varepsilon = \tilde{\varepsilon}$ and $\omega = \tilde{\omega}$ where $\tilde{\varepsilon}$ and $\tilde{\omega}$ are determined by formulas

$$\left.\begin{aligned}
\tilde{\varepsilon} &= 1 - \left(1 - \frac{R}{a}\right)^2, \\[4pt]
5\,\mathrm{Cos}\,2\tilde{\omega} - 1 &= \frac{(1 - \varepsilon_0)\,(5\,\mathrm{Cos}\,2\omega_0 - 1)}{1 - \tilde{\varepsilon}}\,.
\end{aligned}\right\} \tag{14}$$

If $(\sin 2\omega_0) < 0$ the local minimum of eccentricity takes place at $\sin 2\omega = 0$. The $e_{\min}$ value is determined from formulas

$$e_{\min}^2 = \frac{e_0^2(5\,\mathrm{Cos}\,2\omega_0 - 1)}{4}\,, \quad \text{if} \quad \mathrm{Cos}\,2\omega_0 > \frac{1}{5}\,,$$

$$e_{\min}^2 = \frac{e_0^2(1 - 5\,\mathrm{Cos}\,2\omega_0)}{6}\,, \quad \text{if} \quad \mathrm{Cos}\,2\omega_0 < \frac{1}{5}\,.$$

Thus, though in the problem considered when the orbital plane inclination to the disturbing body orbit plane equals to $90°$ the temporary decrease of eccentricity is possible under certain conditions, in the last analysis eccentricity becomes so large (at the constant semiaxis of the orbit) that the fall of the satellite onto the central body takes place.

This case is interesting as a limit one for sufficiently large inclinations.

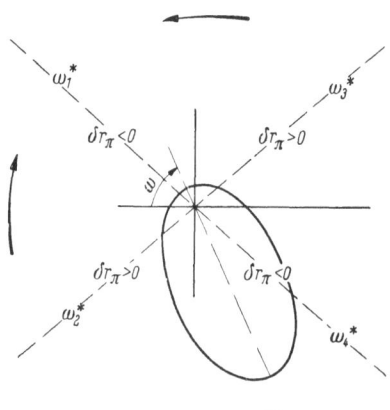

Fig. 5. Evolution of the orbit at $i = 90°$.

In the vicinity of $i_0 = 90°$ there is a continuous dependence of e_{max} on i_0.

In Figure 5 the orbit evolution for the case $i_0 = 90°$ is illustrated. The sign of the change of the pericentre height is marked in the quadrants of the coordinate system. Besides, dashed lines are shown which denote special values of the argument of the latitude of the perigee ω_1^*, ω_2^*, ω_3^*, ω_4^*. The arrows between dashed lines indicate the direction of the ω change depending on the perigee position.

The dependence of ω on the time t for the orbit with the inclination equal to $90°$ is determined from the formulas

$$
\begin{aligned}
t &= \frac{2\pi}{\sqrt{f\,M}}\, a^{3/2}\, N\,, \\[2mm]
N &= \frac{a_N}{A}[\delta\, F(\varphi,\, \varkappa^2) - \delta_0\, F(\varphi_0,\, \varkappa^2)]\,,
\end{aligned}
\qquad (15)
$$

where A is according to (3), F is an elliptical integral of the first kind

$$\delta = 1 \quad \text{if} \quad \text{Sin}\,2\omega > 0; \qquad \delta = -1 \quad \text{if} \quad \text{Sin}\,2\omega < 0,$$

$$\delta_0 = 1 \quad \text{if} \quad \text{Sin}\,2\omega_0 > 0; \qquad \delta_0 = -1 \quad \text{if} \quad \text{Sin}\,2\omega_0 < 0.$$

Values φ_0, φ, $\varkappa^2$ and a_N are determined from different formulas depending on the sign of the integral

$$c = e_0^2(5\,\text{Cos}\,2\omega_0 - 1).$$

	$c > 0$	$c < 0$
B	$\dfrac{c+1}{5}$	$\dfrac{c+1}{5}$
a_N	$2\left[\dfrac{4}{5}(1+B)\right]^{-\frac{1}{2}}$	$2\left[\dfrac{6}{5}(1-B)\right]^{-\frac{1}{2}}$
$\text{Sin}\,\varphi_0$	$\left[\dfrac{(1+B)\,(1-\text{Cos}\,2\omega_0)}{(1-B)\,(1+\text{Cos}\,2\omega_0)}\right]^{\frac{1}{2}}$	$\left[\dfrac{(1-B)\,(1+\text{Cos}\,2\omega_0)}{(1+B)\,(1-\text{Cos}\,2\omega_0)}\right]^{\frac{1}{2}}$
$\text{Sin}\,\varphi$	$\left[\dfrac{(1+B)\,(1-\text{Cos}\,2\omega)}{(1-B)\,(1+\text{Cos}\,2\omega)}\right]^{\frac{1}{2}}$	$\left[\dfrac{(1-B)\,(1+\text{Cos}\,2\omega)}{(1+B)\,(1-\text{Cos}\,2\omega)}\right]^{\frac{1}{2}}$
$\varkappa^2$	$\dfrac{3}{2}\dfrac{(1-B)}{(1+B)}$	$\dfrac{2}{3}\dfrac{(1+B)}{(1-B)}$

The lifetime of the satellite as a function of the initial data is determined from formula (15) at $\omega = \tilde{\omega}$ where $\tilde{\omega}$ is according to (14).

In cases when the initial value $\omega_0 = \omega_1^*$ or $\omega = \omega_4^*$, the satellite lifetime is calculated according to the elementary formula:

$$t = \frac{2\pi}{\sqrt{f\,M}}\, a^{3/2}\, N\,, \qquad (16)$$

where

$$N = \frac{5}{\sqrt{24}} \frac{1}{A} \ln\left[\frac{(1 + \sqrt{1 - e_0^2})(1 - \sqrt{1 - \tilde{e}^2})}{(1 - \sqrt{1 - e_0^2})(1 + \sqrt{1 - \tilde{e}^2})} \right],$$

$$\tilde{e} = 1 - \frac{R}{a}.$$

3. The influence of the noncentricity of the gravitational field

The correctness of the conclusion about the fall of the satellite onto the central body (see the previous section) at the satellite inclination to the disturbing body orbit plane close to 90° was checked by numerical integration of the precise set of equations for several orbits of artificial satellites of the Moon.

However, practical validity of this approximated conclusion, generally speaking, can be doubted by the fact of the existence of satellites of Uranus. Satellites of Uranus rotate about the equator of the planet. The inclination of the equator of Uranus to its orbit is 98°, i.e. orbital planes of satellites of Uranus are inclined to the orbital plane of the disturbing body — the Sun — at an angle close to 90°.

We shall show that the stationary eccentricity of these satellites orbits within the framework of the system of Eq. (2) can be explained if disturbances at the expense of the Uranus gravitational field oblateness are taken into account.

Somewhat idealizing the problem let us assume that in system (2) $I = 90°$, $i_0 = 90°$. In this case the set of equations (2) is transformed as follows

$$\left.\begin{array}{l} \dfrac{da}{dn} = 0; \quad \dfrac{di}{dn}; \quad \dfrac{d\Omega}{dn} = 0, \\[2mm] \dfrac{d\varepsilon}{dn} = -(1 - \varepsilon)\,\varepsilon^{1/2}\,\mathrm{Sin}\,2\omega, \\[2mm] \dfrac{d\omega}{dn} = \varepsilon^{1/2}\left(\dfrac{2}{5} - \mathrm{Sin}^2\omega\right) + \dfrac{\beta}{\varepsilon^2}. \end{array}\right\} \tag{17}$$

After similar substitution integral (4) is written in the form

$$(1 - \varepsilon)\left(\frac{2}{5} - \mathrm{Sin}^2\omega\right) + \frac{2}{3}\,\frac{\beta}{\varepsilon^{3/2}} = c. \tag{18}$$

The charakteristic behaviour of integral curves in the plane (ω, ε) for the case $0 < \beta < \dfrac{3}{5}$ is illustrated in Figure 6.

Arrows show the direction of the evolution. In the figure dashed curves are indicated: the vertical tangents curve AOB whose equation is of the following form

$$\varepsilon = \left(\frac{\beta}{\mathrm{Sin}^2\omega - \dfrac{2}{5}}\right)^{2/5} \tag{19}$$

and special values for the case $\beta = 0$, $\omega = \omega_1^*$ and $\omega = \omega_3^*$.

Point 0 represents a stationary solution

$$\varepsilon = \varepsilon_0 = \left(\frac{5}{3}\beta\right)^{2/5}, \qquad \omega = \omega_0 = \frac{\pi}{2}.$$

If $\beta \to 0$, curve (19) tends to fill the whole rectangle which is limited by dashed vertical lines, and the minimum value ε tends to zero, i.e. the value of the eccentricity maximum in the process of evolution approaches unity, and the fall of the satellite onto the central body is realized.

If $\beta > \frac{3}{5}$, there will be no libration point 0 in the plane (ε, ω).

In the case $\beta \neq 0$ maximum and minimum values of the eccentricity are determined from the equations:

At a) $\beta > \frac{3}{5}$ and b) $\beta < \frac{3}{5}$, $c > \frac{2}{3}\beta$,

$$\left.\begin{aligned}
\frac{2}{5}e_{\min}^2 + \frac{2}{3}\frac{\beta}{(1 - e_{\min}^2)^{3/2}} &= c, \\
-\frac{3}{5}e_{\max}^2 + \frac{2}{3}\frac{\beta}{(1 - e_{\max}^2)^{3/2}} &= c.
\end{aligned}\right\} \tag{20}$$

At c) $\beta < \frac{3}{5}$, $c < \frac{2}{3}\beta$ the extremal values of the eccentricity are found as two roots of the equation

$$-\frac{3}{5}e_{\text{extr}}^2 + \frac{2}{3}\frac{\beta}{(1 - e_{\text{extr}}^2)^{3/2}} = c. \tag{21}$$

For Oberon, a satellite of Uranus, which is the farthest from the planet and wich is maximally disturbed by the Sun, it is possible to obtain an approximated estimate for the value β.

Having supposed that the gravitational potential of the field of Uranus can be approximated by formula (1) we obtain after the substitution of approximated values of all the parameters:

$$\beta \approx 91.$$

For large values β and small values of the eccentricity the amplitude of oscillations according to (20) will be approximately as follows:

$$e_{\max} - e_{\min} \approx \frac{1}{2\beta}e_{\min}$$

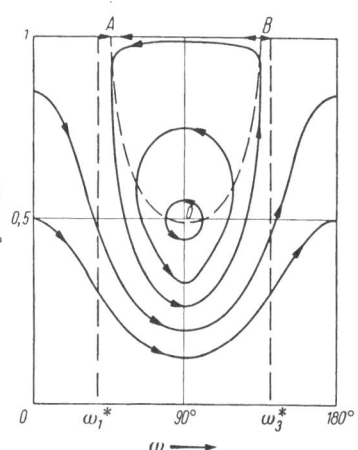

Fig. 6. The typical field of integral curves in the plane (ε, ω) at $i = 90°$, $I = 90°$, $\beta < \frac{3}{5}$.

i.e. for the satellite of Uranus ($e \approx 0.001$) the long-period oscillations of the eccentricity can be of the order of 10^{-5}.

The considerable change of the character of evolution, when the noncentricity of the main field is taken into account, is not typical of satellites of the solar system planets.

As one more example where the oblateness does not influence the qualitative conclusions obtained from equations at $\beta = 0$ we consider the

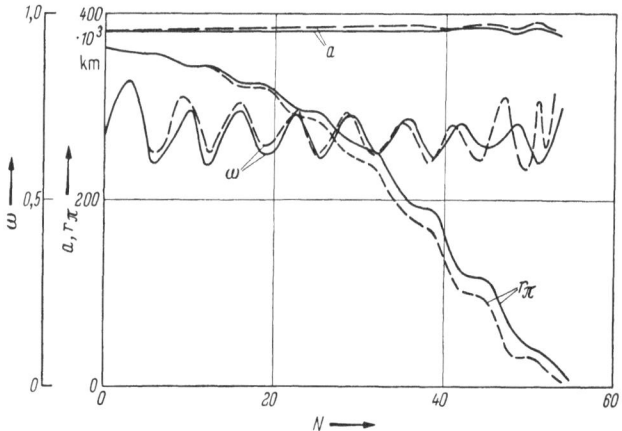

Fig. 7. The orbital element variation a, r, ω in a precise calculation (solid lines) and in an approximated calculation according to the method of single averaging (dashed lines).

orbit evolution of an Earth satellite, whose semiaxis and eccentricity are equal to the semiaxis and the eccentricity of the Moon's orbit, respectively.

The satellite orbital inclination with respect to the plane of the ecliptic was taken equal to $90°$.

For the satellite considered the parameter $\beta = 0.2 \times 10^{-4}$. If the initial value is $\omega_0 = \omega_1^* = \dfrac{1}{2}$ arc $\cos \dfrac{1}{5}$, the lifetime of such satellite at $\beta = 0$ can be estimated from formula (16).

Estimates from formula (16) have shown that such satellite could made only 52 circuits, i.e. it would exist on the orbit for about four years.

Exact estimates of the lifetime of such a "Moon" were obtained by numerical intergration of the set of differential equations of the combined motion of the Earth, Sun and the satellite with the noncentricity of the Earth's gravitational field taken into account. The evolution of the orbital elements depending on the order number of circuits is presented in Figure 7 and Figure 8 by solid lines. The minimum distance at a given circuit r_π was calculated (instead of the orbit eccentricity).

For comparison the same dependences obtained by means of approximated differential equations averaged once with respect to the satellite revolution [2] are indicated by dashed lines.

In a precise solution the minimum distance of the satellite orbit becomes less than the Earth radius after 55 revolutions. The lifetime was 54 revolutions when calculations were made according to the equations averaged once.

This example illustrates the efficiency of the approximated formulas not only in a qualitative, but also in a quantitative aspect.

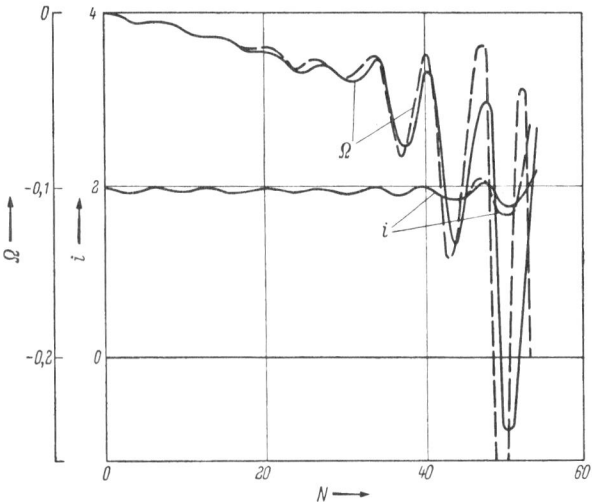

Fig. 8. The orbital element variation i, Ω in a precise calculation (solid lines) and in an approximated calculation according to the method of single averaging.

References

[1] MOISSEYEV, N. D.: a) On Some Basic Approximated Schemes of Celestial Mechanics Obtained by Averaging the Limited Circular Three-Point Problem. b): On the Averaged Variants of the Spatial Limited Circular Three-Point Problem. "Proceedings of the P. K. Sternberg Astronomical Institute", vol. XV, Book 1, 1945.
[2] LIDOV, M. L.: Evolution of Orbits of Planetary Artificial Satellites Under the Influence of Gravitational Disturbances of Outer Bodies. Artificial Earth Satellites, The U.S.S.R. Academy of Sciences Publishing House, vol. 8, 1961.

Discussion

During the discussion of the above paper, Dr. I. I. SHAPIRO did express the following remarks:

The work of Drs. MUSEN and LIDOV may have important cosmological implications: the solar system in the past may have more closely resembled a three-dimensional complex of orbiting bodies. Perhaps formerly existing, unstable satellites could provide explanations for such anomalies as the tilted axis of the earth's rotation, craters on the moon, and the so-called "missing" planet. The last may have existed and have been destroyed by interaction with a large satellite in a highly inclined orbit. Similarly, a former satellite of the earth may have imparted sufficient angular momentum to cause rotation about an inclined axis. It may also be possible that some craters on the moon were created by impacts with the remains of an unstable lunar satellite, broken up after passing within ROCHE's limit.

The effect of terrestrial radiation pressure on satellite orbits

By

Stanley P. Wyatt

University of Illinois Observatory, Urbana, Ill., U.S.A.

Abstract. The earth returns all incident solar radiation to space. About 60 percent of the total is delayed infra-red radiation; the remainder is scattered immediately outward and constitutes the albedo component. The terrestrial radiation exerts a perturbing force on earth satellites, and the question considered here is whether this force causes appreciable secular changes in the dynamical orbital elements of satellites. A variety of approximate models of the radiation are constructed and the secular perturbations of orbital period and eccentricity are calculated. The preliminary conclusions are that the infra-red radiation and the specular part of the albedo radiation cause only negligible effects. The diffuse part of the albedo radiation, on the other hand, appears in a general way to affect the dynamical elements. The secular perturbations are about 10 or 15 percent as large as those arising from the force of direct sunlight. In a general way their effects tend to imitate those of direct solar radiation pressure. Thus far it has not been possible to find clear evidence of terrestrial pressure in the orbital data on individual satellites because of uncertainties in their average A/m ratios and in their own reflection characteristics.

1. Introduction

The mechanical force of direct sunlight on earth satellites can produce major changes in their orbital elements. Several authors have recently examined the secular perturbations arising from this cause (PARKINSON, JONES, and SHAPIRO, 1960; MUSEN, BRYANT, and BAILIE, 1960; MUSEN, 1960; GEYLING, 1960; SHAPIRO and JONES, 1960; KOZAI, 1961; WYATT, 1961; ZADUNAISKY, SHAPIRO, and JONES, 1961; BRYANT, 1961 a; SHAPIRO and JONES, 1961). On the whole, comparison of prediction with observation has been satisfactory (see references above and MUHLEMAN et al., 1960; JASTROW and BRYANT, 1960; BRYANT, 1961 b). The push of solar radiation is ordinarily greater than atmospheric drag at heights above some 1000 km, and it has been increasingly recognized over the past two years that it is necessary to assess and eliminate the perturbing effects of radiation pressure before utilizing observed ·period changes to deduce atmospheric densities at great heights.

The general picture of the influence of sunlight has emerged only recently as a consequence of the work cited above and perhaps of other studies I have not seen. The picture is well illustrated by the Echo I balloon (Satellite 1960 Iota One), with its high orbit and large A/m ratio. Most of the time Echo spends a fraction of each revolution in the earth's shadow and the solar pressure is then cut off. Both the period and eccentricity then change secularly. During a minor part of its life Echo is in sunshine all around the orbit; on these occasions e continues to vary secularly but P remains unchanged. The sign of the secular changes in P and e on any given date depends on the orientation of the orbit with respect to the solar direction.

In the long run the period and eccentricity oscillate around mean values. In particular, the eccentricity of Echo's orbit oscillates around zero, with the argument of perigee changing abruptly every time e passes through zero. For orbits of low or moderate inclination, the total period of oscillation is governed by the length of the day as seen from the perigee point. If a close satellite has a small inclination, the eastward motion of its perigee point outstrips the eastward motion of the sun on the ecliptic, and perigee gains a full lap on the sun every 40 days. Thus the perigee point sees the sun rise in the east and set in the west, and the secular changes of P and e reverse sign at perigee noon and again 20 days later at perigee midnight. At higher inclinations the right ascension of perigee advances more slowly. Vanguard I (Satellite 1958 Beta Two) has an inclination of $34°$ and the secular changes due to radiation pressure require about 2.5 years for a complete oscillation. It is possible to choose a critical inclination, near $43°$ for close satellites, such that the right ascension of perigee moves eastward once a year. In such a case the local solar time at perigee remains statistically constant for a very long time. Under such conditions of resonance the force of sunlight continually perturbs the orbit in the same sense. Depending on the initial conditions the eccentricity will either increase monotonically or else it will decrease and go through zero and then increase monotonically. In either case the perigee will drive toward the ground, and the satellite is doomed to destruction.

Thus the role of solar radiation in changing the dynamical elements of a high satellite is a rather major one, and one is led to inquire whether other radiative forces can produce observable effects. As viewed from an earth satellite the only light source that is at all comparable with the sun is the earth itself; all other objects are very feeble by comparison. The present paper tries to assess the expected effect of terrestrial radiation on the motions of close satellites. Shapiro and Jones accounted for specular reflection in their early computer programs and estimated that it produces perturbations only about one percent as large as those caused by direct solar pressure (SHAPIRO and JONES, 1960). More recently a mixture of specular and diffuse reflection has been included in the programs (ZADUNAISKY, SHAPIRO, and JONES, 1961). CUNNINGHAM has set down the general equations for the energy incident on a satellite due to diffuse reflection by the earth and has integrated them on a computer (CUNNINGHAM, 1961). I have not seen any published work that deals analytically with the expected orbital perturbations arising from terrestrial radiation.

In the sequel we shall first set down some quantitative results for direct solar radiation pressure, for the sake of later comparison. Then we shall estimate crudely the relative magnitude of terrestrial and solar radiation pressure. Finally we shall calculate perturbations arising from various simplified models of the terrestrial radiation. Our attention is restricted to secular effects on the period (or semi-major axis) and the eccentricity.

2. Direct solar radiation pressure

For reference purposes it is convenient first to summarize the effects of sunlight itself. A satellite of average physical cross-section A and mass m at distance $r_\odot = 1$ a.u. from the sun intercepts energy at the rate $L_\odot A/4\pi r_\odot^2$, where $L_\odot$ is the total power output of the sun. Thus the momentum gained per unit time, or repulsive force, is of amount $L_\odot A/4\pi c r_\odot^2$, where c is the velocity of light. If the incident energy is reflected specularly or is absorbed and re-emitted isotropically, virtually no net momentum is carried away. We therefore assume that the radial acceleration has a magnitude

$$f = L_\odot (A/m)/4\pi c r_\odot^2 . \tag{1}$$

We also assume for simplicity that during the few hours a satellite spends revolving once around the earth the vector $\vec{f}$ is a constant relative to the satellite's orbit. Within the framework of these assumptions it is found that the secular acceleration, $\Delta P/P = \dot{P}_s$, is zero when the satellite does not penetrate the earth's shadow during one revolution. The secular rate of change of eccentricity on such occasions is

$$\frac{\Delta e}{P} = \dot{e}_s = \frac{3L_\odot (A/m)\sqrt{q(1+e)}\sin i' \sin\beta}{8\pi c r_\odot^2 \sqrt{GM_\oplus}} . \tag{2}$$

Here G is the constant of gravitation, $M_\oplus$ the mass of the earth, and q the distance from the earth's center to the satellite perigee. The angles i' and β are shown in Figure 1. The xy-plane coincides with the orbit plane of the satellite, Q is the direction of perigee, and P the instantaneous position of the satellite. The direction of the sun is S, inclined by an angle $i' = \widehat{ZS}$ from the orbit normal. The direction J defines the direction of the x-axis and is the intersection of the orbit plane and a perpendicular plane which contains the sun. The instantaneous true anomaly of the satellite is $\theta = \widehat{QP}$; and we define the angle $\beta \equiv \widehat{JQ}$. The results quoted above for the secular perturbations are found on integrating the instantaneous time rate of change of the elements around the orbit.

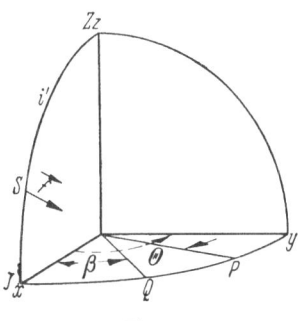

Fig. 1.

For numerical calculations it is convenient to describe the physical characteristics of the satellite itself by a dimensionless quantity D_s such that $(A/m) = D_s \ \mathrm{cm^2/gm}$, and also to express the ratio of perigee distance to the earth's equatorial radius by another dimensionless quantity $K \equiv q/R_\oplus \geq 1$. On substituting constants, the secular rates of period and eccentricity when there is no shadowing are

$$\dot{P}_s = 0,$$

$$\dot{e}_s = \nu \, D_s \, \sqrt{K(1+e)} \, \sin i' \, \sin \beta. \tag{3}$$

The constant $\nu = (1 \ \mathrm{cm^2/gm}) \times (3 L_\odot \sqrt{R_\oplus})/(8 \pi \, c \, r_\odot^2 \, \sqrt{G M_\oplus}) = 8.68 \times 10^{-11} \ \mathrm{sec^{-1}}$.

When a satellite dips into the earth's shadow during a part of every revolution the calculation of the secular perturbations is extremely complicated. The general solution has not been obtained, although several special cases have been solved (WYATT, 1961). The maximum secular period change occurs when the full force of sunlight lies in the orbit plane $(i = 90°)$ and probably at or near the point where the asymmetry is a maximum $(\beta = 90°$ or $270°)$. When $i' = 90°$ and $\beta = 90°$ the secular acceleration can be given explicitly and is

$$\dot{P}_s = - \mu D_s \, U(K, e), \tag{4}$$

$$U(K, e) = \frac{K \sqrt{1+e}}{(1-e)} \{ \sqrt{(K^2-1) + e(K^2+1) + 2Ke}$$
$$- \sqrt{(K^2-1) + e(K^2+1) - 2Ke} \},$$

the period decreasing with the time. Here D_s and K are defined as before, and μ is a dimensionless collection of constants, with

$$\mu = (1 \ \mathrm{cm^2/gm}) \times (3 L_\odot \, R_\oplus^2)/(4 \pi \, c \, r_\odot^2 \, G M_\oplus) = 1.40 \times 10^{-7}.$$

For $i' = 90°$ and $\beta = 270°$ the effect is equal and opposite and the period increases with time. Table 1 gives numerical values of $U(K, e)$ for several values of K and e.

Table 1. *Selected Values of* $U(K, e)$

K	$e = 0.00$	0.10	0.20	0.30	0.40
1.00	0.00	0.74	1.22	1.78	2.49
1.10	0.00	0.45	0.88	1.39	2.04
1.20	0.00	0.41	0.85	1.37	2.05
1.30	0.00	0.41	0.86	1.40	2.11
1.40	0.00	0.41	0.88	1.45	2.19

The secular change in the eccentricity for these special cases can also be given, but the formula is excessively complicated and we shall use Eq. (3) for $\dot{e}_s$ as the reference result for the effect of solar radiation pressure on the orbital eccentricity. It may be pointed out, however, that for the limiting case of $e = 0$ and $K = 1$ a satellite is in the shadow half the time and the general expression reduces to

$$\dot{e}_s = \frac{1}{2} \nu D_s \sin i' \sin \beta, \tag{5}$$

or one-half the value predicted by Eq. (3).

Numerically, the maximum secular effects for a balloon with $D_s \cong 100$ in an orbit with $K = 1.2$ and $e = 0.2$ are approximately $\dot{P}_s \cong \mp 1.40 \times 10^{-7} \times 10^2 \times 0.85 \cong \mp 1.2 \times 10^{-5} \cong \mp 0.017$ min/day and $\dot{e}_s \gtrsim \pm 0.5 \times 8.68 \times 10^{-11} \text{ sec}^{-1} \times 10^2 \cong \pm 0.0004$/day.

3. Terrestrial radiation

For equilibrium to obtain, the earth must return the same amount of radiation to space that it receives from the sun, and therefore the average total power output of the earth is $L_{\odot} \pi R_{\oplus}^2 / 4 \pi r_{\odot}^2 = L_{\odot} R_{\oplus}^2 / 4 r_{\odot}^2$ erg/sec. The mean flow outward over a sphere of radius r concentric with the center of the earth is therefore $L_{\odot} R_{\oplus}^2 / 16 \pi r_{\odot}^2 r^2$. For a very rough preliminary estimate of the magnitude of the perturbing acceleration, assume all photons leave the earth radially. At r the average acceleration due to terrestrial radiation would then have a magnitude of

$$f' = \frac{L_{\odot} (A/m) R_{\oplus}^2}{16 \pi c r_{\odot}^2 r^2}. \tag{6}$$

Dividing by Eq. (1), the ratio of the magnitudes is

$$\frac{f'}{f} = \frac{R_{\oplus}^2}{4 r^2} \lesssim 0.25. \tag{7}$$

At first sight, therefore, one might guess that perturbations due to terrestrial radiation are about one-fourth of those due to direct sunlight. Actually, if the situation were as simple as suggested by Eq. (6) the perturbing force would be radial and inverse-square. The satellite would therefore move in a two-body orbit with a very slightly reduced effective gravity. Numerically, the ratio of Eq. (6) to the gravitational acceleration of the satellite by the earth is $-\mu \, D_s/12 \cong -1.2 \times 10^{-8} \, D_s$. Thus even for a balloon like Echo the upward thrust by terrestrial radiation is only about one-millionth of the downward pull of gravitation.

More realistically, we can divide the radiant energy output of the earth into two parts: the infra-red component (IR) and the albedo component (A). Various investigators have measured earthlight on the moon and have found that the earth's mean albedo, γ, is between 0.35 and 0.40, and probably closer to 0.40. The value of γ is of course a weighted average; the chief contribution is made by highly reflecting clouds that cover about half of the planet at any one time, and lesser contributions are made by bright snow-covered areas, darker land areas, and the very dark oceans. Thus a fraction γ of the solar radiation incident on the earth is scattered immediately back to space, while a fraction $1 - \gamma \cong 0.60$ is absorbed by air and clouds and ground. The latter component is ultimately returned to space as infra-red radiation, and heat balance requires that the mean infra-red power output of the earth is $L_\odot (1 - \gamma) \, R_\oplus^2/4 r_\odot^2$ erg/sec.

4. Infra-red radiation pressure

Although satellites are now collecting much information on terrestrial infra-red radiation, it seems premature at the present time to select a realistic model of its distribution over the earth and its variations with time. For studying the effect of this radiation on satellite orbits, however, it is certainly reasonable to ignore longitudinal variations. The observed secular perturbations are derived from observations made over a day or more, and in this interval a satellite crosses the equator at many different longitudes. Day-and-night effect are more difficult to assess because during one or two days all crossings of a given latitude (for example northbound crossings of the equator) occur at very nearly the same local solar time. In this preliminary work we shall ignore diurnal variations in the infra-red output at any place on the earth on the grounds that diurnal temperature variations are only a few percent of the mean temperature. Seasonal changes in the total rate of infra-red energy output alternate only 3 percent around the mean because of the small eccentricity of the earth's orbit, and we shall neglect them here also. Although we average over longitude and time, it is necessary to take some account of variations in the infra-red component with latitude.

Model IR 1

As a first attempt, let us assume that the infra-red radiation by the earth is not only independent of longitude and time but also independent of latitude. We may imagine this radiation either to be emitted isotropically from a point source at the center of the earth, or else to be emitted equally by every unit area at the surface and in accordance with LAMBERT's law. In the first case each outbound photon is moving radially when it strikes the satellite. The perturbing acceleration is radial and of amount

$$f' = \frac{L_{\odot}\,(A/m)\,(1 - \gamma)\,R_{\oplus}^2}{16\,\pi\,c\,r_{\odot}^2\,r^2}. \tag{8}$$

In the second case the upward flow of radiation through unit area in the direction inclined by an angle z from the vertical is proportional to $\cos z$, where z is the zenith distance. The total radiation from a thin ring at the surface that is incident on a satellite of average cross-section A is

$$\frac{L_{\odot}\,A(1 - \gamma)\,R_{\oplus}^2\,\cos z\,\sin\theta\,d\theta}{8\,\pi\,r_{\odot}^2\,s^2}, \tag{9}$$

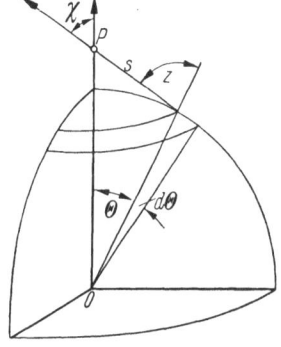

Fig. 2.

where s is the distance from the ring to the satellite. In Figure 2 the satellite is at point P at a distance $r = OP$ from the center of the earth. By symmetry the transverse component of the perturbing acceleration due to the ring is zero, while the radial component is given by

$$\frac{L_{\odot}\,(A/m)\,(1 - \gamma)\,R_{\oplus}^2\,\cos z\,\cos\chi\,\sin\theta\,d\theta}{8\,\pi\,c\,r_{\odot}^2\,s^2}. \tag{10}$$

Integration over the cap of earth visible to the satellite gives a total radial acceleration exactly equal to Eq. (8). Thus with LAMBERT-type infra-red emission that is independent of latitude, longitude, and time, the perturbing force is radial and inverse-square, with a strength about 60 percent as large as that given by Eq. (6). And if Model IR 1 were correct the infra-red radiation would cause no perturbations in the motion of earth satellites, but only a tiny reduction in effective gravity.

Model IR 2

Here we assume that the infra-red radiation is independent of longitude and time, but decreases as the latitude increases. Model IR 1 ignored latitude effects while the present model probably overestimates them. To keep the integration tractable we here assume that the radiation is proportional to the annual insolation on an earth whose obliquity is zero. Thus we have a dependence on the cosine of the latitude. In particular we set the transverse perturbing acceleration equal to zero and the radial perturbing acceleration equal to $C \cos \delta / r^2$, where C is a constant and $\delta \equiv \widehat{FP}$ is the instantaneous declination of the satellite as shown in Figure 3. Also in the figure $\alpha_N \equiv \widehat{VN}$, $\alpha \equiv \widehat{VF}$, and $\omega + \theta \equiv \widehat{NP}$. With this choice for the perturbing acceleration the secular change in the period is given by

$$\dot{P}_s = \frac{3 C e}{G M_\oplus (1 - e^2)} \int\limits_0^{2\pi} \cos \delta \sin \theta \, d\theta$$

$$= \frac{3 C e}{G M_\oplus (1 - e^2)} \int\limits_0^{2\pi} \left| \sqrt{1 - \sin^2 i \sin^2(\omega + \theta)} \right| \sin \theta \, d\theta = 0. \quad (11)$$

And similarly for the eccentricity,

$$\dot{e}_s = \frac{C(1 - e)^{3/2}}{2\pi G M_\oplus q^{3/2}} \int\limits_0^{2\pi} \cos \delta \sin \theta \, d\theta = 0. \quad (12)$$

Thus with the assumptions as given, the secular perturbations of P and e are zero. A defect of this model, however, is that the infra-red photons are assumed to move radially from a point source at the earth's center in proportion to the cosine of the latitude. If instead we employ LAMBERT's rule at the earth's surface and adhere to the dependence on $\cos \delta$, a satellite will be hit by more infra-red photons from the brighter low-latitude part of the visible ground than from the dimmer poleward area. Thus a small horizontal force pushes the satellite northward in

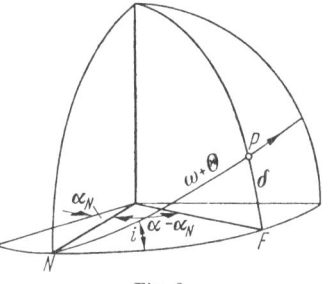

Fig. 3.

the northern hemisphere and southward in the southern. Intuitively the effect of such a brightness gradient on a close satellite would seem to be very small but I have not checked it quantitatively.

Model IR 3

A somewhat more general model of the infra-red radiation may be devised to account for the fact that the poles receive a finite amount of solar radiation and that atmospheric heat transport reduces the sharpness of dependence of infra-red radiation on latitude. Let us set the transverse perturbing acceleration equal to zero, as before, and set the radial component equal to $(C_1 + C_2 \cos \delta)/r^2$. Here C_1 and C_2 are adjustable constants subject only to the requirement that the total infra-red power output of our point source is $L_\odot (1 - \gamma) \, R_\oplus^2/4 r_\odot^2$ erg/sec. On integration around the orbit we again find that $\dot{P}_s = \dot{e}_s = 0$.

At this point we leave the infra-red problem, with the preliminary conclusion that the orbital effects are zero or small. It should be remembered, however, that more realistic models may reveal finite and observable perturbations on the period and eccentricity.

5. Albedo radiation pressure

The reflected component of the terrestrial radiation is decidely non-isotropic and we may anticipate that it can give rise to secular changes in the orbital period and eccentricity of a close earth satellite. We deal first with diffuse radiation (D) and at the end we consider specular reflection (S). For the diffuse models that follow, we assume that the mean cloud cover of the earth is distributed at random over all latitudes and longitudes and that diurnal variations are negligible. We therefore employ a constant mean albedo, γ, and the total albedo power output of the earth is $L_\odot \, \gamma \, R_\oplus^2/4 r_\odot^2$. Again it should be stressed that a satellite samples many longitudes during each day as the earth spins under the orbit, and it therefore appears resonable to average over longitude. Systematic latitude and time effects may exist, but in the present work we assume they are small enough to be neglected.

Model AD 1

The first and simplest model for the albedo component adopts an inverse-square outward radial force while the satellite is over the illuminated hemisphere of the earth and no perturbing force for the rest of the time. For this case the radial acceleration is given by

$$
\left.
\begin{aligned}
f' = R &= \frac{L_\odot \, (A/m) \, \gamma \, R_\oplus^2}{8\pi \, c \, r_\odot^2 \, r^2}, && 0 \leq z_\odot \leq \frac{\pi}{2}, \\
R &= 0, && \frac{\pi}{2} \leq z_\odot \leq \pi.
\end{aligned}
\right\}
\tag{13}
$$

Here $z_\odot$ is the zenith distance of the sun as seen by the satellite. Integration around the orbit reveals that the secular perturbations are

$$
\left.
\begin{aligned}
\dot{P}_s &= - \frac{\mu \, \gamma \, D_s e \sin \beta}{(1 - e^2)}, \\
\dot{e}_s &= - \frac{\nu \, \gamma \, D_s (1 - e)^{3/2} \sin \beta}{3\pi \, K^{3/2}}.
\end{aligned}
\right\}
\tag{14}
$$

The constants μ and ν are the same as those employed in the results for direct solar radiation pressure.

Model AD 2

A somewhat preferable model uses a radial inverse-square perturbing force that varies as $\cos z_\odot$ over the illuminated hemisphere and vanishes over the dark hemisphere. From Figure 1, $\cos z_\odot = \cos \widehat{SP} = \sin i' \cos(\beta + \theta)$, and the radial acceleration becomes

$$f' = R = \frac{L_\odot (A/m) \gamma R_\oplus^2 \sin i' \cos(\beta + \theta)}{4 \pi c\, r_\odot^2\, r^2}, \quad 0 \leq \beta + \theta \leq \frac{\pi}{2},$$

$$R = 0, \qquad\qquad\qquad \frac{\pi}{2} \leq \beta + \theta \leq \pi. \tag{15}$$

The secular perturbations are found to be

$$\dot{P}_s = - \frac{\pi \mu \gamma D_s e \sin i' \sin\beta}{2(1 - e^2)},$$

$$\dot{e}_s = - \frac{\nu \gamma D_s (1 - e)^{3/2} \sin i' \sin\beta}{6 K^{3/2}}. \tag{16}$$

These results resemble those for the effects of direct sunlight. For example on an occasion when $i' = 90°$ and $\beta = 90°$ or $270°$, the ratio of $\dot{P}_s$ for model AD 2 to that given by Eq. (4) is $\pi \gamma e/2(1 - e^2)\, U(K, e)$. If a satellite is moving with $K = 1.2$ and $e = 0.2$, for example, $U(K, e) = 0.85$ from Table 1, and with $\gamma = 0.40$ the ratio becomes 0.15. Division of $\dot{e}_s$ for Model AD 2 by that given in Eq. (3) for the case of no shadowing yields a ratio of $-\gamma(1 - e)^{3/2}/6 K^2 (1 + e)^{1/2} \leq -0.07$. Thus with this model perturbations are about ten percent of those arising from direct sunlight.

Model AD 3

In principle it would be interesting to compute the perturbations arising from diffuse reflection by LAMBERT's law from an earth with a fixed albedo, γ. The calculations, however, are formidable, and it appears impossible without recourse to numerical techniques to obtain the components of the acceleration for a close satellite. The perturbing force goes to zero, of course, in the earth's shadow, and it is chiefly radial when the entire cap visible to the satellite is in sunlight. But in the intermediate zones a part of the earth's cap is bright and the rest is in darkness. In general there is an outward radial force and an orthogonal component in the plane containing sun, center of earth, and satellite.

From a very great distance one sees the earth as a point source and its albedo radiation then produces a pure radial acceleration that is proportional to the well-known LAMBERT phase function. The perturbing acceleration on such an occasion is

$$f' = R = \frac{L_\odot (A/m) \gamma R_\oplus^2}{6 \pi^2 c\, r_\odot^2\, r^2} [(\pi - z_\odot) \cos z_\odot + \sin z_\odot], \tag{17}$$

where $z_\odot$ is the angle between sun and satellite as seen from the center of the earth. Providing $K \gtrsim 100$ the phase angle is very closely equal to the zenith distance of the sun as seen from the satellite. This choice of perturbing acceleration is invalid for close satellites, of course, and we shall not consider it further here.

The general perturbing acceleration due to diffuse LAMBERT reflection can also be found in the limiting case when we bring the satellite arbitrarily close to the earth's surface or, alternatively, let the radius of the earth become infinite. The result reduces to the special case of a satellite moving above an infinite plane, with the sun at zenith distance $z_\odot$. Each unit area of the surface emits total energy at the rate $L_\odot \gamma \cos z_\odot / 4 \pi r_\odot^2$. By symmetry the transverse component of the perturbing acceleration is zero, while the radial component is given by

$$f' = R = \frac{L_\odot (A/m) \gamma \cos z_\odot}{4 \pi c r_\odot^2}, \qquad 0 \leq z_\odot \leq \frac{\pi}{2}, \left.\begin{array}{c} \\ \\ \\ \end{array}\right\}$$
$$R = 0, \qquad\qquad\qquad \frac{\pi}{2} \leq z_\odot \leq \pi, \qquad (18)$$

and is independent of the height of the satellite. Expressed as power series in the eccentricity the secular perturbations with this model are

$$\dot{P}_s = -\frac{\pi}{2} \mu \gamma D_s K^2 e \sin i' \sin \beta \left[1 + e \left(2 - \frac{8 e \cos \beta}{3 \pi} \right) + \cdots \right], \left.\begin{array}{c} \\ \\ \end{array}\right\}$$
$$\dot{e}_s = -\frac{1}{6} \nu \gamma D_s \sqrt{K} \sin i' \sin \beta \left[1 + e \left(\frac{1}{2} - \frac{8 e \cos \beta}{3 \pi} \right) + \cdots \right]. \qquad (19)$$

The leading terms in these formulae are rather similar to the result for Model AD 2; each being larger than its AD 2 counterpart by a factor of about K^2. But the present formulae probably overestimate the perturbations because the flat-earth approximation fails to account for the decrease in radiation pressure with height above ground.

Model AD 4

DANJON's measures of earthlight on the moon indicate that the earth's phase function is more peaked than predicted by the LAMBERT law. At large distances the empirical phase function can be moderately well represented over the well-observed parts of the curve by $(1 - \cos z_\odot)^2$. Let us therefore adopt a simplified model in which the albedo emission leaves the earth radially. The normalized perturbing acceleration for this case is

$$f' = R = \frac{3 L_\odot (A/m) \gamma R_\oplus^2 (1 - \cos z_\odot)^2}{64 \pi c r_\odot^2 r^2}. \qquad (20)$$

With this model the force of the albedo radiation is finite at all points except when the sun is in the nadir, whereas with the others the force is cut off when the sun is more than 90° from the satellite's zenith. The present model, like the others, is defective in the sense that the adopted acceleration is fully radial, the orthogonal component being assumed to be zero. Integration around the orbit shows that for Model AD 4 we have

$$\dot{P}_s = -\frac{3\pi\mu\gamma D_s e \sin i' \sin\beta}{8(1-e^2)},$$

$$\dot{e}_s = -\frac{\nu\gamma D_s(1-e)^{3/2}\sin i' \sin\beta}{8K^{3/2}}. \tag{21}$$

These perturbations amount to 3/4 of those given by Model AD 2 and thus suggest that the orbital effects are not very sensitive to the specific choice of the radial disturbing force.

6. Albedo radiation pressure — The specular component

The only times a satellite is exposed to pure specular reflection are those occasions when it is moving over a calm water surface and sees a reflected image of the solar disk. Although the oceans cover 71 percent of the globe, it is only very rarely that a patch of ocean is perfectly calm. Nearly always it is wavy and rippled, and the solar image is seen to be smeared out over a rather large angular area. Pure specular reflection is more likely to be experienced over small bodies of water that are less disturbed by the wind. Viewed from infinity, the specular image of a vertical sun comes from a circle with radius 15 km. From heights of 1000 km and 500 km, the radii of the reflecting surfaces are only 4 km and 2 km respectively. Therefore relatively small lakes can image the entire solar disk. However, the increased efficiency of small lakes is offset by the small fraction of the earth's surface that they occupy. It appears very likely that the secular effect of specular radiation pressure is small.

To attempt a quantitative evaluation, let us first derive an expression for the reflected energy flow at a point above the earth's surface. We assume that the sun is a point source and that a fraction γ' of the incident photons are reflected into space, independently of the angle of incidence. We ignore the FRESNEL law of reflection for two

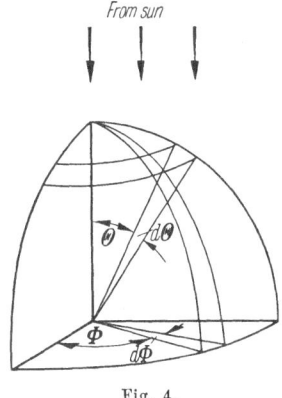

Fig. 4.

reasons: first, it is mathematically intractable and, secondly, the increased reflectivity at large angles of incidence is more or less compensated by the increasing air path of the radiation.

The energy per unit time that is incident on an element of surface with coordinates in the interval θ to $\theta + d\theta$ and ϕ to $\phi + d\phi$ is $L_\odot R_\oplus^2 \sin\theta \cos\theta \, d\theta \, d\phi / 4\pi r_\odot^2$, as can be seen from Figure 4. The fraction γ' is reflected and the outgoing beam is confined between the two vertical planes ϕ and $\phi + d\phi$ and the two slant planes inclined by 2θ and $2(\theta + d\theta)$, as shown in Figure 5. If the satellite is at point P at a distance $s = TP$ from the point of reflection, it sees this radiation pro-

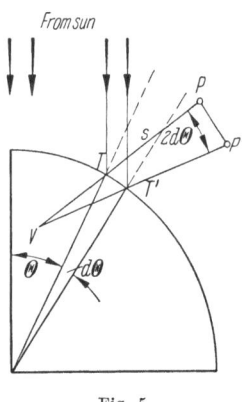

Fig. 5.

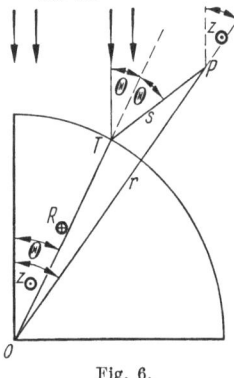

Fig. 6.

ceeding from the point V. On arrival in the vicinity of P this radiation crosses normally a rectangular area of dimensions PP' by $r \sin z_\odot \, d\phi$. From Figure 5,

$$PP' = 2d\theta(s + VT) = 2d\theta\left(s + \frac{R_\oplus \cos\theta}{2}\right)$$
$$= (2s + R_\oplus \cos\theta)\, d\theta. \tag{22}$$

The cross-section at P is therefore of area $r \sin z_\odot (2s + R_\oplus \cos\theta)\, d\theta \, d\phi$, and hence the energy incident on unit area of the satellite from the direction V is given by

$$\frac{L_\odot \gamma' R_\oplus^2 \sin\theta \cos\theta}{4\pi r_\odot^2 \, r \sin z_\odot (2s + R_\oplus \cos\theta)}. \tag{23}$$

From Figure 6 it may be found that $r \sin z_\odot = \sin\theta[R_\oplus + 2s \cos\theta]$, and thus the perturbing acceleration is given by

$$f' = \frac{L_\odot (A/m)\gamma' R_\oplus^2 \cos\theta}{4\pi c \, r_\odot^2 (R_\oplus + 2s \cos\theta)(R_\oplus \cos\theta + 2s)}. \tag{24}$$

Although this expression is complicated, it can be easily degenerated to two special cases.

Model AS 1

First let the satellite recede arbitrarily far from the earth so that $s \gg R_\oplus$ and $s \to r$. For this case, Eq. (24) reduces to

$$f' = \frac{L_\odot (A/m)\, \gamma'\, R_\oplus^2}{16\pi\, c\, r_\odot^2\, r^2}, \tag{25}$$

which agrees with the well-known fact that a specular sphere reflects isotropically if seen from a sufficiently great distance. Here again we encounter a radial inverse-square law for the perturbing acceleration and thus Model AS 1 causes no secular effects on the orbital elements.

Model AS 2

For the other degenerate case we let the satellite come arbitrarily close to the earth so that $s \to 0$ and $\theta \to z_\odot$. Under these conditions, except at $\theta = \pi/2$, Eq. (24) becomes

$$f' = \frac{L_\odot (A/m)\, \gamma'}{4\pi\, c\, r_\odot^2}, \tag{26}$$

a formula that is identical with the acceleration produced by reflection from a flat earth. To evaluate the special case for grazing incidence, set $\theta = \pi/2 - \varepsilon$ in Eq. (24) and evaluate the limit of f' as $s \to 0$. The result is identical with Eq. (26). For $z_\odot > \pi/2$, of course, no reflected image is seen. The situation is analogous to that of Model IR 1, where a perturbing force acts when the zenith distance of the sun is less than $90°$ and vanishes the rest of the time.

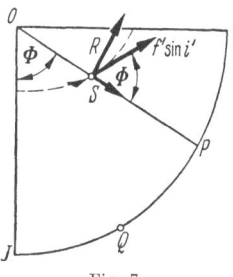

Fig. 7.

Unlike Model IR 1, however, the force here is non-radial. Returning to Figure 5, the reflected radiation is incident on the satellite along the line VP, and in the limiting case as $s \to 0$, we have $P \to T$, and the angle $VTO = z_\odot$. The component of the total perturbing acceleration that lies in the orbit plane is $f' \sin i'$. Figure 7 shows Figure 1 in plan view and also includes the components of the disturbing acceleration in the orbit plane. They are

$$\left. \begin{array}{l} R = f'\, \sin i'\, \cos\phi, \\ S = f'\, \sin i'\, \sin\phi, \end{array} \right\} \tag{27}$$

where $\phi \equiv \beta + \theta$. The perturbations, expressed as power series in the eccentricity, are found to be

$$\left. \begin{array}{l} \dot P_s = -\pi\, \mu\, \gamma'\, D_s\, K^2 e \sin i'\, \sin\beta \left[1 + e\left(2 - \dfrac{8\cos\beta}{3\pi} \right) + \cdots \right], \\[2ex] \dot e_s = \dfrac{1}{6}\, \nu\, \gamma'\, D_s\, \sqrt{K}\, \sin i'\, \sin\beta \left[1 + e\left(\dfrac{1}{2} - \dfrac{32\cos\beta}{3\pi} \right) + \cdots \right]. \end{array} \right\} \tag{28}$$

For small eccentricities, the magnitude of these perturbations differs by a factor of order $\pm (\gamma'/\gamma)\, K^2$ from those predicted by the diffuse Models AD 2 and AD 4. The factor K^2 arises from the flat-earth approximation and thus Eq. (28) probably overestimates the effect of specular reflection. With an index of refraction of 4/3 for water, the reflectivity at normal incidence is 0.02. If we set $\gamma' = 0.02$, then $(\gamma'/\gamma) \cong 0.05$. This model therefore suggests that pure specular reflection modifies the orbital period and eccentricity by minor amounts.

Model AS 3

Approximate account of the decline in the perturbing force with height above ground can be taken by adopting perturbing accelerations that are reduced from those of Model AS 2 by the factor $(R_\oplus^2/r^2)$. Integration then shows that

$$\left.\begin{aligned}
\dot{P}_s &= 0, \\
\dot{e}_s &= \frac{\nu\,\gamma'\, D_s \sin i' \sin\beta}{6\,K^{3/2}} \left[1 - e\left(\frac{3}{2} + \frac{8\cos\beta}{3\pi} \right) + \cdots \right].
\end{aligned}\right\} \quad (29)$$

For this model there is no secular effect in the period, while the effect on the eccentricity is reduced by a factor of about $-(\gamma'/\gamma)$ below that predicted by the diffuse models. It appears that the dynamical orbital elements are altered very little when specular reflection acts everywhere on the sunlit part of the earth. And because a satellite probably sees a specular image of the sun in calm waters during only an extremely limited portion of its circuit, it seems that we can afford to ignore specular radiation pressure entirely and restrict our attention to the diffuse albedo component.

7. Conclusions

Terrestrial radiation causes perturbations in the period and eccentricity of a close satellite. The infra-red component appears to contribute very little to changes of the orbital elements, largely because of the general symmetry we have assumed in the IR models. Future data from satellites should allow us to decide better how the infra-red radiation of the earth varies with latitude and time and whether it is necessary to construct more realistic models.

The albedo output of the earth may be considered as partly diffuse and partly specular. The specular models worked out here suggest that the imaged sun has only a small effect on the orbital elements. This result depends in part on the legitimacy of using the low reflectivity of water for normal incidence. The use of a larger effective γ', however, is more than offset by the small fraction of the terrestrial water surface that is calm enough to give pure specular reflection.

The diffuse part of the albedo energy output causes larger changes of period and eccentricity than do the other components of terrestrial radiation. On the other hand, the pressure of diffuse radiation is relatively small when compared with that of direct sunlight. As a preliminary estimate we suggest that earthlight produces secular perturbations that average around 10 percent of those due to the sun itself. Physically, there are several reasons for this reduction. First, the non-isotropic albedo component of the earth's radiation is only about 40 percent of the whole. Second, its effectiveness is diluted with increasing height above ground. Third, a close satellite sees the earth's radiation coming from a large solid angle and a large fraction of the horizontal momentum of the incident photons is cancelled.

The diffuse models show that the period decreases when β lies in the first two quadrants and therefore the local solar time at perigee is after noon and between noon and midnight. If we replace the continuous perturbing force by an upward impulse once each revolution at satellite noon, then on these occasions the orbital velocity is reduced. And therefore the semi-major axis and period decrease. The effect of such an impulse on the eccentricity acts in the same sense; the orbit becomes rounder. When β lies in either the third or fourth quadrants and therefore the local time at perigee is between midnight and noon, the effects are opposite in sign. An instantaneous impulse at satellite noon then increases the orbital velocity and therefore increases a and P. The eccentricity increases too.

The sign of $\dot{P}_s$ agrees with that due to direct solar pressure, while the sign of $\dot{e}_s$ is opposite to that of direct sunlight. Thus the two effects reinforce each other in changing the period, but counteract each other in changing the eccentricity. Let us adopt Model AD 2 for numerical esatimates and consider the relative importance of diffuse radiation to sunlight itself. The period is not changed by either force when $e = 0$ or in the symmetric case when $\beta = 0°$ or $180°$. When the satellite is in sunshine all around the orbit the period is unchanged by direct sunlight, but it may be affected by earthlight. In general the effect will be small, however, because i' is small on such occasions. When $\beta = 90°$ or $270°$ the ratio of Eq. (16) for $\dot{P}_s$ to Eq. (4) lies in the range 0.13 to 0.16 in the entire domain of Table 1 for which $K \geq 1.10$. The ratio of the perturbations in the eccentricity ranges from -0.07 for $K = 1$ and $e = 1$ to -0.01 for $K = 1.4$ and $e = 0.4$, on occasions when there is no shadowing. The relative effect of diffuse earthlight will be twice as large when Eq. (5) is applicable and the earth's shadow reduces the effect of direct sunlight on the change of eccentricity. In general, it appears that the diffuse component of the earth's albedo radiation may alter the orbits of earth satellites by small but detectable amounts.

It is a pleasure to acknowledge the help of Mr. RONALD A. SCHORN of the University of Illinois Observatory, who has been working with me on this and related topics. The investigation reported here was carried out in part under a grant from the United States Air Force.

References

BRYANT, R. W. (1961)a: NASA Technical Note, D-1063; (1961)b: NASA Technical Note, D-1124, and J. Geophys. Res., **66**, 3066.
CUNNINGHAM, F. G. (1961): NASA Technical Note, D-1099.
GEYLING, F. T. (1960): J. Franklin Inst., **269**, 375.
JASTROW, R., and R. BRYANT (1960): J. Geophys. Res., **65**, 3512.
KOZAI, Y. (1961): Smithsonian Astrophys. Observ. Spec. Report, No. 56.
MUHLEMAN, D. G., et al. (1960): Science, **132**, 1487.
MUSEN, P. (1960): J. Geophys. Res., **65**, 1391.
MUSEN, P., R. BRYANT, and A. BAILIE (1960): Science, **131**, 935.
PARKINSON, R. W., H. M. JONES, and I. I. SHAPIRO (1960): Science, **131**, 920.
SHAPIRO, I. I., and H. M. JONES (1960): Science, **132**, 1484; (1961): Science, 134, 973.
WYATT, S. P. (1961): Smithsonian Astrophys. Observ. Spec. Report, No. 60.
ZADUNAISKY, P. E., I. I. SHAPIRO, and H. M. JONES (1961): Smithsonian Astrophys. Observ. Spec. Report, No. 61.

The general relativity "force" on a satellite

By

G. C. McVittie

University of Illinois Observatory, Urbana, Illinois, U.S.A.

Mention should be made in this symposium of the fact that the motion of an infinitesimal satellite about a spherical earth, predicted by general relativity, differs from the motion under the Newtonian inverse square law of gravitation. The general relativity motion may be treated by classical methods in the following way: — Let r be the distance of the satellite from the centre of the (spherical) earth and let θ be an angle mesured from a fixed direction in the plane of the orbit. If M is the mass of the earth, G the constant of gravitation and c the velocity of light, we write $m = G\,M/c^2$. It is then known [1] that the relativistic motion possesses an integral of areas, namely,

$$h = r^2 \frac{d\theta}{dt}, \tag{1}$$

where t is the proper-time of the satellite. The equation of the orbit is also known and, with $u = 1/r$, it is

$$\frac{d^2 u}{d\theta^2} + u = \frac{m\,c^2}{h^2} + 3\,m\,u^2. \tag{2}$$

For any artificial satellite of the earth, it can easily be shown that the second term on the right is far smaller than the first.

It is a known result [2] of classical mechanics that the orbit whose equation is

$$\frac{d^2 u}{d\theta^2} + u = -\frac{F(u)}{h^2\,u^2}, \tag{3}$$

is obtained when the radial force per unit mass on the moving particle is $F(u)$. Moreover the integral of areas (1) then also exists provided that t is interpreted as the absolute Newtonian time. Therefore the Eqs. (1) and (2) would arise under a classical central force

$$F(u) = -m\,c^2\,u^2 - 3\,m\,h^2\,u^4. \tag{4}$$

This means that there is a small perturbing force, additional to the inverse square attraction, which has a *radial* component only. If R is the perturbing function, the only non-zero space derivative of R is

therefore
$$\frac{\partial R}{\partial r} = -3\,m\,h^2/r^4. \tag{5}$$

The presence of h^2 in this expression prevents us from integrating the formula with respect to r and thus finding R. But it is not necessary to do this; let $(x,\,y,\,z)$ be fixed rectangular coordinates with origin at the centre of the earth. Then

$$\frac{\partial R}{\partial x} = \frac{x}{r}\,\frac{\partial R}{\partial r}\,,$$

and similarly for the $\dot{y}$ and z derivatives. Since the motion is in a plane, R involves only the three elements a, e, χ of the undisturbed elliptic orbit [3]. If σ denote any one of these elements, we have

$$\frac{\partial R}{\partial \sigma} = \left(\sum_{x,\,y,\,z} \frac{x}{r}\,\frac{\partial x}{\partial \sigma}\right) \frac{\partial R}{\partial r} = \frac{\partial r}{\partial \sigma}\,\frac{\partial R}{\partial r}\,. \tag{6}$$

Also [4]

$$\frac{\partial r}{\partial \sigma} = \frac{\partial a}{\partial \sigma}\,\frac{1-e^2}{1+e\cos f} - \frac{\partial e}{\partial \sigma}\,a\cos f + \frac{\partial \chi}{\partial \sigma}\,\frac{a\,e\sin f}{(1-e^2)^{1/2}}\,, \tag{7}$$

where f is the true anomaly in the unperturbed elliptic orbit.

We now apply the standard planetary equations [5] for the variations of the elements. The orbit being plane, the inclination is irrelevant and the variation of the argument of perigee, ω, reduces to

$$\frac{d\omega}{dt} = \frac{(1-e^2)^{1/2}}{n\,a^2\,e}\,\frac{\partial R}{\partial e}\,,$$

where $n = h\,a^{-2}(1-e^2)^{-1/2}$ is the mean motion. Replacing the time by the true anomaly through (1) with $df/dt \equiv d\theta/dt$, to the order to which we are working, and using (5), (6) and (7) also, we have

$$\frac{\partial \omega}{df} = \frac{3\,m}{a\,e(1-e^2)}\,\cos f\,(1 + e\cos f)^2. \tag{8}$$

We define an anomalistic revolution of the satellite as one in which f increases by 2π. If $\Delta\omega$ is the corresponding change of ω, it is easy to show by integrating (8) with respect to f over 0 to 2π that

$$\Delta\omega = \frac{6\pi\,m}{a(1-e^2)} = \frac{6\pi\,m}{p}\,, \tag{9}$$

where p is the semi-latus rectum of the orbit. This is the general relativity result for the motion of the perigee. It has thus been found by the use of classical perturbation theory instead of by the customary method ound in textbooks on general relativity [6]. Numerical values of $\Delta\omega$ have been calculated by LA PAZ [7] and need not be repeated here.

The classical perturbation method having thus given the change of ω to the first order in m/p, we are encouraged to proceed and to try to find the alteration in the period. We calculate the change of the semi-major axis, a, which is given by

$$\frac{da}{dt} = \frac{2}{na}\frac{\partial R}{\partial \chi}. \tag{10}$$

Proceeding as before, the change in a for an increase of f by 2π is found to be

$$\Delta a = -\frac{6m\,h^2(1-e^2)\,e}{n^2\,p^4}\int\limits_0^{2\pi}(1+e\cos f)^2\sin f\,df = 0.$$

Thus if the period were defined, to the first order in m/p, by KEPLER's Third Law, namely, $P = 2\pi\,a^{3/2}(m\,c^2)^{-1/2}$, it would follow that $\Delta P = 0$ also. This procedure therefore is unable to show how the perturbed period depends on m/p. That it does so can be demonstrated by the use of DARWIN's method [8] for the exact solution of Eq. (2). It is of interest to observe that this equation is of the same mathematical form as that for the orbit of a satellite moving in the equatorial plane of an oblate earth [9]. Therefore DARWIN's method can also be applied to the solution of this second problem. DARWIN shows that the approximately elliptic orbit is obtainable by assuming that $du/d\theta = 0$ at the perigee and apogee points defined by $u = (1+e)/p$ and $u = (1-e)/p$. The first integral of (2) is then

$$\left(\frac{du}{d\theta}\right)^2 = 2m\,(u-C)\left(u-\frac{1-e}{p}\right)\left(u-\frac{1+e}{p}\right), \tag{11}$$

where

$$C = \frac{1}{2m}-\frac{2}{p},$$

$$\frac{1}{h^2} = \frac{1}{(m\,c^2)\,p}\left\{1-\frac{m}{p}(3+e^2)\right\}.$$

To the first order in m/p, the second equation is

$$\frac{1}{h} = \frac{1}{(m\,c^2)^{1/2}\,p^{1/2}}\left\{1-\frac{1}{2}\frac{m}{p}(3+e^2)\right\}. \tag{12}$$

The orbit is $u = (1+e\cos f')/p$ where f' is DARWIN's relativistic anomaly, and then (11) reduces to

$$\left(\frac{df'}{d\theta}\right)^2 = 1 - \frac{2m}{p}(3+e\cos f'). \tag{13}$$

The calculations are simplified by the introduction of the relativistic eccentric anomaly E' where

$$(1+e\cos f')\,(1-e\cos E') = 1 - e^2. \tag{14}$$

A period P is defined by integrating (1) over a revolution of the satellite and it may be expressed as

$$P/p^2 = \frac{1}{h} \int_{\theta_0}^{\theta_1} \frac{d\theta}{(1 + e \cos f')^2} ,$$

where θ_0 and θ_1 refer to the limits of θ corresponding to a revolution. It is sufficient to work to the order m/p. After some calculation, in which (12), (13) and (14) are used, we obtain

$$P = \frac{a^{3/2}}{(m\,c^2)^{1/2}} \left[E' - e \sin E' + \frac{m}{p} \left\{ \frac{3}{2}(1 - e^2) E' - e \sin E' \right\} \right]_{E_0'}^{E_1'}$$

since a may still be defined in terms of p by $p = a(1 - e^2)$. The anomalistic period, P_a, is found at once by taking $E_0' = 0$, $E_1' = 2\pi$; it is the period from perigee to perigee and amounts to

$$P_a = \frac{2\pi a^{3/2}}{(m\,c^2)^{1/2}} \left\{ 1 + \frac{3}{2} \frac{m}{p}(1 - e^2) \right\}. \tag{15}$$

The sidereal period is more difficult to calculate for it corresponds to an increase θ from 0 to 2π. However, by the use of (13) to the first order in m/p, and of (14) I find that

$$E_1' = E_0' + 2\pi - \frac{6m\pi}{p} \frac{(1 - e^2)^{1/2}}{1 + e \cos\omega} ,$$

where E_0' corresponds to $\theta = 0$ and so to $f' = -\omega$. Thus by (14)

$$1 - e \cos E_0' = \frac{1 - e^2}{1 + e \cos\omega} .$$

Using these results the sidereal period, P_s, correct to the order m/p, is

$$P_s = \frac{2\pi a^{3/2}}{(m\,c^2)^{1/2}} \left\{ 1 - \frac{3m}{p} \left(\frac{(1 - e^2)^{3/2}}{(1 + e \cos\omega)^2} - \frac{1 - e^2}{2} \right) \right\}. \tag{16}$$

It is unfortunate that the equation (10), combined with KEPLER's Third Law, gave no clue to the terms in m/p in the anomalistic period, P_a, of Eq. (15). The Eq. (10) did, in fact, make a true statement, namely, that the semi-major axis a and the anomalistic period P_a were constants of the perturbed motion just as they are of the unperturbed. As to the sidereal period given by Eq. (16), we observe that it undergoes periodic changes as the position of perigee alters relative to the initial line $\theta = 0$. A corresponding result is known from the theory of equatorial orbits about an oblate earth. Both examples therefore illustrate the fact that purely radial perturbations can produce oscillations in the value of the sidereal period of a satellite.

I am indebted to Sir CHARLES DARWIN and to Dr. D. G. KING-HELE for valuable criticisms of this note.

References

[1] McVITTIE, G. C.: General Relativity and Cosmology, London: Chapman and Hall, (1956) p. 85.

[2] WHITTAKER, E. T.: Analytical Dynamics, Cambridge: University Press, 3 rd. ed. (1927) Sec. 47.

[3] Notations as in W. M. SMART: Celestial Mechanics, London: Longmans, Green (1953) p. 21.

[4] SMART, W. M.: loc. cit., p. 220.

[5] SMART, W. M.: loc. cit., p. 69.

[6] McVITTIE, G. C.: loc. cit., p. 87.

[7] LA PAZ, L.: Publ. Astron. Soc. Pacific, 66, 13 (1954).

[8] DARWIN, C. G.: Proc. Roy. Soc., Lond., A, 263, 39 (1961).

[9] KING-HELE, D. G.: Proc. Roy. Soc., Lond., A, 247, 49 (1958), Eq. (80).

Determination of orbits based on visual and photographical observations

By

A. S. Sochilina

Institute of Theoretical Astronomy, Leningrad, U.S.S.R.

At the Institute for Theoretical Astronomy a preliminary determination of orbits of Earth's artificial satellites based on visual photographical observations is performed.

To compute preliminary orbits of satellites the series of visual observations at intervals of 10–15 days are used. Precise orbital elements are received using photographical observations at short intervals of 1–2 days.

Optical observations of satellites in the U.S.S.R. may be divided into three groups with respect to their accuracy.

Visual observations made by means of tubes AT-1 have a precision of $0.°5$ in position. Photographical observations made by means of small cameras have a precision of $0.°1$. At last photographical observations made by means of camera HAΦA-3 have a precission about $0.'1–0.'2$. In the last case accuracy in time is nearly $0.^s005–0.^s01$. All these kinds of observations are used for the determination of satellites' orbits at the Institute for Theoretical Astronomy. Orbits obtained from visual observations are used for computing of ephemeris and also as preliminary ones for computation of precise orbits. The determination of orbits from visual observations has been performed for all Soviet satellites.

For improvement the elements the modified method by ECKERT-BROUWER is used. The modification consists in using a moving system of coordinates with axis OX directed towards the ascending node of the orbit.

Then the formulae for improvement of elements are expressed as follows

$$
\begin{bmatrix} \varrho \cos\delta\, \Delta\alpha \\ \varrho\, \Delta\delta \end{bmatrix} = \begin{bmatrix} \Delta x \\ \Delta y \\ \Delta z \end{bmatrix} \begin{bmatrix} -\sin(\alpha-\Omega), & \cos(\alpha-\Omega), & 0 \\ -\cos(\alpha-\Omega)\sin\delta, & -\sin(\alpha-\Omega)\sin\delta, & \cos\delta \end{bmatrix},
$$

$$
\begin{bmatrix} \Delta x \\ \Delta y \\ \Delta z \end{bmatrix} = \begin{bmatrix} \Delta\Omega \\ \Delta i \\ \Delta M \\ \Delta\omega \\ \Delta\varphi \\ \Delta n \end{bmatrix} \begin{bmatrix} -y, & 0, & \dfrac{\dot{x}}{n}, & -r\sin u, & A_x, & \dfrac{\dot{x}}{n}(t-t_0)-\dfrac{2}{3n}x \\[2mm] x, & -z, & \dfrac{\dot{y}}{n}, & r\,\cos i\cos u, & A_y, & \dfrac{\dot{y}}{n}(t-t_0)-\dfrac{2}{3n}y \\[2mm] 0, & y, & \dfrac{\dot{z}}{n}, & r\,\sin i\cos u, & A_z, & \dfrac{\dot{z}}{n}(t-t_0)-\dfrac{2}{3n}z \end{bmatrix},
$$

$$x = r \cos u, \qquad \frac{\dot{x}}{n} = -\frac{a}{\cos\varphi}\,[\sin u + \sin\varphi \sin\omega],$$

$$y = r \cos i \sin u, \qquad \frac{\dot{y}}{n} = \frac{a \cos i}{\cos\varphi}[\cos u + \sin\varphi \cos\omega],$$

$$y = r \sin i \sin u, \qquad \frac{\dot{z}}{n} = \frac{a \sin i}{\cos\varphi}[\cos u + \sin\varphi \cos\omega],$$

$$\left.\begin{array}{l} A_x = -a[\sin u \sin E + \cos\varphi \cos\omega], \\ A_y = a \cos i [\cos u \sin E - \cos\varphi \sin\omega], \\ A_z = a \sin i [\cos u \sin E - \cos\varphi \sin\omega]. \end{array}\right\} \qquad (*)$$

The topocentric right ascension and declination (α, δ) are computed from

$$x - X = \varrho \cos(\alpha - \Omega) \cos\delta, \qquad X = R \cos\varphi' \cos(s - \Omega),$$

$$y - Y = \varrho \sin(\alpha - \Omega) \cos\delta, \qquad Y = R \cos\varphi' \sin(s - \Omega),$$

$$z - Z = \varrho \sin\delta, \qquad\qquad Z = R \sin\varphi'.$$

The following notations are adopted in these formulae:

Ω longitude of acsending node,
i inclination,
M mean anomaly,
φ angle of eccentricity $(e = \sin\varphi)$,
n mean motion,
a major semi-axis,
u argument of latitude,
s local sidereal time,
R geocentric distance of the observatory,
φ' geocentric latitude of the observatory.

For an orbit with small eccentucity elements $M_0 + \omega$, $e \sin\omega$, $e \cos\omega$ are used and the formulae $(*)$ are accordingly changed.

To compute the influence atmospheric drag the elements n' and n'' are introduced into the expression of mean anomaly

$$M = M_0 + n(t - t_0) + n'(t - t_0)^2 + n''(t - t_0)^3.$$

The influence of the Earth's oblateness is computed according to the theory by PROSKURIN and BATRAKOV (1960) [2]. Orbits based on visual observations are determined on intervals of 10–15 days, which permits to receive more precise values of n' and n''. But due to air-drag the O–C increase. Therefore orbits based on photographical observations are computed at short intervals of time, the elements n' and n'' being computed beforehand. In this case the O–C attain the value $1'-3'$. This accuracy is rather satisfactory. The obtained orbit can be improved only after taking into account tesseral harmonics and more precise coordinates of the observating station.

The results of a preliminary improvement of the elements of satellite 1958 δ_1 have been published in Bulletin of Astronomical Council (1960) [1].

These elements can be represented by the following polynomials

$$\Omega = (173.°421 \pm 0.°030) - (2.°6964 \pm 0.°0015)\,(t - t_0) - $$
$$-0.°1461 \times 10^{-2}(t - t_0)^2 - 0.°467 \times 10^{-5}(t - t_0)^3,$$

$$i = (65.°144 \pm 0.°002) - 0.°125 \times 10^{-2}(t - t_0),$$

$$\omega = (30.°346 \pm 0.°082) - 0.°3868\,(t - t_0) - 0.°168 \times 10^{-3}(t - t_0)^2,$$

$$\varphi = (5.°450 \pm 0.°039) - 0.°01507\,(t - t_0) - 0.°1029 \times 10^{-3}(t - t_0)^2,$$

$$n = 5044.18 \pm 0.°01 \quad \text{for moment } t_0,$$

where t_0 means 1958 july 31.0. From changes of these elements we have determined the Earth's oblateness which is equal to $\varepsilon = \dfrac{I}{297.17 \pm 0.08}$.

References

[1] BATRAKOV, JU. V., and A. S. SOCHILINA (1960): Elements of the soviet satellite 1958 δ_1 and the value of the Earth's oblateness. Bull. Stations of Optical observations of Artificial Satellites, N 7 (17).
[2] PROSKURIN, V. F., and JU. V. BATRAKOV (1960): Perturbations in the motion of artificial satellites due to the Earth's slattening. Bull. ITA, **7**, 7 (90).

Satellite orbit analyses for geodetic purposes

By

W. M. Kaula

Theoretical Division, Goddard Space Flight Center
National Aeronautics and Space Administration. Greenbelt, Maryland, U.S.A.

Abstract. The difficulties in obtaining accurate determinations of tracking station positions and longitudinal variations of the earth's gravitational field are not problems of mechanics, but rather of data analysis: non-uniform distribution of observations, inadequacy of the mathematical model of the atmosphere, and similarity of effects of different parameters on the same orbit.

Because the characteristic periocities of the geodetic effects fall in a different part of the spectrum of orbital variations from drag effects, the geodetic effects should, in theory, be determinable even though appreciably smaller. Given a statistical model for drag effects, a generalization of linear autoregression theory can be developed, which indicates, however, that in the computation arrays of dimensions on the order of the number of observations must be manipulated. Hence either the number of observations treated must be limited or the rigorous treatment modified. The second alternative has been applied in analyzing BAKER-NUNN camera observations of satellites 1959α-1, 1959η-1, 1960ι-2, in which various empirical methods are applied in place of allowing for the correlation between different observations. Some preliminary results obtained from 1960ι-2 are presented.

I. Introduction

The geometrical [1, 2] and gravitational [2, 3] problems in analyzing close satellite orbits for geodetic parameters beyond the oblateness — station positions and the higher harmonics of the gravitational field — are relatively straight forward, since the effects involved are small enough to be treated as first-order perturbations. Since there are numerous independent parameters, of which several have similar effects on a given orbit, two factors — the non-uniform distribution of observations and the inadequacy of the atmospheric model — require attention to some statistical problems to attain fullest geodetic exploitation of any accurate satellite observations. This paper has two main parts: (1) development of the appropriate statistical theory and (2) numerical results obtained thus far from simpler approximate methods.

II. Theory of non-uniformly distributed observations

We start from the generalized linear autoregression of prediction theory [4, 5], which states that, given $x(t)$ over and interval $A < t < B$, the optimum prediction (in a least-squares sense) $E\{x(z)\}$, for $z > B$, can be expressed as

$$E\{x(z)\} = \int_A^B w(z, t)\, x(t)\, dt,\qquad(1)$$

where $w(z, t)$ is the solution of

$$K(z, s) = \int_A^B w(z, t)\, K(t, s)\, dt,\quad A < s < B.\qquad(2)$$

$K(u, v)$ is the covariance between $x(u)$ and $x(v)$.

The modifications of (1) and (2) appropriate to our problem:

(a) The variable x is a 6-dimensional vector — position and momentum of a satellite — so make the replacements:

$$x(z) \to x_f(z),\quad w(z, t) \to w_{fh}(z, t),\quad K(z, s) \to K_{fg}(z, s).\qquad(3)$$

(b) Observations are not continuous over the interval A, B, but are in several disconnected bits, so make the replacements:

$$\int_A^B \cdots dt \to \sum_k \int_{a_k}^{b_k} \cdots dt = \int_A^B I(t) \cdots dt,\qquad(4)$$

where minimum $a_k \geqq A$, maximum $b_k \leqq B$, $I(t) = 1$, $a_k < t < b_k$, and $I(t) = 0$, $b_k < t < a_{k+1}$ for any k.

(c) The $x_h(t)$ are not observed, but rather a linear transform $y_j(t)$ of lesser dimension (e.g. 2 for photos, 1 for range or range-rate), so make the replacements

$$
\left.
\begin{aligned}
x_h(t) \to y_j(t),\quad w_{fh}(z, t) &\to w_{fh}(z, t)\frac{\partial y_j(t)}{\partial x_h(t)} = v_{fj}(z, t),\\[4pt]
K_{fg}(z, s) &\to K_{fg}(z, s)\frac{\partial y_c(s)}{\partial x_g(s)} = M_{fc}(z, s),\\[4pt]
K_{hg}(t, s) &\to \frac{\partial y_j(t)}{\partial x_h(x)}\, K_{hg}(t, s)\frac{\partial y_c(s)}{\partial x_g(s)} = L_{jc}(t, s).
\end{aligned}
\right\}\qquad(5)
$$

In (5), and hereafter, the rule of summation over repeated subscripts in a product applies.

(d) The observations $y_j(t)$ may be affected by errors in certain parameters (e.g., datum position), so make the replacement

$$y_j(t) \to \hat{y}_j(t) - \frac{\partial y_j(t)}{\partial q_m}\, E\{q_m\},\qquad(6)$$

where $E\{q_m\}$ is the correction to q_m.

(e) We may not be interested in the $x_f(z)$, but rather certain parameters p_l of which they are functions (e.g., departures from secularly changing Keplerian elements are functions of parameters of the gravitational field and the atmospheric model), so make the replacement

$$E\{x_f(z)\} \to \frac{\partial x_f(z)}{\partial p_l}\, E\{p_l\}.\qquad(7)$$

Putting (3) thru (7) in (1) and (2), we get

$$\frac{\partial x_f(z)}{\partial p_l} E\{p_l\} = \int_A^B I(t)\, v_{fj}(z,\,t) \left[\hat{y}_j(t) - \frac{\partial y_j(t)}{\partial q_m} E\{q_m\} \right] dt. \qquad (8)$$

$$M_{fc}(z,\,s) = \int_A^B I(t)\, v_{fj}(z,\,t)\, L_{jc}(t,\,s)\, dt. \qquad (9)$$

Over a finite interval of time, $x_f(z)$ which are departures from a model of secularly changing Keplerian elements (or a model taking into account any major periodic variations, such as those due to J_2, as well) can be represented by a FOURIER expansion

$$x_f(z) = \Re \sum_{n=0}^{\infty} \alpha_{fn} \exp [i\,\lambda_n\, z]. \qquad (10)$$

The α_{fn} are complex; $\Re$ denotes the real part, and i is $(-1)^{1/2}$. In any practical application, we are concerned only with the finite interval $A < z < B$, so

$$\lambda_n = \frac{2\,\pi\, n}{B - A}. \qquad (11)$$

Also, in any practical case, a finite upper limit can be taken for n; for satellite orbits, certainly one such that $1/\lambda_n > 20$ minutes.

Apply the FOURIER transformation $\int_A^B \cdots e^{i\lambda_n z}\, dz$ to (8) and (9):

$$\frac{\partial \alpha_{fn}}{\partial p_l} E\{p_l\} = \int_A^B I(t)\, \Omega_{fjn}(t) \left[\hat{y}_j(t) - \frac{\partial y_j(t)}{\partial q_m} E\{q_m\} \right] dt. \qquad (12)$$

$$\varrho_{fcn}(s) = \int_A^B I(t)\, \Omega_{fjn}(t)\, L_{jc}(t,\,s)\, dt, \qquad (13)$$

where $\varrho_{fcn}(s)$ and $\Omega_{fjn}(t)$ are the transformation of $M_{fc}(z,\,s)$ and $v_{fj}(z,t)$ respectively.

The advantage of the FOURIER representation in (12) is apparent. For the effect of any gravitational harmonic, which has a discrete spectrum, most of the $\partial \alpha_{fn}/\partial p_l$ are zero, while in the equations where they are non-zero, the $\partial \alpha_{fn}/\partial p_l$ due to drag affects are negligibly small.

To solve (13) for the $\Omega_{fjn}(t)$ to use in (12), there are needed statistical models for the gravitational and drag effects on the orbit. Given these models, the $L_{jc}(t,\,s)$ can be constructed from the $K_{hg}(t,\,s)$ by (5), while to obtain the $\varrho_{fcn}(s)$ the FOURIER transform with respect to z must be applied to $K_{fg}(z,\,s)$ before applying the post-multiplication $\partial y_c(s)/\partial x_g(s)$.

The statistical model for the gravitational part of K_{fg} is obtainable from the autocovariance analysis of terrestrial gravimetry [6], the orders-of-magnitude of which have been confirmed by the zonal harmonics thus far obtained from satellite orbits [7].

For the drag part of K_{fg} the appropriate data are less sure, mainly due to the considerable variation from one year to another and from one altitude to another. Such statistical analyses of drag as have so far been carried out [8, 9] have been on a limited variety of orbital specifications.

III. Numerical analysis of actual orbits

The aforedescribed theoretical development has not been applied in numerical computations because it has not yet been worked out to what extent there can be modified the computational manipulation of arrays of dimension comparable to the number of observations, as is implied by Eq. (13). Instead, it has been attempted to avoid allowing for covariance between different observations by various devices, of more-or-less intuitive justification, as described in [9, 10]:

1. Inclusion in the reference model to the maximum extent possible of the various longterm effects: zonal harmonics, luni-solar perturbations, radiation pressure, and drag by physical or empirical models.

2. Weighting of observations inversely as their density with respect to the angles critical for the determination of the parameters of interest.

3. Use of the covariance matrix of parameters as a means of combining results of different satellites, or of satellite with terrestrial results.

4. Increasing the variance of observations and their timing over that which can be reasonably ascribable to the observations and timing themselves.

5. Use of arbitrary polynomials.

From the viewpoint of the more rigorous theory discussed in Sec. II, devices 1. and 5. attempt to reduce the segment of the spectrum which must be considered; device 2. is an approximation of what should be obtained for $\Omega_{fjn}(t)$; device 3. extends the set of observations used beyond those described by $I(t)$, $A < t < B$, in the modified form of the solution obtained by the added observations; and device 4. in effect arbitrarily shifts the higher frequency (of the order 0.1 to 1.0 cycle/day) part of the drag spectrum to a band of much higher frequency than that of the observations.

These methods are being applied to BAKER-NUNN camera observations of satellites 1959α-1, 1959η, and recently, 1960ι2. These analyses are being made for those high frequency (daily, semi-daily, etc.) effects which should be expected to cause perturbations of more than $\pm 10^{-5}$ or ± 70 meters: the harmonics of the gravity field up to $n, m = 4$,

2, and the station positions for the cameras located in Australia, Japan, Argentina, and Hawaii. The atmospheric model of JACCHIA [11] is applied to the two Vanguard orbits (1959α-1 and 1959η), but none is used for the high, dense Echo I rocket case (1962ι 2). Orbital arcs of about

Table 1. *Coefficients for normalized spherical harmonics[1] obtained from satellite 1960 ι-2 ($h_\pi = 1500$ km, $i = 47^0$, $e = .01$, $A/m = 0.2$ cm^2/gm) Sept.—Dec. 1960. Five orbital arcs of 23 days each Multiply by scaling factor 10^{-6}*

No. Obs.	$\bar{C}_{21}$	$\bar{S}_{21}$	$\bar{C}_{22}$	$\bar{S}_{22}$	$\bar{C}_{31}$	$\bar{S}_{31}$	$\bar{C}_{41}$	$\bar{S}_{41}$
151	0.17	−0.13	1.72	−1.46	1.82	0.00	0.09	0.16
179	−0.05	0.39	1.10	−2.91	1.43	−1.44	−0.68	1.13
126	−0.13	−0.01	2.37	−1.57	0.34	0.12	−0.71	0.09
113	−0.03	0.12	2.55	−0.96	1.76	0.03	0.03	0.64
110	0.26	0.25	1.44	−1.67	2.26	−0.39	−0.35	0.42
679	0.04	0.13	1.77	−1.81	1.50	−0.41	−0.34	0.52

[1] Functions Y_{nm} such that $\int\limits_{\text{sphere}} Y_{nm}^2 \, d\sigma = 4\pi$.

20 days are used, and the arbitrarily polynomials included go to t^4 in the mean anomaly plus secular terms for the inclination and eccentricity. To absorb some of the high frequency drag effect, the timing standard deviation is assumed to be ± 0.007 seconds. The angle-weighting used

Table 2. *Datum shift for Woomera, Australia, obtained from satellite 1960 ι-2 Sept.—Dec. 1960. Units of earth Radii Multiply by Scaling Factor 10^{-6}*

No. Obs. from Woomera	dU	dV	dW
30	−17.1	−16.3	5.2
25	3.3	39.6	3.8
9	2.4	− 9.9	0.1
20	− 4.8	−19.4	4.0
26	−18.3	−12.2	1.8
110	− 8.9	− 2.7	3.4

(U, V, W axes toward lat., long. of $0°$, $0°$; $0°$, $90°$ E; $90°$ N respectively.)

is with respect to the angle (node-GST), integral multiples of which appear as the arguments of the principal tesseral harmonic effects. The observational residuals obtained from the solution for each satellite are on the order of $\pm 10''$ in direction and $\pm 0.03^s$ time. The tests of agreement between solutions from different orbital arcs and of a negligibly small answer for the "impossible" harmonic J_{21} have as yet been satisfied only by the analysis of 1960 ι 2, for which preliminary results are shown for some gravitational coefficients in Table 1 and for the position of Woomera, Australia (which had the most observations) in Table 2. Since these arcs happen to fall across the events of October and November 1960, appreciable improvement is expected.

References

[1] VEIS, G.: Geodetic uses of satellites. Smithsonian Contrib. to Astrophys. **3**, 95—161 (1960).

[2] KAULA, W. M.: Analysis of gravitational and geometric aspects of geodetic utilization of satellites. Roy. Astr. Soc. Geophys. J. **5**, 104—133 (1961).

[3] GROVES, G. V.: Motion of a satellite in the earth's gravitational field. Proc. Roy. Astr. Soc. **254 A**, 48—65 (1960).

[4] BARTLETT, M. S.: "An introduction to stochastic processes." Cambridge Univ. Press (1956).

[5] PARZEN, E.: An approach to time series analysis. Ann. Math. Stat. **32**, 951—988 (1961).

[6] KAULA, W. M.: Statistical and harmonic analysis of gravity. J. Geophys. Res. **65**, 2401—2421 (1959).

[7] KOZAI, Y.: The potential of the earth derived from satellite motions. Proc. IUTAM Sym. Dynamics of Satellites, Paris (1962).

[8] MOE, K.: The errors in orbital predictions for artificial earth satellites. Pres. Amer. Geophys. Un. 43rd. Ann. Meeting, Washington (1962).

[9] KAULA, W. M.: Analysis of satellite observations for longitudinal variations of the gravitational field. Space Research II, Proc. 2nd Int. Space Sci. Sym., 360—372 (1961).

[10] KAULA, W. M.: Satellite orbit analysis for geodetic purposes. Proc. Conf. Gen. & Prac. Problems in Th. Astron., Moscow (1961).

[11] JACCHIA, L. G.: A variable atmospheric density model from satellite observations. J. Geophys. Res. **65**, 2775—2782 (1960).

The contraction of satellite orbits under the action of air drag, allowing for the variation of scale height with altitude

By

D. G. King-Hele

Royal Aircraft Establishment, Farnborough, England

Abstract. Equations are given which specify the contraction of satellite orbits under the influence of air drag in an atmosphere with a constant 'density scale height' H, and it is shown how these equations are to be modified when H varies with height. Some examples are given of values of H determined by means of these equations from the observed changes in satellite orbits.

1. Introduction

One of the chief perturbations to the orbits of most artificial satellites is the effect of air drag, acting mainly at heights near that of perigee. The effect of air drag is to make the orbit contract and become more nearly circular. This contraction can be completely specified [1, 2] during the whole of the satellite's life, if we assume (1) that the atmosphere is spherically symmetrical, and (2) that the air density ϱ varies exponentially with height y in the height-band of about 100 km above perigee where drag is important, i.e.

$$\varrho \propto \exp\left(-\frac{y}{H}\right), \tag{1}$$

where H is assumed constant. The quantity H, which may be defined strictly as

$$H = -\frac{\varrho}{d\varrho/dy} \tag{2}$$

is often called the 'density scale height'.

In reality the atmosphere is not spherically symmetrical, but slightly oblate, and H is not constant. The effect of atmospheric oblateness has already been investigated [3, 4]. The present paper reports some new work on the effects of the variation of H with height, and its application to the practical problem of determining values of H from the observed changes in satellite orbits.

14*

2. Results for constant H

For an orbit contracting under the influence of air drag in a spherically symmetrical atmosphere with constant H, the variation of perigee distance r_p and orbital period T with eccentricity e are, for $0.02 < e < 0.2$, given by the following equations [1]:

$$(r_{p0} - r_p)_{H \text{ const}} = \frac{1}{2} H \left\{ \left(1 - \frac{5H}{2a_0} \right) \ln \frac{e_0}{e} - \right.$$
$$\left. - (e_0 - e) \left(1 - \frac{3H}{4a_0} \frac{1 + e_0}{e \, e_0} - \frac{e + e_0}{2} \right) \right\} + 0 \left(H e^3, \frac{H^3}{a^2 e} \right) \qquad (3)$$

$$\left(\frac{T}{T_0} \right)_{H \text{ const}} = \left(\frac{1 - e_0}{1 - e} \right)^{3/2} \left[1 - \frac{3H}{4 r_{p0}} \left\{ \ln \frac{e_0 (1 + e)}{e (1 + e_0)} + \frac{3H}{4 a_0} \left(\frac{1}{e} - \frac{1}{e_0} \right) \right\} \right] +$$
$$+ 0 (6 \times 10^{-4}), \qquad (4)$$

where suffix o denotes initial values, and a is the semimajor axis. The variations of r_p and T with e are shown in Figures 1 and 2. These equa-

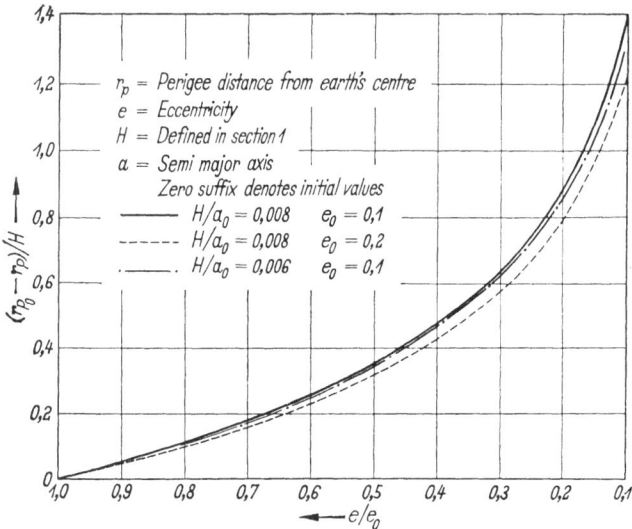

Fig. 1. Variation of perigee distance, r_p, with eccentricity, e, when $0.02 < e < 0.2$.

tions are independent of fluctuations in air density and of day-to-day changes in the effective cross-sectional area of the satellite. The variation of eccentricity with time t, again for $0.02 < e < 0.2$, is given by

$$\left(\frac{e}{e_0} \right)_{H \text{ const}} = \sqrt{1 - \frac{t}{t_L}} \left\{ 1 - \frac{e_0}{6} \left(1 - \frac{19}{3} e_0 \right) \left(1 - \sqrt{1 - \frac{t}{t_L}} \right) - \right.$$
$$\left. - \frac{29}{288} e_0^2 \frac{t}{t_L} + \frac{3H}{16 a_0} \ln \left(1 - \frac{t}{t_L} \right) + 0 \left(e^3, \frac{H^2}{a^2 e} \right) \right\}, \qquad (5)$$

where t_L is the total lifetime. Despite its apparent complication, Eq. (5) gives a variation of e^2 with t which is virtually a straight line, since the term in curly brackets never differs from 1 by a factor of more than 1.006. The delightful simplicity of this theoretical result is, however, not always preserved in practice, since the equation is affected by day-to-day variations in air density.

Corresponding equations are available [1, 2] for $e < 0.02$ and for $e > 0.2$. Since the range of values $o < e < 0.2$ is the most useful in

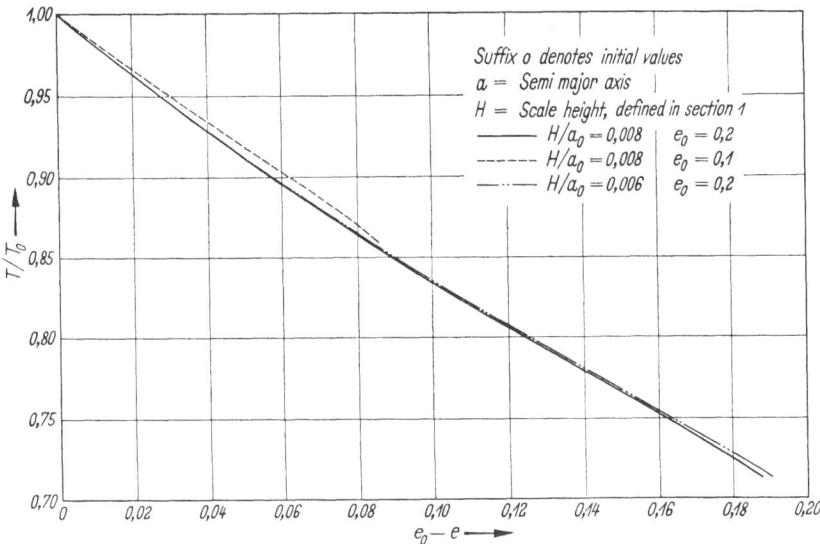

Fig. 2. Variation of orbital period T with eccentricity e, as given by Eq. (4).

practice, only two other equations will be quoted here, those for r_p and T when $e < 0.02$, which are, in terms of $z = a\,e\,/H$,

$$r_{p1} - r_p = H\left[\left(1 - \frac{3H}{a_1}\right)\ln\frac{z_1 I_1(z_1)}{z I_1(z)} + \right.$$
$$\left. + \frac{2H}{a_1}\left\{\frac{z_1 I_0(z_1)}{I_1(z_1)} - \frac{z I_0(z)}{I_1(z)}\right\} - (z_1 - z)\right] + 0(a\,e^3), \qquad (6)$$

$$\frac{T}{T_1} = 1 - \frac{3H}{2a_1}\ln\frac{z_1 I_1(z_1)}{z I_1(z)} + 0(10^{-4}), \qquad (7)$$

where suffix 1 denotes initial values, and I_0 and I_1 are the BESSEL functions of the first kind and imaginary argument, of order zero and 1 respectively. Eq. (7) applies for $z < 3$, which, since $a_1/H \triangleq 150$, is roughly equivalent to $e < 0.02$.

3. Modifications required when H varies with altitude

3.1 The variation of H. In reality the value of the quantity H defined in Eq. (2) does not remain constant, but tends to increase with altitude in the relevant range of heights (150–800 km). The variation

of H with height in the Cospar International Reference Atmosphere [5] is shown in Figure 3. It should be emphasized that the values in Figure 3 represent an average between day and night, and apply for fairly high solar activity, that in 1959 for example. Also it is probable

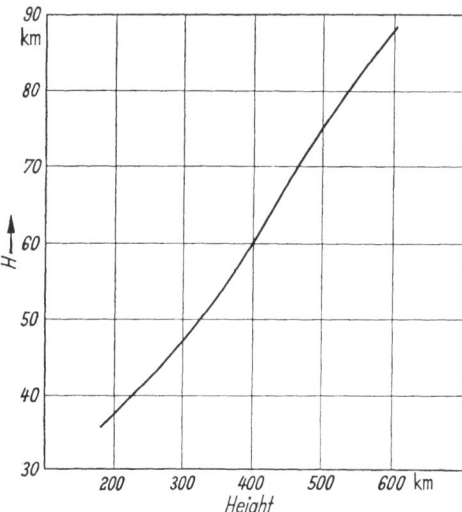

that the variation of H with height y is not monotonic but has a maximum and minimum between 200 and 300 km [6, 7]. Nevertheless, Figure 3 provides us with a useful general picture of the variation of H with height.

From Figure 3, the best simple approximation is to assume that H varies linearly with y in the height-band of about 100 km above perigee height, so that

$$H = H_p + \mu(y - y_p), \quad (8)$$

Fig. 3. Values of 'density scale height' H from Cospar International Reference Atmosphere.

where suffix p denotes values at perigee, and μ is a constant, which would have a value of about 0.12 if Figure 3 were correct. We have developed the theory [8] for variable H on the assumption that H varies in the manner indicated by (8) and that μ can take any value between -0.2 and 0.2.

3.2 Results for variable H (Ref. [8]). When H varies with height in the manner indicated by (8), Eqs. (3)–(7) are no longer valid. They can be amended in two different ways. The first possible amendment is to specify some constant mean value of H, e.g. the value H_2, say, at a height H_{p0} above initial perigee [so that, by (8), $H_2 = H_{p0}(1 + \mu)$], and then to express $r_{p0} - r_p$ in terms of H_2 and an extra term in μ. This process gives, in place of (3), the equation

$$\left(\frac{r_{p0} - r_p}{H_2}\right)_{H\,\text{variable}} = \left(\frac{r_{p0} - r_p}{H}\right)_{H\,\text{const}} + \frac{1}{4}\mu\left[\left(1 + e_0 + e - \frac{3H_2}{2a_0 e}\right) \times\right.$$

$$\times \ln\frac{e_0}{e} - \frac{1}{2}\left(1 + \frac{3}{2}\mu\right)\left(\ln\frac{e_0}{e}\right)^2 -$$

$$\left. - (e_0 - e)\left\{1 - \frac{3H_2}{a_0 e}\left(\frac{1}{e_0} - \frac{1}{2}\right)\right\}\right] +$$

$$+ 0\left(\frac{\mu a e^3}{6H}, \frac{9\mu^2}{32}\ln\frac{x_0}{x}\right)$$

$$= \left(\frac{r_{p0} - r_p}{H}\right)_{H\,\text{const}} + \Omega \quad (9)$$

say, where H is to be replaced by H_2 in the first term on the r.h.s., which is given by (3), or Figure 1. Values of Ω are plotted in Figure 4, which shows that Ω is small, and frequently of the same order as the error-terms in (9).

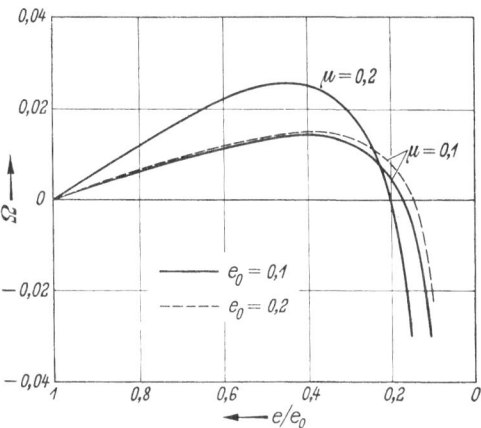

Fig. 4. Variation of Ω, the correction to $(r_{p0} - r_p)/H$, when H varies with height.

The second and perhaps better way of amending Eq. (3) is to find the value of H, H_3 say, such that Eq. (3) can be used unchanged, except that H_3 is written in place of H. In general H_3 will be dependent on e and e_0: but if we take H_3 as the value of H at height αH_{p0} above the mean perigee height [which is itself at height $(y_{p0} - y_p)/2$ below initial perigee], so that

$$H_3 = H_{p0}\left\{1 - \frac{1}{2}\mu\left(\frac{y_{p0} - y_p}{H_{p0}}\right) + \mu\alpha\right\},$$

then it turns out that $\alpha = 3/2$ for all $e > 0.02$.

When $e < 0.02$, the situation is not so simple, since it is obvious that when $e = 0$ the best height at which to evaluate H will be at the mean perigee height, so that $\alpha = 0$ for $e = 0$. For $e < 0.02$, therefore, α must vary with e. For $0 < e < 0.02$, or rather for $0 < z < 3$, the appropriate value of α is found to be approximately as shown in Figure 5, where α has been plotted against the mean value of z, $\bar{z} = (z_1 + z)/2$.

Thus, to sum up, Eqs. (3), (4), (6) and (7) can be used unchanged if H is taken as H_3, the value of

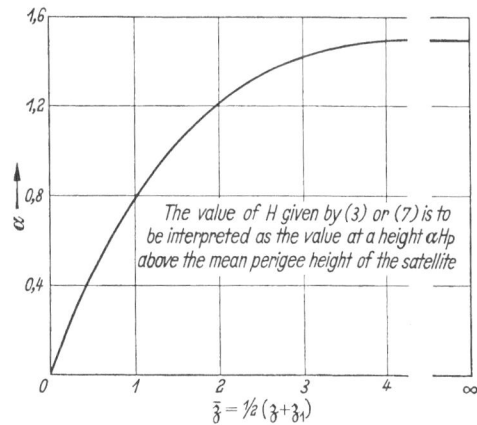

The value of H given by (3) or (7) is to be interpreted as the value at a height αH_p above the mean perigee height of the satellite

Fig. 5. Assumed variation of α with $\bar{z}$.

the variable H at a height αH_{p0} above the mean perigee height during the relevant time-interval, where $\alpha = 3/2$ for $e > 0.02$, and α decreases to zero as $e \to 0$, as shown in Figure 5.

In the Eq. (5) for e/e_0 it is not appropriate to seek the value of H which keeps the equation unchanged, because e/e_0 does not depend

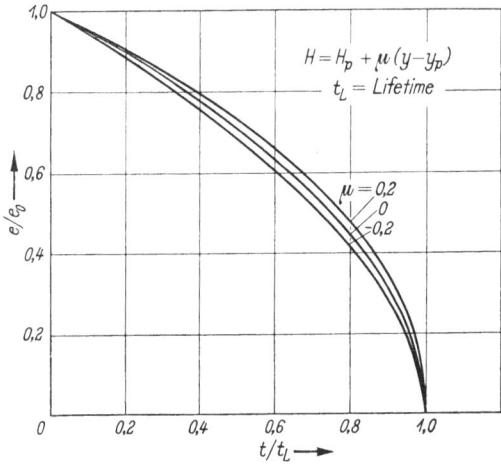

Fig. 6. Variation of eccentricity e with time when H varies with height y.

primarily upon H. Instead we express the effect of variable H by evaluating the extra term in μ which has to be added in (5). We find

$$\left(\frac{e}{e_0}\right)_{H \text{ variable}} = \left(\frac{e}{e_0}\right)_{H \text{ const}} \left\{1 + \frac{7\mu}{16} \ln \frac{1}{\sqrt{1 - t/t_L}} + 0(\mu^2)\right\}.$$

e/e_0 is plotted against t/t_L in Figure 6, for various values of μ: the maximum change in e/e_0 due to variation in H, with μ at its maximum of 0.2, is 0.03.

3.3 Discussion. It may at first seem rather surprising that in calculating the change in perigee height it is necessary to evaluate H at a height as great as $3H/2$ above perigee (for $e > 0.02$). At this height the drag has fallen to less than $1/4$ of its value at perigee and might be expected to have little influence on the orbit. This paradox can be explained by considering the simplest possible theoretical model for the effect of drag — concentrating the drag into an impulse at perigee. Under this over-simple assumption the apogee height would decrease and the perigee height would remain constant. Thus it is only the drag at points away from perigee, i.e. appreciably higher than perigee, which has any effect in reducing the perigee height. Seen in this light, the need to evaluate H at a height $3H/2$ above perigee seems more reasonable.

The effects of variations in H with height, for eccentricities greater than about 0.05, were first investigated in 1959 by GROVES [9], who was primarily concerned with the effects over one revolution. The need to evaluate H at a height $3H/2$ above perigee is deducible from one of his equations, though Groves himself did not interpret his results in this manner.

It should be emphasized that only in the equations for perigee height and period does H have to be evaluated at $3H/2$ above perigee (for $e > 0.02$): in other equations lower altitudes are appropriate. The formula for air density at perigee in terms of $\dot{T}$, for example, is best modified to allow for variations in H by evaluating H at a height of $3H/4$ above perigee [8] (for $e > 0.02$), the lower height being natural because air drag at heights right down to perigee is effective in making the orbit contract.

4. Practical application in determining H from orbital changes

If we know the orbital elements of a satellite, in particular the perigee distance and eccentricity, initially and at a later date, and if we correct the perigee height to allow for other relevant effects (such as those due to the odd harmonics in the earth's gravitational field), Eq. (3) provides a powerful method of determining the mean value of H during the time interval between the initial and later date. Similarly Eq. (7) provides an excellent method for finding H from the changes in orbital period for small-eccentricity orbits. With the constant-H theory, however, there was considerable doubt about the height to which the values of H referred, and there was a tendency to assume that they applied at the perigee height or perhaps $H/2$ above. With the

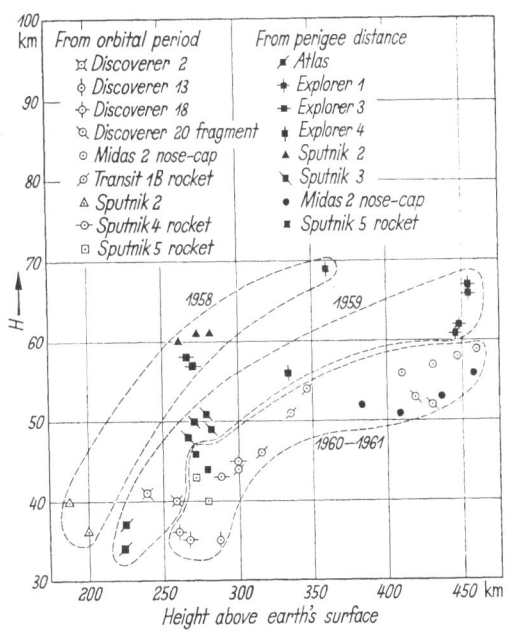

Fig. 7. Values of density scale height H obtained from the changes in perigee distance and orbital period of various satellites.

variable-H theory this uncertainty disappears. The interpretation of the variable-H theory which has been advocated in this paper has the further advantage that the value of μ does not need to be known.

Recently, using the results of the variable-H theory, we determined values of H from either (3) or (7), utilizing all the 14 satellites found to be suitable for this purpose. We obtained [10] 44 values of H (all 'average day/night' values), which are plotted in Figure 7. Although the orbital information from which they were derived was not always of satisfactory accuracy, the results prove to be consistent and, when

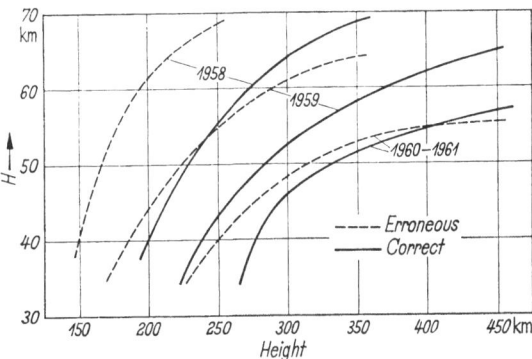

grouped according to year, as shown by the broken lines in Figure 7, they reveal clearly that (1) H decreased as solar activity declined between 1958 and 1961 and (2) the increase of H with height becomes much less rapid at heights above 350 km.

Fig. 8. Variation of H with height obtained from Fig. 7 (unbroken lines). The broken lines give erroneous values obtained by assuming that H refers to average perigee height.

Figure 8 shows mean curves drawn through the points of Figure 7 (unbroken lines), together with the erroneous values of H which would be obtained on the assumption that they applied at the mean perigee height rather than $\alpha\ H$ above (broken lines). Some of the differences are very large: for example the erroneous value for 1958 at 200 km is 62 km and the correct value 40 km. This shows that allowing for the variation of H with height is no minor refinement to the theory, but is of considerable practical significance in obtaining geophysical information from the observed changes in satellite orbits.

References

[1] KING-HELE, D. G., G. E. COOK, and D. M. C. WALKER: Proc. Roy. Soc., Lond., A, **257**, 224—249 (1960).
[2] KING-HELE, D. G.: Proc. Roy. Soc., Lond., A, **267**, 541—557 (1962).
[3] PARKYN, D. G.: J. Geophys. Res. **65**, 9—17 (1960).
[4] KING-HELE, D. G., G. E. COOK, and D. M. C. WALKER: Proc. Roy. Soc., Lond., A, **264**, 88—121 (1961).
[5] KALLMANN-BIJL, H. K. et al.: CIRA 1961. North-Holland Publ. Co., Amsterdam (1961).
[6] KING-HELE, D. G., and D. M. C. WALKER: Space Research II, North-Holland Publ. Co., Amsterdam (1961), pp. 918—957.
[7] GROVES, G. V.: Proc. Roy. Soc. Lond., A, **263**, 212—216 (1961).
[8] KING-HELE, D. G., and G. E. COOK: Ministry of Aviation Report (1962) (to be published).
[9] GROVES, G. V.: Proc. Roy. Soc., Lond., A, **252**, 16—34 (1959).
[10] KING-HELE, D. G., and J. M. REES: Proc. Roy. Soc., Lond., A **270**, 562—587 (1962).

The libration of a satellite on an elliptic orbit

By

V. V. Beletsky

The U.S.S.R. Academy of Sciences, Moscow, U.S.S.R.

The paper [1] has given analysis of a problem on oscillations of a satellite about the centre of mass under the influence of gravitational moments. In the present paper oscillations on an elliptic orbit are considered in more detail. Attention is focused on resonance effects.

1. Forced small oscillations of a satellite on an elliptic orbit

As was shown in [1], the orbit ellipticity causes forced oscillations of a satellite in the orbit plane. These oscillations exist at all, specifically, zero initial data.

The equation of plane oscillations has the form

$$(1 + e\cos\theta)\frac{d^2\delta}{d\theta^2} - 2e\sin\theta\frac{d\delta}{d\theta} + \omega^2\sin\delta = 4e\sin\theta; \qquad \delta = 2\vartheta \qquad (1.1)$$

where ϑ is the angle between one of the body major central axes of inertia which lie in the orbit plane, and the orbit radius-vector $\omega^2 = 3\dfrac{A-C}{B}$; A, B, C are the main central moments of inertia of a satellite; e is the eccentricity, θ is the true anomaly.

For the case of small oscillations the linearization of Eq. (1.1) and substitution of

$$\vartheta = \frac{z}{1 + e\cos\theta} \qquad (1.2)$$

lead to the equation

$$\frac{d^2z}{d\theta^2} + \left\{\frac{\omega^2 + e\cos\theta}{1 + \cos\theta}\right\} z = 2e\sin\theta. \qquad (1.3)$$

Forced eccentricity oscillations on an elliptic orbit are described by the formula which was given in [1] without the derivation

$$\vartheta_e = \frac{2e\sin\theta}{1 + e\cos\theta}\left\{\frac{1}{\omega^2 - 1} + \frac{e}{\omega^2 - 4}\cos\theta\right\} \approx$$

$$\approx \frac{2e}{\omega^2 - 1}\sin\theta + \frac{3e^2}{(\omega^2 - 4)(\omega^2 - 1)}\sin2\theta. \qquad (1.4)$$

In this paragraph the derivation of formula (1.4) is described.

Let us seek partial solution of nonhomogeneous Eq. (1.3) in the form

$$z_e = 2e \sin\theta \; \varphi. \tag{1.5}$$

Then for φ we have the equation

$$(1 - x^2)(1 + e\,x)\,\varphi'' - 3(1 + e\,x)\,x\,\varphi' + (\omega^2 - 1)\,\varphi = 1 + e\,x, \tag{1.6}$$

where the primes denote derivatives with respect to the argument

$$x = \cos\theta. \tag{1.7}$$

Partial solution of Eq. (1.6) can be sought in the form of series with respect to the small parameter e.

The series will converge according to POINCARE's theory for sufficiently small values of e.

$$\varphi = \varphi_0 + e\,\varphi_1 + e^2\,\varphi_2 + \cdots + e^n\,\varphi_n + \cdots. \tag{1.8}$$

Having substituted in (1.6) and equated the terms at equal powers n we obtain the system of equations:

$$\left. \begin{aligned} &(1 - x^2)\,\varphi_0'' - 3x\,\varphi_0' + (\omega^2 - 1)\,\varphi_0 = 1, \\ &(1 - x^2)\,\varphi_1'' - 3x\,\varphi_1' + (\omega^2 - 1)\,\varphi_1 = x - x(1 - x^2)\,\varphi_0'' + 3x^2\,\varphi_0', \\ &(1 - x^2)\,\varphi_2'' - 3x\,\varphi_2' + (\omega^2 - 1)\,\varphi_2 = -x(1 - x^2)\,\varphi_1'' + 3x^2\,\varphi_1', \\ &\cdots\cdots\cdots\cdots\cdots\cdots\cdots\cdots\cdots\cdots\cdots\cdots\cdots\cdots \\ &(1 - x^2)\,\varphi_n'' - 3x\,\varphi_n' + (\omega^2 - 1)\,\varphi_n = -x(1 - x^2)\,\varphi_{n-1}'' + 3x^2\,\varphi_{n-1}'. \end{aligned} \right\} \tag{1.9}$$

From this system we found successively partial solutions for φ_n which are polinomials of the nth power

$$\varphi_0 = \frac{1}{\omega^2 - 1}; \qquad \varphi_1 = \frac{1}{\omega^2 - 4}\,x; \quad \ldots; \tag{1.10}$$

$$\varphi_{n-1} = \sum_{m=0}^{m=n-1} a_m\,x^m; \qquad \varphi_n = \sum_{m=0}^{m=n} b_m\,x^m. \tag{1.11}$$

Here to avoid confusion the coefficients of the $n - 1^{\text{th}}$ polynomial are denoted by a_m, and coefficients of the nth polynomial by b_m. To determine them we obtain recurrent ratios:

$$\left. \begin{aligned} &b_m = \frac{m^2 - 1}{\omega^2 - (m+1)^2}\,a_{m-1} \quad \text{for the case} \quad m = n \text{ and } m = n - 1 \\ &(m+2)(m+1)\,b_{m+2} + [\omega^2 - (m+1)^2]\,b_m \\ &\qquad\qquad = (m^2 - 1)\,a_{m-1} - m(m+1)\,a_{m+1} \\ &\text{for the case} \quad m = 0, 1, 2, \ldots, n - 2. \end{aligned} \right\} \tag{1.12}$$

We see that two elder terms n^{th} and $n - 1^{\text{th}}$ of the nth polynomial are determined from (1.12) from the corresponding ($n - 1^{\text{th}}$ and $n - 2^{\text{th}}$) elder terms of the $n - 1^{\text{th}}$ polynomial. Subsequent terms of the nth polinomial are determined from (1.12) from the just determined elder terms of this polinomial and from the beforehand known terms ($n - 3^{\text{th}}$ and $n - 1^{\text{th}}$) of the previous, the $n - 1^{\text{th}}$, polinomial.

For instance, the coefficients of the polinomial

$$\varphi_2 = b_0 + b_1 x + b_2 x^2$$

are determined through the coefficients of the polinomial

$$\varphi_1 = a_0 + a_1 x; \qquad a_0 = 0; \qquad a_1 = \frac{1}{\omega^2 - 4}$$

in the following way. From (1.12) we have

$$b_2 = \frac{3}{\omega^2 - 9} \frac{1}{\omega^2 - 4}; \qquad b_1 = 0$$

and from (1.12), taking into account that $a_{-1} \equiv 0$ we obtain

$$2 b_2 + (\omega^2 - 1) b_0 = 0, \qquad b_0 = \frac{6}{(\omega^2 - 9)(\omega^2 - 4)(\omega^2 - 1)}.$$

The constructed solution (1.5), (1.8), (1.10)–(1.12) is of interest as an example of precise partial solution of nonhomogeneous Eq. (1.3) of HILL's type. For considered concrete case of the satellite oscillations on an elliptic orbit it is reasonable to consider only the first terms of this solution, namely, terms (1.10) which due to (1.7), (1.5) and (1.2) give formula (1.4) for forced oscillations. Terms with e^n, $n \geq 3$ can be left without consideration since precise solution of the linear equation satisfies actual forced oscillations, i.e. the adequate solution of nonlinear equation (1.1) only with an accuracy to e^3.

At $\omega^2 \approx 1$ solution (1.4) is not adequate (the resonance case). Oscillations in the range of the resonance value of the parameter will be considered below. Physically there will not be some other resonance in forced oscillations since $\omega^2 = 3\dfrac{A - C}{B} < 3$ and, therefore, the second term in braces of formula (1.4) is always limited.

2. Plane small oscillations of a satellite on an elliptic orbit

The BOGOLYUBOV-KRYLOV method [2] can be used for determining the general solution of linearized Eq. (1.3) of small plane oscillations of a satellite on an elliptic orbit. For this purpose let us write Eq. (1.3) in the form

$$z'' + \omega^2 z = (\omega^2 - 1) \{e \cos\theta - e^2 \cos^2\theta + e^3 \cos^3\theta + \cdots\} z + 2e \sin\theta.$$

Let us consider

$$\omega^2 = 3\frac{A - C}{B} > 0. \tag{2.1}$$

The solution of Eq. (2.1) at $e = 0$ will be

$$z = a \cos\psi; \qquad \psi = \omega t + \psi_0.$$

At $e \neq 0$ the solution should be sought in the form

$$z = a \cos \psi + e \, u_1(a, \, \psi, \, \theta) + e^2 \, u_2(a, \, \psi, \, \theta) + e^3 \, u_3(a, \, \psi, \, \theta) + \cdots \quad (2.2)$$

It should be mentioned that

$$\left. \begin{aligned} \frac{da}{d\theta} &= e \, A_1(a) + e^2 \, A_2(a) + e^3 \, A_3(a) \ldots \\ \frac{d\psi}{dt} &= \omega + e \, B_1(a) + e^2 \, B_2(a) + e^3 \, B_3(a) \ldots \end{aligned} \right\} \quad (2.3)$$

Using the well known algorithm of the Bogolyubov-Krylov method [2] it is not difficult to find

$$A_1 = A_2 = A_3 = 0, \quad (2.4)$$

$$B_1 = B_3 = 0; \qquad B_2 = \frac{3}{4} \frac{\omega \, (\omega^2 - 1)}{4 \omega^2 - 1}, \quad (2.5)$$

$$u_1 = \frac{2}{\omega^2 - 1} \sin \theta + \frac{a(\omega^2 - 1)}{2} \left\{ \frac{1}{2\omega - 1} \cos(\psi - \theta) - \right.$$
$$\left. - \frac{1}{2\omega + 1} \cos(\psi + \theta) \right\}, \quad (2.6)$$

$$u_2 = \frac{1}{\omega^2 - 4} \sin 2\theta + \frac{a}{16} \left\{ \frac{\omega(\omega + 1)(\omega - 2)}{2\omega - 1} \cos(\psi - 2\theta) - \right.$$
$$\left. - \frac{\omega(\omega - 1)(\omega + 2)}{2\omega + 1} \cos(\psi + 2\theta) \right\}. \quad (2.7)$$

There is no use to consider the term u_3 for the reasons indicated in the previous paragraph.

It is supposed in the solution that $\omega^2 \neq 1$. At $\omega^2 = 1$ Eq. (1.3) is turned into equation with constant coefficients and is integrated at once.

In the solution (2.2) to (2.7) forced eccentricity oscillations are contained in the same form as in § 1 where they were obtained by another method. The resonance values of the parameter ω^2 in forced oscillations lie in the region close to 1. Besides, as evident from (2.5) to (2.7), there is a region (and only one) of parametric resonance

$$\omega^2 = \frac{1}{4} \, .$$

In the parametric resonance region the amplitude of oscillations will increase to some limit value. These steady resonance oscillations can be determined by considering nonlinear Eq. (1.1). To obtain a more general formula for forced oscillations than (1.4) namely, suitable both in the range of the resonance value of frequency and outside this range, one should refer also to nonlinear Eq. (1.1).

Consideration of the indicated oscillations will be done in the paragraphs below.

3. Forced finite oscillations of a satellite

Forced (eccentricity) oscillations will be sought in the form of the FOURIER series with respect to θ. Let us consider at first the first harmonic of this series since precisely in the first harmonic the resonance effect is observed. Let us assume that

$$\delta = a\cos\psi; \quad \psi = \theta + \psi_0. \tag{3.1}$$

The amplitude a and phase ψ_0 should be determined.

Having substituted (3.1) into Eq. (1.1) and taken into account the well known expansion

$$\sin(a\cos\psi) = 2\sum_{n=0}^{\infty}(-1)^n I_{2n+1}(a)\cos(2n+1)\psi, \tag{3.2}$$

where $I_k(a)$ is the BESSEL function of the first kind, we obtain (by equating the coefficients at equal harmonics) the equations:

The free term

$$\frac{1}{2}\,e\,a\,\cos\psi_0 = 0.$$

Coefficient at $\cos\theta$:

$$(2\omega^2 I_1 - a)\cos\psi_0 = 0.$$

Coefficient at $\sin\theta$:

$$\{-2\omega^2 I_1(a) + a\}\sin\psi_0 = 4e. \tag{3.3}$$

We see that $\cos\psi_0 = 0$ which corresponds to two values of phases $\psi_0 = \dfrac{\pi}{2}$ and $\psi_0 = -\dfrac{\pi}{2}$.

For the first case we have from Eq. (3.3)

$$\delta = -a\sin\theta, \tag{3.4}$$

$$\omega^2 = \frac{a - 4e}{2 I_1(a)}. \tag{3.4'}$$

For the second case we have

$$\delta = a\sin\theta, \tag{3.5}$$

$$\omega^2 = \frac{a + 4e}{2 I_1(a)}. \tag{3.5'}$$

Formulas (3.4), (3.4') and (3.5), (3.5') describe the first harmonic of forced oscillations. The connection of the amplitude a of the first harmonic with the natural frequency ω^2 and the eccentricity e is given by formulas (3.4') and (3.5'). These formulas enable us to simply calculate the dependence $\omega^2(a, e)$ and, therefore, to determine the dependence $a(\omega^2, e)$. The example of such calculation is given in Figure 1 for the value $e = 0.01$. In this figure instead of the amplitude "a" of the oscillations of the angle δ the amplitude α of oscillations of the angle ϑ

is considered (Let us remind that $\vartheta = \delta/2$). We see that for the above indicated value of the eccentricity the amplitude of oscillations at the most unfavourable frequency $\omega^2 = 1.12$ reaches $31°$ while at the frequencies far from the resonance value the amplitude of oscillations is of the order of $\sim 1°$.

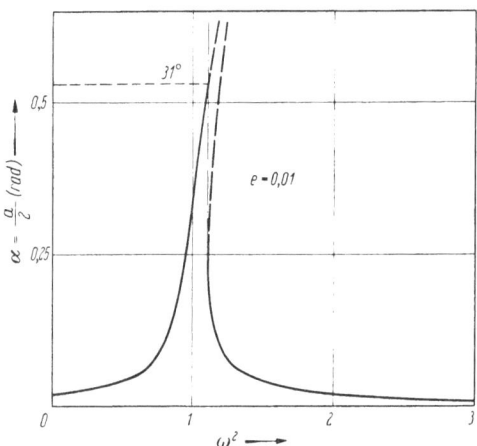

Fig. 1. Amplitude of oscillations of the angle ϑ.

Let us note that for the case of small amplitudes when $I_I(a) \sim a/2$ formulas (3.4) and (3.5) give

$$\frac{\delta}{2} = \vartheta = \frac{2e}{\omega^2 - 1}\sin\theta$$

which coincides with the first term of formula (1.4) obtained earlier. Let us now seek forced oscillations with their second harmonic taken into account.

Previous analysis of the linear and nonlinear problem makes it possible to conclude that these oscillations should be sought in the form

$$\delta = \pm\, a \sin\theta + a_1 \sin 2\theta. \tag{3.6}$$

Let us take into account the expansion

$$\left.\begin{aligned}
\sin(\alpha \sin\theta) &= 2 \sum_{\varkappa=0}^{\infty} I_{2\varkappa+1}(\alpha) \sin(2\varkappa + 1)\,\theta, \\
\cos(\alpha \sin\theta) &= I_0(\alpha) + 2 \sum_{\varkappa=1}^{\infty} I_{2\varkappa}(\alpha) \cos 2\varkappa\,\theta.
\end{aligned}\right\} \tag{3.7}$$

Let us at first consider (3.6) with the sign $+$ in the first term.

Having equated coefficients at equal harmonics we obtain (for determination of a and a_1) the set of transcendental equations:

$$\left.\begin{aligned}
-a - 4e &= -2\omega^2\{I_1(a)\,I_0(a_1) + \sum_{m=1}^{\infty} I_{2m}(a_1)\,[I_{4m+1}(a) - I_{4m-1}(a)]\} \\
-\frac{3}{2}\,ae - 4a_1 &= -2\omega^2 \sum_{\varkappa=0}^{\infty} I_{2\varkappa+1}(a_1)\,[I_{4\varkappa}(a) - I_{4\varkappa+1}(a)].
\end{aligned}\right\} \tag{3.8}$$

Having substituted "a" by "$-a$" we obtain the case with the sign "$-$" in the first term (3.6). From analysis of the linear problem it is seen that the amplitude a_1 of the second harmonic is proportional to the amplitude a of the first harmonic multiplied by the eccentricity. Therefore for small eccentricities a_1 will be a value small even at the resonance a values. For smal a_1 we have the following

$$I_0(a_1) \sim 1; \quad I_1(a_1) \sim \frac{a_1}{2}; \quad I_p(a_1) \approx 0 \quad (p \geq 2).$$

Then system (3.8) turns into the system (with $\pm a$ taken into account):

$$\left. \begin{array}{r} \pm a + 4e = \pm 2\omega^2 I_1(a), \\ \pm \dfrac{3}{2} a\,e + 4a_1 = \omega^2 I_0(a)\,a_1. \end{array} \right\} \tag{3.9}$$

The first of Eq. (3.9) gives the previous result (3.4) and (3.5), and the second equation determines a_1 through a, ω^2 and e:

$$a_1 = \pm \frac{3\,a\,e}{2\,[\omega^2 I_0(a) - 4]}. \tag{3.10}$$

Sign $+$ should be taken when a is determined by formula (3.5), sign $-$ should be taken when a is determined by formula (3.4). At small e formula (3.10) gives

$$a_1 = \frac{4e}{\omega^2 - 1} \frac{3}{2} e \frac{1}{\omega^2 - 4} =$$

$$= \frac{3 \cdot 2\,e^2}{(\omega^2 - 1)(\omega^2 - 4)}. \tag{3.11}$$

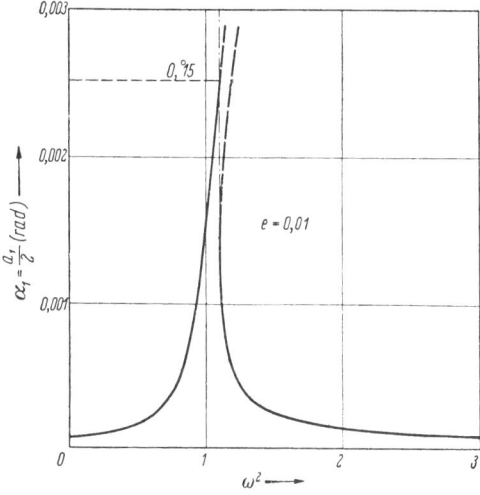

Fig. 2. Amplitude of the second harmonic.

Since the amplitude of the second harmonic of oscillations of the nutation angle is equal to $a/2$, we see that we have again obtained the result coinciding with (1.4).

The dependence $a_1(\omega^2)$ calculated from formulas (3.10), (3.4), (3.5) is shown in Figure 2 for the case $e = 0.01$. We see that the amplitude a_1 is very small (on the order of $0.°3$). Therefore for small eccentricities it is enough to limit oneself with the first harmonic of oscillations at the analysis of forced oscillations.

4. Steady plane oscillations in the range of parametric resonance

In the case of the frequency value close to the resonance one the amplitude of oscillations at parametric resonance will increase from the original value to some steady value, and the frequency of oscillations will tend to resonance frequency. The phase of oscillations also has some limit value. As a result of this, some oscillations with a finite amplitude will become steady. In this paragraph we shall consider such steady oscillations and determine to a first approximation the region of resonance values of natural frequency ω.

In § 2 it was shown that $\omega = 1/2$ is the frequency resonance value. We shall seek the first harmonic of steady resonance oscillations in the region of parametric resonance in the form.

$$\delta = a \cos\left(\frac{1}{2}\theta + \psi_0\right). \quad (4.1)$$

Having substituted into Eq. (1.1) and used calculations similar to those of the previous paragraph we obtain solutions:

$$1.\ \sin\psi_0 = 0,\qquad \omega^2 = a\frac{\left(\frac{1}{4} - \frac{3}{8}e\right)}{2\,I_1(a)}, \quad (4.2)$$

$$2.\ \cos\psi_0 = 0,\qquad \omega^2 = \frac{a\left(\frac{1}{4} + \frac{3}{8}e\right)}{2\,I_1(a)}. \quad (4.3)$$

Dependence $a(\omega)$ of the amplitude a on the natural frequency ω is obtained by means of calculations according to formulas (4.2), (4.3) and is unique in the region

$$\frac{1}{4} - \frac{3}{8}e \le \omega^2 \le \frac{1}{4} + \frac{3}{8}e. \quad (4.4)$$

In this region only branch (4.2) is contained (Figure 3). Region (4.4) determines the region of unsteady oscillations. No matter how small will be the initial amplitude of oscillations a_0 it increases to the finite value a determined by formula (4.2) if only the value of the natural frequency ω of oscillations satisfies inequality (4.4). If ω^2 lies outside the region (4.4), oscillations with small initial amplitudes will remain small when there is an unlimited increase of θ and will be described by formulas of § 2.

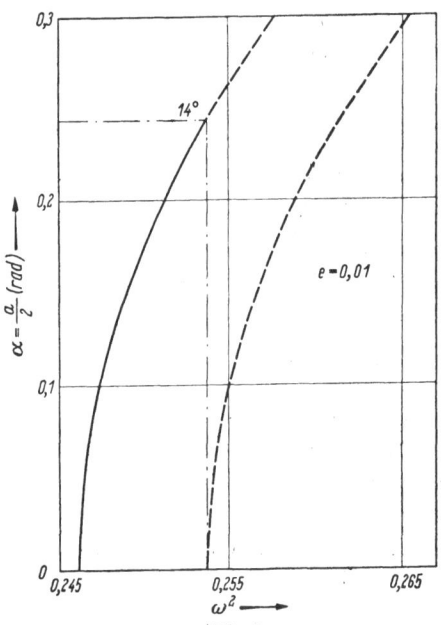

Fig. 3.

As evident from Figure 3, the maximum amplitude of resonance oscillations for $e = 0.01$ equals $a/2\ \max = 14°$, i.e. is approximately by two times lower than the maximum amplitude of forced oscillations.

A more precise definition of the instability region (4.4) can be obtained by considering higher harmonics in resonance oscillations which will give terms proportional to e^2 in (4.4). For sufficiently small e these corrections are small and we can confine ourselves with the above analysis.

It is also not difficult to consider the interaction of parametric excitation and forced oscillations. This consideration has shown that such interaction is small, which is accounted for by noncoincidence of the regions of resonance frequences in parametric resonance and in forced oscillations.

5. Resonance values of parameters in spatial oscillations

Equations of small spatial oscillations on an elliptic orbit consist of already analyzed Eq. (1.3) for the pitch angle ϑ and of two equations for roll angles $\bar{\gamma}$ and yaw angles $\bar{\alpha}$ [1]

$$\ddot{\bar{\gamma}} + (3\zeta + \omega^2)\frac{B-C}{A}\bar{\gamma} - \omega\left(1 + \frac{(C-B)}{A}\right)\dot{\bar{\alpha}} - \dot{\omega}\bar{\alpha} = 0, \quad (5.1)$$

$$\ddot{\bar{\alpha}} + \omega^2\frac{B-A}{C}\bar{\alpha} - \omega\left(-1 + \frac{B-A}{C}\right)\dot{\bar{\gamma}} + \dot{\omega}\bar{\gamma} = 0, \quad (5.2)$$

where

$$\omega = \dot{\theta} \qquad \zeta = \frac{g_0 R_0^2}{\varrho^3}; \qquad \omega = \frac{R_0\sqrt{g_0 P_0}}{\varrho^2}; \qquad \varrho = \frac{P_0}{1 + e\cos\theta}. \quad (5.3)$$

g_0 is the acceleration of the gravity force at the distance R_0 from the Earth's centre; P_0, e are the focal parameter and eccentricity of the satellite orbit, θ is the true anomaly.

To turn to Eqs. (5.1) and (5.2). Let us transform

$$\bar{\gamma} = \frac{\gamma}{1 + e\cos\theta}; \qquad \bar{\alpha} = \frac{\alpha}{1 + e\cos\theta} \quad (5.3)$$

and turn to the new independent variable θ.

The Eqs. (5.1) and (5.2) take the form

$$\left.\begin{aligned}
\gamma'' + \gamma\,\frac{1}{1 + e\cos\theta}\{e\cos\theta + a_1(4 + e\cos\theta)\} - \\
- a_2\alpha' + \frac{e\sin\theta}{1 + e\cos\theta}a_3\alpha = 0, \\
\alpha'' + \alpha\frac{1}{1 + e\cos\theta}\{e\cos\theta + c_1(1 + e\cos\theta)\} + \\
+ c_2\gamma' - \frac{e\sin\theta}{1 + e\cos\theta}c_3\gamma = 0,
\end{aligned}\right\} \quad (5.4)$$

where there are designations

$$\left.\begin{aligned}
a_1 = \frac{B-C}{A}; \quad a_2 = \frac{A+C-B}{A}; \quad a_3 = \frac{B+A-C}{A}; \\
c_1 = \frac{B-A}{C}; \quad c_2 = \frac{A+C-B}{C}; \quad c_3 = \frac{B+C-A}{C}.
\end{aligned}\right\} \quad (5.5)$$

Variable coefficients of Eq. (5.4) for small values of e can be written with an accuracy to the first power of e, and Eq. (5.4) be presented in the form

$$\left.\begin{aligned}
\gamma'' + 4a_1\gamma - a_2\alpha' &= -e\{\gamma\cos\theta(1 - 3a_1) + a_3\alpha\sin\theta\} = e\,F_\gamma, \\
\alpha'' + c_1\alpha + c_2\gamma' &= -e\{\alpha\cos\theta - c_3\gamma\sin\theta\} = e\,F_\alpha.
\end{aligned}\right\} \quad (5.6)$$

At the $e = 0$ we obtain the solution on a circular orbit

$$\left.\begin{aligned}
\gamma &= A_i\sin\psi_i, \\
\alpha &= K_i A_i\cos\psi_i, \\
\psi_i &= \lambda_i\frac{\theta}{\omega_0} + \varkappa_i, \quad \theta = \omega_0 t.
\end{aligned}\right\} \quad (5.7)$$

For the sake of brevity, the sign of summation with respect to $i = 1.2$ is dropped.

Constants K_i, λ_i are determined according to formulas (3.7) and (3.8) of the article (1). $A_i, \varkappa_i$ are integration constants determined by the initial data. For the sake of brevity let us denote λ_i/ω_0 simply by λ_i Eq. (5.6) have solution (5.7) to a zero approximation. To obtain a first approximation, let us substitute (5.7) into right-hand sides of Eq. (5.6) and find partial solution of the nonhomogeneous system obtained. Then the sum of this partial solution and of solution (5.7) gives the sought solution of Eq. (5.6). Substitution of (5.7) into right-hand sides of (5.6) will give

$$\left.\begin{aligned}
F_\gamma &= \Gamma_{i,\delta}\sin(\psi_i + \delta\theta), \\
F_\alpha &= B_{i,\delta}\cos(\psi_i + \delta\theta),
\end{aligned}\right\} \quad (5.8)$$

where

$$\left.\begin{aligned}
\Gamma_{i,\delta} &= \frac{A_i}{2}[(3a_1 - 1) - \delta a_3 K_i], \\
B_{i,\delta} &= -\frac{A_i}{2}[K_i + \delta c_3].
\end{aligned}\right\} \quad (5.9)$$

The partial solution of system (5.6) can be sought in the form

$$\left.\begin{aligned}
\gamma &= e\,M_{i,\delta}\cdot\sin(\psi_i + \delta\theta), \\
\alpha &= e\,N_{i,\delta}\cdot\cos(\psi_i + \delta\theta).
\end{aligned}\right\} \quad (5.10)$$

In (5.8) and (5.10) the sign of twofold summation with respect to $i = 1,2$ and $\delta = -1, +1$ is dropped for the sake of brevity.

Having substituted (5.10) into the left-hand sides of (5.6) and having taken into account (5.8), we obtain (equating coefficients at equal harmonics) a system of eight equations for determining constants $M_{i,\delta}$

and $N_{i,\delta}$

$$M_{i,\delta}\{4a_1 - (\lambda_i + \delta)^2\} + N_{i,\delta} a_2(\lambda_i + \delta) = \Gamma_{i,\delta};$$
$$M_{i,\delta} c_2(\lambda_i + \delta) + N_{i,\delta}\{c_1 - (\lambda_i + \delta)^2\} = B_{i,\delta};$$
$$i = 1,2; \qquad \delta = -1, +1. \tag{5.11}$$

All pairs of coefficients $M_{i,\delta}$; $N_{i,\delta}$ are easily determined from the system of two corresponding linear algebraic equations (5.11). The sum of solutions of (5.7) and (5.10) describes transversal oscillations on an elliptic orbit [with (5.3) taken into account]. Determinants of system (5.11) have the form

$$D = \begin{vmatrix} 4a_1 - (\lambda_i + \delta)^2 & a_2(\lambda_i + \delta) \\ c_2(\lambda_i + \delta) & c_1 - (\lambda_i + \delta)^2 \end{vmatrix} \tag{5.12}$$
$$i = 1,2; \qquad \delta = -1, +1.$$

Values of the system parameters at which determinant D turns to zero are resonance ones. If parameters of the system are close to resonance values, parametric resonance appears in transversal oscillations. Equation $D = 0$ has the form

$$x^4 - x^2(c_1 + 4a_1 + a_2 c_2) + 4a_1 c_1 = 0, \tag{5.13}$$

where

$$x^2 = (\lambda_i + \delta)^2. \tag{5.14}$$

But solutions of Eq. (5.13) with respect to x^2 are

$$x^2 = \lambda_j^2, \qquad j = 1.2.$$

Therefore resonance values of parameters satisfy one of the relationships:

$$(\lambda_\varkappa \pm 1)^2 = \lambda_j^2, \quad \varkappa = 1,2, \quad j = 1,2. \tag{5.15}$$

Investigating ratios (5.15) one should take into account the fact that according to the physical sense of the moments of inertia and due to stability conditions imposed on them, the inequalities take place $\qquad 0 < \lambda_2 < 1 < \lambda_1 < 2.$

It turns out that (5.15) is satisfied
1. For external values of λ_1 and λ_2 when the three-axis ellipsoid of the body inertia degenerates into a rotations ellipsoid or a spheroid. Consideration of these limit cases can be ignored.
2. For the value $\lambda_1 = 1/2$ which has no physical sense.

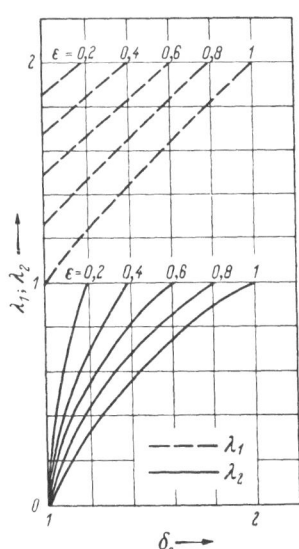

Fig. 4. Diagrams of $\lambda_i(\delta_0, \varepsilon)$.

3. For the value

$$\lambda_2 = 1/2. \tag{5.16}$$

The closeness of the frequency λ_2 to the value (5.16) is the only condition of the appearance of parametric resonance in transversal oscillations of a satellite on an elliptic orbit.

In Figure 4 dependencies $\lambda_i(\varepsilon, \delta_0)$ are shown where $\varepsilon = C/A$, $\delta_0 = B/A$. In the considered region of stable oscillations $B > A > C$ and in a physical sense of moments of inertia $B + A > C$; $B + C > A$; $C + A > B$.

References

[1] BELETSKY, V. V.: Artificial Earth Satellites, Vol. 3. The USSR Academy of Sciences Publishing House, 1959.

[2] BOGOLYUBOV, N. N., and YU. A. MITROPOLSKY: Asymptotic Methods in the Theory of Nonlinear Oscillations, Moscow. The State Publishing House for Physical and Mathematical Literature, 1958.

Variation problem of transfer between two points in a central field

By

V. N. Lebedev and B. N. Rumyantsev

Computation Centre of the U.S.S.R. Academy of Sciences, Moscow

This paper examines a plane problem of minimum time orbital transfer in a central gravitational field between two points located at predetermined distances from the centre of attraction. It is assumed that the engine of the spaceship develops an acceleration constant in magnitude. The use of the LAGRANGE method leads to the solution of a boundary problem for the system of differential equations. Calculations have been made for two classes of problems: transfer between two circular orbits and transfer from a certain point to the predetermined circular orbit. One of the problems of this kind (transfer between the orbits of the Earth and Mars) was considered in [1] where the author employed MAYER's method. In distinction to [1], the method employed in the present work do not lead to conditions of transversality which makes it possible to investigate a broader class of problems.

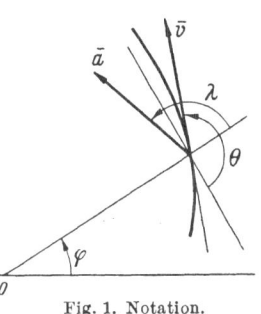

Fig. 1. Notation.

1. The travel of the spaceship in the central field is described by the following system of equations in the polar coordinate system:

$$\frac{dv}{dt} = -\frac{\sin\theta}{r^2} + a\sin(\theta - \lambda), \tag{1.1}$$

$$\frac{d\theta}{dt} = \left(v - \frac{1}{rv}\right)\frac{\cos\theta}{r} + \frac{a\cos(\theta - \lambda)}{v}, \tag{1.2}$$

$$\frac{dr}{dt} = v\sin\theta, \tag{1.3}$$

$$\frac{d\varphi}{dt} = -\frac{v\cos\theta}{r}, \tag{1.4}$$

where v is the velocity modulus (see Figure 1), r-distance from the centre of attraction O, θ-angle between the perpendicular to the radius vector and the velocity vector, φ-polar angle, a-thrust acceleration (ratio between the force of gravity to the mass of the spaceship), λ-angle between the radius vector and the acceleration vector, t-time. All the variables are non-dimensional. The corresponding dimensional variables marked by an asterisk can be obtained from the following formulas:

$$r^* = r\, r_0^*, \qquad v^* = v \sqrt{\frac{\mu}{r_0^*}}, \qquad t^* = t\, r_0^* \sqrt{\frac{r_0^*}{\mu}}, \qquad a^* = a\, \frac{\mu}{r_0^{*2}} \quad (1.5)$$

here r_0^* is the initial radius, $\mu = 13.29 \times 10^{10}\,\mathrm{km^3/sec^2}$ the Gauss constant for the Sun.

Transfer should be effected in minimum time from the point where

$$r = 1, \qquad v = v_0, \qquad \theta = \theta_0 \tag{1.6}$$

to the point where

$$r = r_1, \qquad v = v_1, \qquad \theta = \theta_1. \tag{1.7}$$

2. Let us note that the Eqs. (1.1) to (1.3) can be solved independently of (1.4) and φ found from (1.4) by quadrature. We shall not consider here the Eq. (1.4). Let us assume r to be an independent variable. Then (1.1) and (1.2) will be reduced to the following form:

$$\frac{dv}{dr} + \frac{1}{r^2\,v} - \frac{a\sin(\theta - \lambda)}{v\sin\theta} \equiv \varphi_1 = 0, \tag{2.1}$$

$$\frac{d\theta}{dr} + \frac{1}{r^2\,v^2\tan\theta} - \frac{1}{r\tan\theta} - \frac{a\cos(\theta - \lambda)}{v^2\sin\theta} \equiv \varphi_2 = 0. \tag{2.2}$$

Integrating the Eq. (1.3) we obtain

$$T = \int_{r_0}^{r_1} \frac{dr}{v\sin\theta} \equiv \int_{r_0}^{r_1} F\,dr. \tag{2.3}$$

Our problem can now be formulated as follows: find the functions v, θ and λ of r that satisfy the Eqs. (2.1) and (2.2) and the boundary conditions (1.6) and (1.7) and reducing the functional (2.3) to minimum.

We have arrived at the formulation of the general Lagrange problem (see [2]). Therefore, following his method we introduce the function

$$H = F + \lambda_1\,\varphi_1 + \lambda_2\,\varphi_2,$$

where λ_1 and λ_2 are some functions of r. The following Euler's equations will hold for the function H:

$$H_{x_i} - \frac{d}{dr} H_{x_i'} = 0, \qquad i = 1, 2, 3 \tag{2.4}$$

where $x_1 = v$, $x_2 = \theta$, $x_3 = \lambda$, the stroke denotes the r derivative. Differentiating in the Eq. (2.4) we reduce them to the form:

$$\frac{d\lambda_1}{dr} = -\frac{1}{v^2 \sin\theta} - \frac{\lambda_1}{r^2 v^2} + \frac{\lambda_1 a \sin(\theta - \lambda)}{v^2 \sin\theta} - \frac{2\lambda_2}{r^2 v^3 \tan\theta} + \frac{2\lambda_2 a \cos(\theta - \lambda)}{v^3 \sin\theta},$$

$$(2.5)$$

$$\frac{d\lambda_2}{dr} = -\frac{\cos\theta}{v \sin^2\theta} - \frac{\lambda_1 a \sin\lambda}{v \sin^2\theta} - \frac{\lambda_2}{r^2 v^2 \sin^2\theta} + \frac{\lambda_2}{r \sin^2\theta} + \frac{\lambda_2 a \cos\lambda}{v^2 \sin^2\theta}, \quad (2.6)$$

$$\lambda_2 = \lambda_1 \frac{v}{\tan(\theta - \lambda)}. \quad (2.7)$$

Let us use (2.7) to eliminate λ_2 after which the Eqs. (2.5) and (2.6) will take the form:

$$\frac{d\lambda_1}{dr} = \frac{a\lambda_1[1 + \cos^2(\theta - \lambda)] - \sin(\theta - \lambda) - \dfrac{\lambda_1}{r^2}[\cos\lambda + \cos\theta \cos(\theta - \lambda)]}{v^2 \sin\theta \sin(\theta - \lambda)}, \quad (2.8)$$

$$\frac{d\lambda}{dr} = \frac{\dfrac{\sin\lambda}{v^2}\left[\dfrac{\sin(\theta - \lambda)}{\lambda_1} - a + \dfrac{\cos\lambda}{r^2}\right] + \dfrac{1}{r}[\sin(\theta - \lambda)\cos(\theta - \lambda) + \sin\theta \cos\theta]}{\sin^2\theta}.$$

$$(2.9)$$

Now, to solve the initial problem we must find the functions v, θ, λ_1 and λ that satisfy the Eqs. (2.1), (2.2), (2.8) and (2.9) and the boundary conditions (1.6) and (1.7). The number of equations is equal to the number of unknown variables and the number of boundary conditions. Hence the problem is definite.

3. For our calculations it is convenient to assume t to be an independent variable since r can change non-monotonously. In this case the Eqs. (2.8) and (2.9) are transformed into:

$$\frac{d\lambda_1}{dt} = \frac{a\lambda_1[1 + \cos^2(\theta - \lambda)] - \sin(\theta - \lambda) - \dfrac{\lambda_1}{r^2}[\cos\lambda + \cos\theta \cos(\theta - \lambda)]}{v \sin(\theta - \lambda)}, \quad (3.1)$$

$$\frac{d\lambda}{dt} = \frac{\dfrac{v}{r}[\sin(\theta - \lambda)\cos(\theta - \lambda) + \sin\theta \cos\theta] - \dfrac{\sin\lambda}{v}\left[a - \dfrac{\sin(\theta - \lambda)}{\lambda_1} - \dfrac{\cos\lambda}{r^2}\right]}{\sin\theta}.$$

$$(3.2)$$

Let us add the Eq. (1.4) which allows us to calculate φ to the system (1.1) to (1.3), (3.1), (3.2). Thus, finally, we must find the solution for the system of Eqs. (1.1) to (1.4), (3.1), (3.2) under the boundary conditions:

$$t = 0, \quad v = v_0, \quad \theta = \theta_0, \quad r = 1, \quad \varphi = 0, \quad (3.3)$$

$$t = t_1, \quad v = v_1, \quad \theta = \theta_1, \quad r = r_1. \quad (3.4)$$

Since the time t_1 is unknown the end of the flight path will be the point where $r = r_1$. This is a boundary value problem—it can be solved by integrating numerically the equations selecting at the left end of the path two parameters: the initial values of λ and λ_1.

Let us consider the case when the travel begins from a circular orbit ($v_0 = 1$, $\theta_0 = \pi$). In this case the denominator in the right side of the Eq. (3.2) is converted to zero. For the derivative λ by t to be definite at the initial point the numerator should also be converted to sero. From this condition we obtain the following relationship at the point $t = 0$:

$$\sin \lambda_0 = a\,\lambda_{10}. \qquad (3.5)$$

However, as distinct from the general case, we do not know here at the initial point the value of $d\lambda/dt$ which should be specified as the second parameter.

4. The system of Eqs. (1.1) to (1.4), (3.1), (3.2) was solved on the BESM-2 computer by the ADAMS method with an automatic selection of the integration step. During calculations the right sides of the Eqs. (3.1)

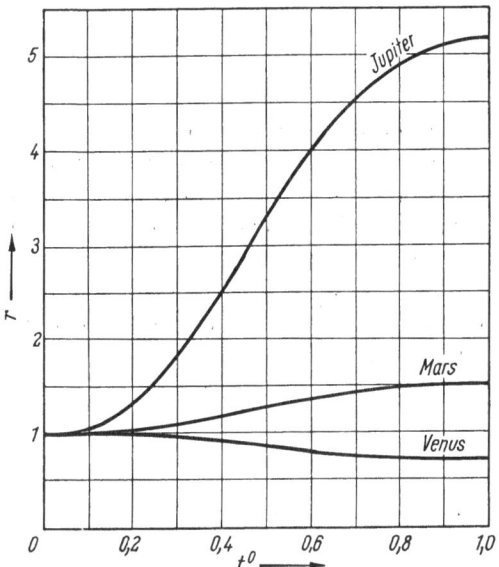

Fig. 2. Time dependence of heliocentric radius.

and (3.2) were reduced at some points to uncertainty of 0/0 type. The programme was so devised that these points were located in the middle of a certain fixed step whose magnitude was selected with a view to a minimum integration error.

Two problems have been examined: transfer between circular orbits and transfer to the predetermined circular orbit from a certain point with an arbitrary initial velocity vector. The latter problem arises when a spaceship returning from a cosmic flight is placed in a circular orbit.

Figure 2 shows a heliocentric radius as a function of time for flights from the Earth to Venus, Mars and Jupiter at $a = 0.1667$ which corresponds to a thrust acceleration of 1 mm/sec². The relative time $t^0 = t/t_1$

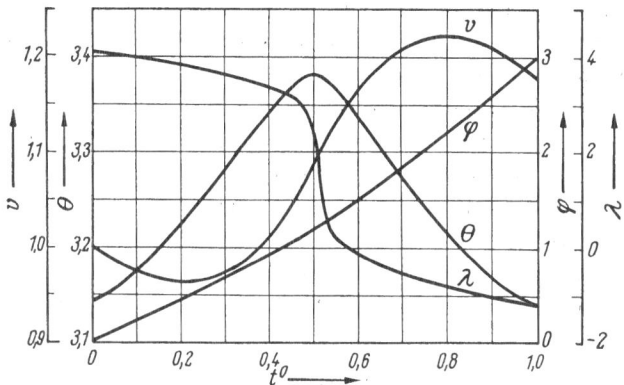

Fig. 3. Time dependence of functions v, θ, φ and λ for the case of travel to Venus's orbit.

is plotted on the x-axis; here t_1 is the time of transfer equal to 2.33 (Venus), 3.25 (Mars) and 9.31 (Jupiter), the corresponding time in days being equal to 135, 189 and 540. The flight angle φ_1 equals 172°, 138° and 185°, respectively. The time of flight in days to the orbits of other

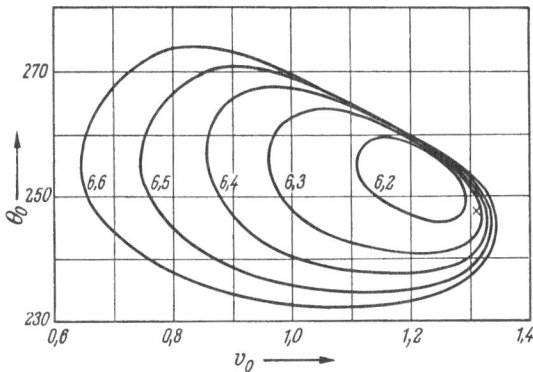

Fig. 4. Curves of equal travel times from a fixed point to a given circular orbit.

planets (except Mercury) is 789 for Saturn, 1.150 for Uranus, 1.480 for Neptune, 1.720 for Pluto. It should be noted that in transfers to radii over 4.21 the function r becomes non-monotonous. For example, during flight to Jupiter r decreases to 0.99999641 at the beginning of the flight path. Figure 3 shows dependence of the functions v, θ, φ and λ on time during flight to Venus.

Figure 4 gives the results of calculations for the second problem: flight time as a function of the initial values of v and θ with the initial radius $r = 5$ and $a = 0.1667$. To decrease the computation time integration was performed in a reverse direction i.e., the initial point was on the predetermined circular orbit of a single radius.

With the magnitude of the initial velocity vector marked with a cross in Figure 4, transfer can be effected from the radius equal to five to the predetermined circular orbit by using a thrust tangential to the flight path. The solution of the variation problem under these initial conditions gives the flight time to an accuracy of three decimal points coinciding with the time at a tangential thrust. This testifies to a high efficiency of this kind of control.

The computations were done with the assistance of N. A. MEISTER.

The authors wish to thank N. N. MOISSEYEV for valuable advice.

References

[1] FAULDERS, C. R.: Minimum Time Steering Programs for Orbital Transfer with Low-Thrust Rockets. Astronaut. Acta 7 (1961) No. 1.
[2] LAVRENTIEV, M. A., and L. A. LYUSTERNIK: A Course of Calculus of Variations, GTTI, 1950.

Observed torque-producing forces acting on satellites

By

Robert J. Naumann

Research Projects Division, Marshall Space Flight Center
Huntsville, Alabama, U.S.A.

Abstract. In addition to the forces that influence the motion of the center of mass of a satellite, various forces are also present which influence the motion about the center of mass. These torque-producing forces may result in a change in satellite orientation that affects the thermal balance, operation of solar cells, various scientific measurements, and drag forces. This latter effect results in a coupling between the equations describing the orientation with those describing the orbital motion. Therefore, strictly speaking, the orientation problem is part of the orbital problem.

To solve the orientation equations it is necessary to determine the nature of the torque-producing forces. This is done by observing the change in orientation of several Explorer satellites where the drag force is sufficiently small so the orbital dependence on orientation is negligible. It was found that permanent magnetic moments in the satellites were the dominant effect responsible for the observed changes in orientation. Gravitational torques are also significant. The changes in orientation of the satellites considered were explained extremely well by these two effects, hence it is concluded that other effects are not significant for similar satellites.

Various approximations are used in this study which greatly reduce the effort required to integrate the orientation equations and do not require the simultaneous solution of the orbital equations. The findings, however, are applicable to the simultaneous solution of both the orbital and orientation sets of equations.

I. Introduction

The most general treatment of satellite motion requires six second-order differential equations to describe the motion. Three equations describe the motion of the center of mass in phase space, and the remaining three equations describe the orientation and rates of three body-fixed axes relative to a space-fixed system. In orbital calculations it is usually assumed that this set of six equations separates into two sets with the orbital set independent of the set describing the orientation. This is not rigorously true if the satellite in question has any asymmetry, since the drag forces will depend on the orientation of the satellite.

It is desirable for several reasons to formulate the full six degree of freedom problem and integrate the equations describing the orientation along with the orbital equations. It may be argued that since the previously mentioned drag dependence is usually small compared to the uncertainties in atmospheric density, such an exercise is largely academic. While this is a valid argument there are many satellite experiments in which it is just as important, or even more important, for the experimenter to have knowledge of the satellite orientation as it is know the position of the center of mass. This is particularly true in the case of the Explorer XI gamma ray astronomy satellite.

In most cases where it is necessary to have orientation information, the approach has been to derive such information from a combination of solar and earth horizon sensors contained in the satellite, or by analyzing the recorded radio signal strength patterns. The latter has been particularly successful with Explorers IV, VII, and XI.

While it has been possible to determine satellite orientation a posteriori, it would be desirable to be able to predict the satellite orientation a priori. Since the external torque-producing forces are coupled closely to the motion of the center of mass, and since the motion of the center of mass depends to a lesser degree on the orientation, it would seem reasonable to consider the entire six degree of freedom formulation in the integration of the equations of motion. A convenient formulation of the full six degree problem has been suggested by LUNDQUIST and NAUMANN [1] in which the orientation of the body-fixed axes is expressed in terms of the EULER quaternion parameters [2]. This formulation has several advantages over the more conventional EULER angle formulation in that it yields symmetrical equations in which no trigonometric functions are involved. These parameters also have certain advantages over the CALEY-KLEIN parameters [3] in that complex numbers are not involved. The disadvantage is that they are not independent and exceed by one the number of degrees of freedom. This difficulty is alleviated by the use of a LAGRANGE undetermined multiplier and the constraint relationship that the sum of the squares of the parameters is unity. Several computer programs based on this formulation have been developed by the Computation Division at the Marshall Space Flight Center and by CUNNINGHAM at the University of California.

In order to use such programs to predict satellite motion about the center of mass, it is necessary to know the nature of the external torques acting on the satellite. Analysis of the motion of several of the Explorer satellites has yielded some knowledge concerning the nature of these torques. The purpose of this paper is to discuss some of the experimental satellite orientation data, and attempt to explain the nature and origin of the torques responsible for the observed motion about the center of mass.

The satellites to be considered are Explorer XI, Explorer IV, Explorer VIII and Explorer VII. This particular order was chosen because

xplorer XI has yielded the most accurate and extensive orientation
iformation. Sufficient information has been obtained to allow the
xtraction of the torques by numerical differentiation, hence, the deter-
iination of significant effects. Much less orientation data are available
ir Explorer IV, but in the light of the Explorer XI analysis, its motion
now fairly well understood. Very little data exist on Explorers VIII
id VII, but again using results from the previous analyses a motion
derived that is consistent with known data.

II. Computational techniques

Before discussing the various satellites individually, it would be
ell to discuss various sources of torques and the various approxi-
iations used in computing their effects.

To begin, it will be assumed that all of the satellites have cylin-
rical symmetry and are initially set spinning about their axes of
ymmetry. The moment of inertia about the axis of symmetry is denoted
y I_3. If a transverse axis has a moment of inertia I_1 less than I_3, the
atellite is said to be stable and will remain spinning about I_3. If I_1 is
reater than I_3, as is the case in Explorer XI and Explorer IV, the
atellite is in a maximum rotational energy configuration and is unstable.
)ue to internal energy dissipation and angular momentum conservation,
he satellite longitudinal axis will precess about the angular momentum
ector in a cone. The half-angle of this precession cone α is related to
he longitudinal spin ω_3 by

$$\cos \alpha = \frac{I_3\,\omega_3}{L}, \tag{1}$$

vhere L is the total angular momentum. As energy is dissipated, $\omega_3 \to 0$,
$\iota \to \pi/2$, and the satellite approaches the minimum energy configur-
ition in which the rotation is a tumbling motion with the axis of sym-
netry rotating in a plane perpendicular to the angular momentum
ector.

1. Gravitational gradient torques. The gravitational torque on a
atellite may be found by expanding the potential energy about the
enter of mass in a TAYLOR series and differentiating with respect to
he generalized angles expressing the orientation of the body-fixed
ixes. The result given by ROBERSON [4] and others [5] is

$$\vec{L}_G = 3 \frac{m\,k^2}{r^3} (I_1 - I_3)\,(\hat{A} \cdot \hat{R})\,(\hat{A} \times \hat{R}), \tag{2}$$

vhere $m\,k^2$ is the gravitational constant; r is the radius vector magnitude;
$\hat{A}$ is a unit vector along the axis of cylindrical symmetry; and $\hat{R}$ is a unit
vector along the radius vector.

Assuming a half-cone angle α, this torque may be time averaged over a precessional or tumble cycle while holding $\hat{R}$ constant. This results in

$$\vec{L}_G = \frac{3}{2}\,\frac{m\,k^2}{r^3}(I_1 - I_3)\,(\hat{R}\cdot\hat{L})\,(\hat{R}\times\hat{L})\,(1 - 3\cos^2\alpha), \qquad (3)$$

where $\hat{L}$ is a unit vector directed along the angular momentum vector.

Assuming a circular orbit, this $\vec{L}_G$ is time averaged over an orbital revolution while holding $\hat{L}$ constant. This is done by expressing $\hat{R}$ in terms of orbital inclination i, right ascension of the ascending node ϕ, and mean anomaly M.

$$\hat{R} = \begin{pmatrix} \cos M \cos\phi - \cos i \sin\phi \sin M \\ \cos M \sin\phi + \cos i \cos\phi \sin M \\ \sin i \sin M \end{pmatrix}. \qquad (4)$$

The unit vector along the orbital angular momentum is

$$\hat{\Omega} = \begin{pmatrix} \sin i \sin\phi \\ -\sin i \cos\phi \\ \cos i \end{pmatrix}. \qquad (5)$$

The time average is found by

$$\vec{L}_G = \frac{3}{4\pi}\,\frac{m\,k^2}{r^3}(I_1 - I_3)\,(1 - 3\cos^2\alpha) \int_0^{2\pi} (\hat{R}\cdot\hat{L})\,(\hat{R}\times\hat{L})\,dM. \qquad (6)$$

Expressing $\hat{R}$ in terms of ϕ, i, and M from Eq. (4), integrating, and expressing the resulting combinations of ϕ and i in terms of components of $\hat{\Omega}$ using Eq. (5), results in

$$\vec{L}_G = K_G(\hat{L}\cdot\hat{\Omega})\,(\hat{L}\times\hat{\Omega})\,(1 - 3\cos^2\alpha), \qquad (7)$$

where

$$K_G = \frac{3}{4}\,\frac{m\,k^2}{r^3}(I_1 - I_3). \qquad (8)$$

2. Magnetic torques. If a satellite has a permanent magnetic moment projection, M_L, on the axis of rotation, an interaction torque will be experienced due to the geomagnetic field. This torque is given by

$$\vec{L}_M = \vec{M}_L \times \vec{B} = M_L(\hat{L}\times\vec{B}). \qquad (9)$$

A component of magnetic moment along the axis of symmetry, M_3, results in $M_L = M_3 \cos\alpha$. Due to rotational symmetry, a transverse component of magnetic moment does not contribute unless $\omega_3 = 0$. If this happens, the M_L is M_1 or the projection of the transverse component on $\hat{L}$.

To faciliate computation, an earth-centered dipole representation of the geomegnetic field is employed,

$$\vec{B} = B_N \begin{pmatrix} -3\,R_x\,R_z \\ -3\,R_y\,B_z \\ 1 - 3\,R_z^2 \end{pmatrix}, \tag{10}$$

where B_N is the scalar field at the equator at r km from the center of the earth. This is given by

$$B_N = \frac{\gamma}{r^3}, \tag{11}$$

where γ is the geomagnetic moment taken as 8.1×10^{25} cgs units.

Assuming a circular orbit, the magnetic torque is time averaged over an orbital revolution in the same manner as before, giving

$$\vec{L}_M = M_L\,B_N \begin{vmatrix} -\dfrac{1}{2}\,L_y + \dfrac{3}{2}\,\Omega_z(\hat{L} \times \hat{\Omega})_x \\[2mm] \dfrac{1}{2}\,L_x + \dfrac{3}{2}\,\Omega_z(\hat{L} \times \hat{\Omega})_y \\[2mm] \dfrac{3}{2}\,\Omega_z(\hat{L} \times \hat{\Omega})_z \end{vmatrix}. \tag{12}$$

The constant $M_L\,B_N$ is termed the magnetic couple.

3. Induced Magnetic torques. For a satellite having a long cylindrical geometry, it is possible that a magnetic moment may be induced that alternates in sign during a tumble cycle [6]. Torques produced by this process would tend to add rather than cancel due to rotational symmetry. The induced moment is

$$\vec{M}_I = \frac{\mu_r - 1}{\mu_0}\,U(\vec{B} \cdot \hat{A})\,\hat{A}, \qquad \text{(rationalized MKS units)} \tag{13}$$

where μ_r is the relative permeability; μ_0 is the permeability of free space; and U is the volume of material in the walls of the cylinder.

Let

$$K_I = \frac{\mu_r - 1}{\mu_0}\,U, \tag{14}$$

then the torque is

$$\vec{L}_I = K_I(\hat{A} \cdot \vec{B})\,(\hat{A} \times \vec{B}). \tag{15}$$

Time averaging over a tumble cycle results in

$$\vec{L}_I = -\frac{1}{2}\,K_I(\hat{L} \cdot \vec{B})\,(\hat{L} \times \vec{B}). \tag{16}$$

Time averaging over an orbital revolution, as before, results in the expressions

$$\vec{L}_I = -\frac{1}{2}\,K_I\,B_N^2(\hat{L} \times \vec{\mathfrak{F}}) \tag{17}$$

where

$$\mathfrak{F}_i = B_{ij}\,L_j; \qquad i = 1, 2, 3; \qquad j = 1, 2, 3$$

where
$$L_1 = L_x, \qquad L_2 = L_y, \qquad L_3 = L_z$$
and

$$B_{ij} = \begin{vmatrix} \dfrac{9}{8}(\Omega_y^2 + 3\,\Omega_x^2\,\Omega_y^2) & -\dfrac{9}{8}\,\Omega_x\,\Omega_y(1 - 3\,\Omega_z^2) & \dfrac{3}{8}\,\Omega_x\,\Omega_z(4 - 9\,\beta^2) \\[2mm] -\dfrac{9}{8}\,\Omega_x\,\Omega_y(1 - 3\,\Omega_z^2) & \dfrac{9}{8}(\Omega_x^2 + 3\,\Omega_y^2\,\Omega_z^2) & \dfrac{3}{8}\,\Omega_y\,\Omega_z(4 - 9\,\beta^2) \\[2mm] \dfrac{3}{8}\,\Omega_x\,\Omega_z(4 - 9\,\beta^2) & \dfrac{3}{8}\,\Omega_y\,\Omega_z(4 - 9\,\beta^2) & 1 - 3\,\beta^3 + \dfrac{27}{8}\,\beta^4 \end{vmatrix}$$

where
$$\beta^2 = \Omega_x^2 + \Omega_y^2.$$

4. Drag torques. The drag force on a satellite is given by

$$\vec{F}_D = -\frac{1}{2}\,C_D\,\varrho\,S\,V\,\vec{V}, \tag{18}$$

where C_D is the drag coefficient usually taken to be 2.0; ϱ is the atmospheric density; and S the area projected normal to the velocity vector $\vec{V}$. The drag torque is

$$\vec{L}_D = \vec{a} \times \vec{F}_D \tag{19}$$

where $\vec{a}$ is the displacement of the center of pressure from the center of mass.

For symmetrical satellites, such as Explorers VII and VIII, the a is for all practical purposes zero and the drag torque is not significant. For asymmetrical satellites, such as Explorer XI, the drag torque is significant only before the satellite has reached the tumbling mode. The first order effects cancel by rotational symmetry after α becomes $\pi/2$.

For an eccentric orbit, such as Explorer XI, the atmospheric density is significant only near perigee. Hence, the $\vec{V}$ is taken as the perigee velocity vector. The time average torque over an orbital revolution is found by

$$\vec{L}_D = -K_D(\hat{L} \times \hat{V}_P) \tag{20}$$
where

$$K_D = \frac{1}{2}\,C_D\,S\,V^2\,a\,\bar{\varrho}$$

and $\bar{\varrho}$ is the time average atmospheric density given by

$$\bar{\varrho} = \frac{1}{2\pi}\int\limits_0^{2\pi} \varrho(r)\,dM.$$

5. Accuracy of approximations. In order to determine the errors introduced in assuming circular orbits and the earth-centered dipole representation for the geomagnetic field, numerical integrations were carried out over actual orbits of Explorer XI, using the Finch and Leaton 48-term LEGENDRE Polynomical representation for the magnetic field [7].

It was found that the gravitational torques computed using the approximate methods differed from those computed by integration over an entire orbit by a maximum of 1%.

The permanent magnetic torques differed by approximately 10% using a value of B_N of .191 gauss computed from Eq. (11). However, it was noted that the values obtained from the numerical integration were always larger than those from the approximate method. Assuming a B_N of .205 gauss reduced this difference to approximately $\pm 3\%$. This is not serious in this analysis since the M_L is not known accurately and the product $B_N M_L$ must be assumed.

The computation of the induced torques suffers somewhat more from the approximations. The maximum observed deviation was 14% of the maximum value, but generally the deviations were no more than $\pm 8\%$ of the maximum value. The spread was such that no apparent improvement could be gained by slightly altering the value of B_N.

III. Explorer XI

The orientation history of Explorer XI was obtained by determining the angle a line from the tracking station to the satellite makes with angular momentum vector by analyzing the recorded radio signal strength variations [8]. Many such angles with their respective line of sight vectors can be determined during a set of successive passes. These data result in an overdetermined set of non-linear equations that are solved by an iterative process to yield the angular momentum unit vector that best fits the observed data in the sense of least squares. The observed orientation is plotted in terms of right ascension and declination in Figure 1.

Beginning at day 35 after 27.0 April 1961, these plots are fitted in segments by polynomials which are then differentiated with respect to time to yield the torques responsible for the motion.

Since the gravitational torques can be computed accurately from the known moments of inertia of the satellite (Table 1), they are subtracted from the observed torques and the results are plotted in Figure 2. This resulting torque must then be due either to a permanent or induced magnetic moment. The induced torques are computed from Eq. (12) assuming a value of -384 dyne-cm for $K_I B_N^2/2$ which was chosen to yield the approximate observed amplitude. It may be seen from Figure 2 that the induced torques have the same general shape as the required torques, but the correlation is not very convincing. A further difficulty

16*

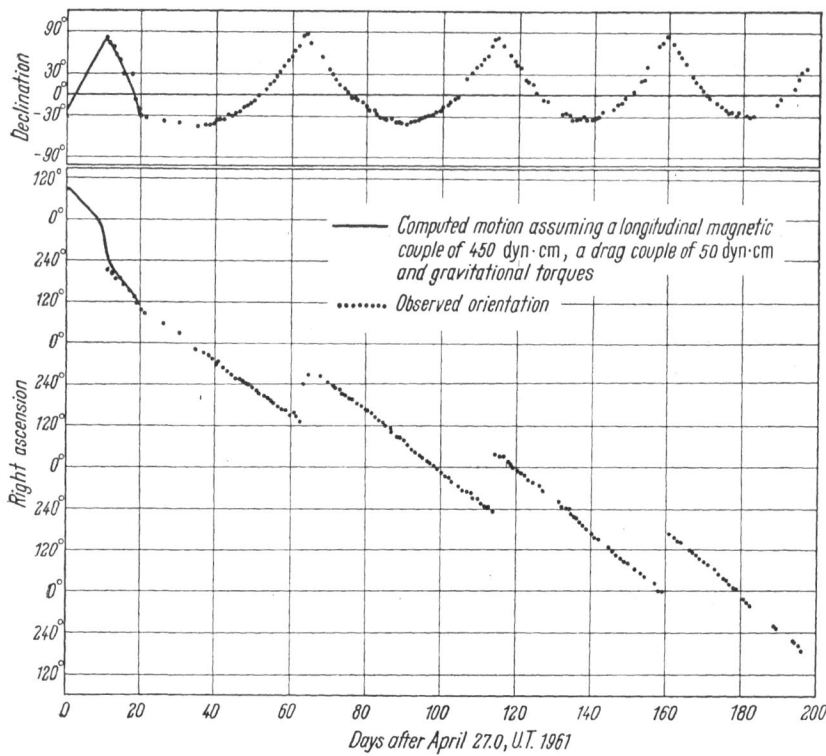

Fig. 1. Observed orientation of the angular momentum vector of Explorer XI.

Table I

Quantity	Explorer XI	Explorer IV	Explorer VII	Explorer VIII
Time Origin	27.0 Apr. 1961	26.0 Jul. 1958	13.0 Oct. 1959	3.0 Nov. 1960
I_1 (gm — cm²)	1.627×10^8	4.752×10^7	1.08×10^7	2.162×10^7
I_3 (gm — cm²)	0.040×10^8	0.056×10^7	2.93×10^7	2.936×10^7
$I_1 - I_3$ (gm — cm²)	1.587×10^8	4.696×10^7	-1.85×10^7	-0.774×10^7
r (km)	7512	7616	7191	6964
K_G (dyne-cm)	111.92	31.78	-14.87	-6.85
$M_1 B_N$ (dyne-cm)	148.14	N/A	N/A	N/A
$M_3 B_N$ (dyne-cm)	450	-3500	75	-200
M_1 (amp-m²)	.7756	N/A	N/A	N/A
M_3 (amp-m²)	2.35	-19.125	.345	$-.837$
ϕ_0 (deg)	253.912	65.81	149.217	16.356
ϕ (deg/day)	-5.0036	-3.6505	-4.1745	-3.3698
i (deg)	28.8	51.0	50.27	49.98
L (gm cm²/sec)	1.276×10^8	4.568×10^7	1.104×10^9	3.073×10^8
ΔL/day	$-0.223 \times 10^{7*}$	-0.0076×10^7	N/A	-0.032×10^7
	$-0.0143 \times 10^{7**}$			

     * For first 20 days before tumbling commenced.
   ** After 26 days when α became $\pi/2$.

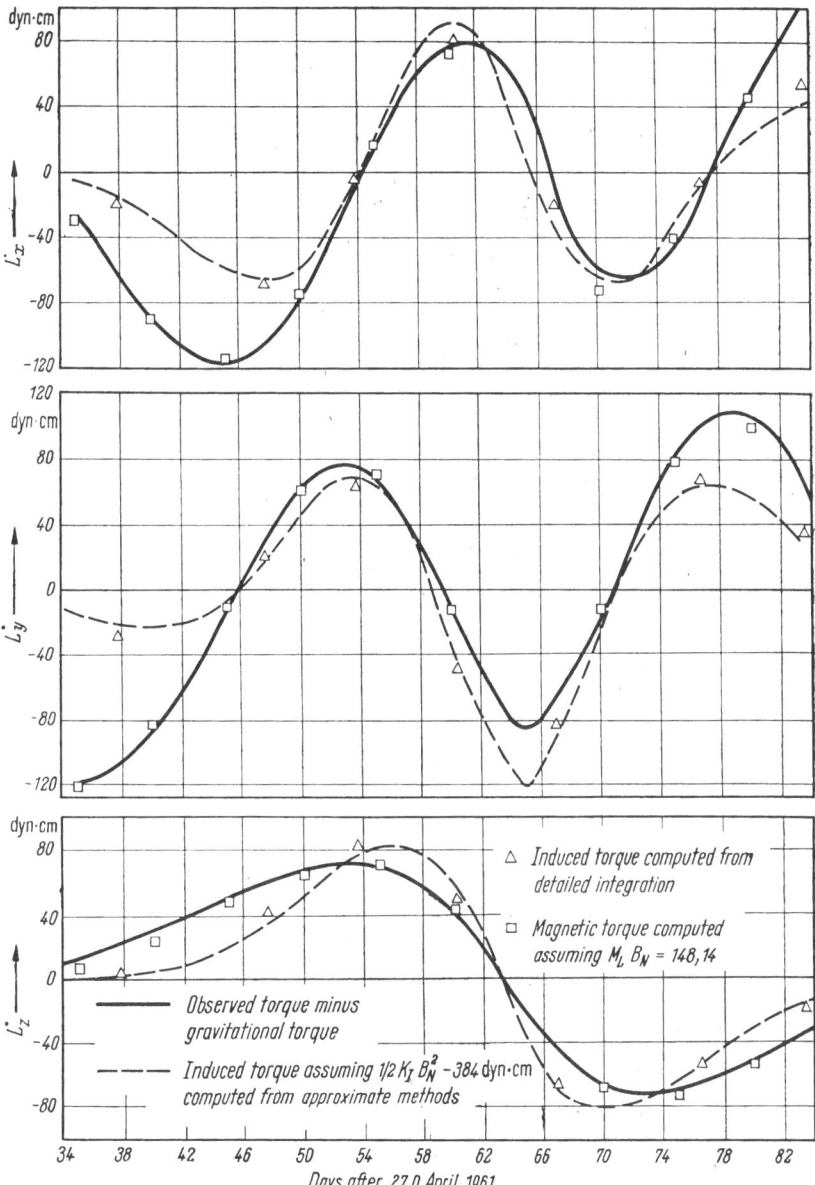

Fig. 2. Analysis of the permanent and induced magnetic torques compared to the difference between the observed torque and gravitational torque for Explorer XI.

arises in the negative sign that must be assumed for K_I since K_I should be positive. Several points are shown that were computed from the detailed integration procedure. It is obvious that the errors introduced in the approximations are small compared to the differences between the induced torques and the required torques.

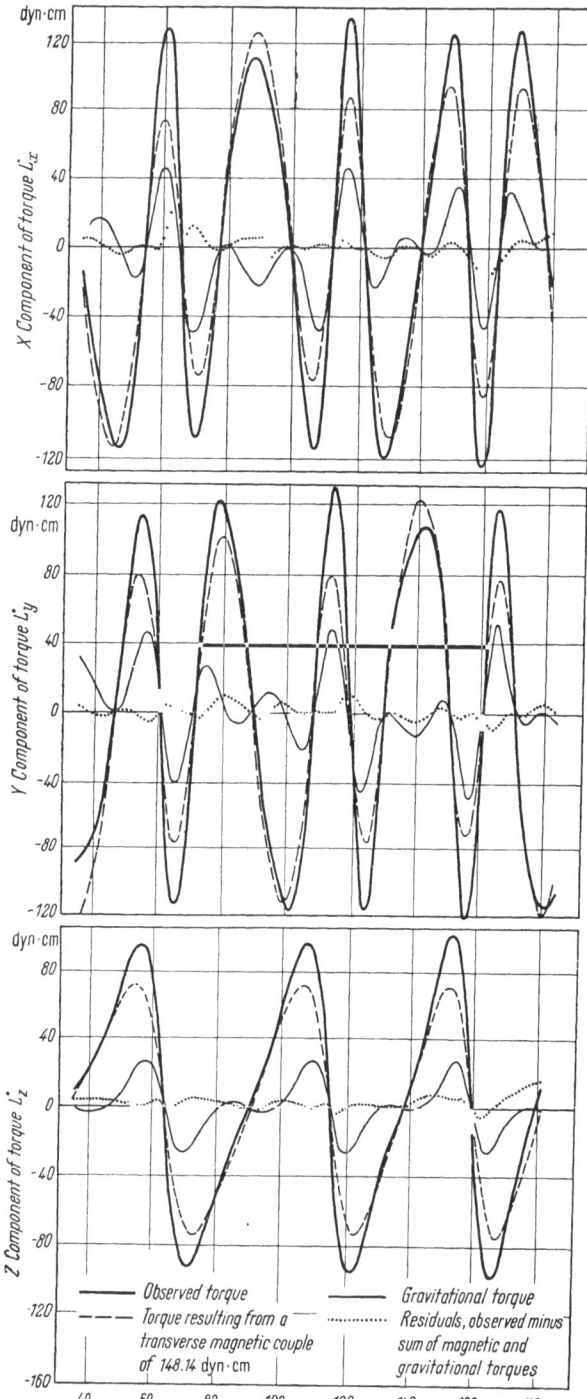

Fig. 3. Analysis of the permanent magnetic and gravitational torque compared to the observed torque for Explorer XI.

On the other hand, the points resulting from an assumed value of 148. 14 dyne-cm for the permanent magnetic couple $M_L B_N$ fall right on the plots of the required torques. Therefore, it is concluded that a permanent transverse magnetic moment is the dominant effect responsible for the observed motion.

The observed torques together with the gravitational and permanent magnetic torques are plotted in Figure 3 for day 35 through day 180. It is seen that the sum of gravitational and magnetic torques equal almost exactly the observed torques, the residuals being generally less than 5 dyne-cm. It is believed that the errors in obtaining the torques from the differentiation of the fitted observations are primarily responsible for these deviations, since the greatest deviations usually occur near the ends of the fitted segments.

It is known from solar cell output data that Explorer XI had no ω_3 after day 26. This probably occurred because one transverse axis has a slightly greater moment of inertia; hence, there is a preferred axis of tumble. Also, a mercury toroidal damper near the tail is quite effective in removing any residual roll about the longitudinal axis. Therefore, it is quite likely that a transverse magnetic moment can have a constant projection on $\hat{L}$.

There is also evidence that a transverse magnetic moment did indeed exist in the attached stainless steel rocket case. Handling difficulties with the loaded rocket casing prevented an accurate measurement of the magnetic moments, but a small hand compass held near the case showed evidence of a transverse as well as a longitudinal magnetic moment.

Although it has been shown that the permanent magnetic moment is the dominant effect, it is still necessary to estimate the magnitude of the induced effect. This is extremely difficult to do in a meaningful manner, since the relative permeability is not well known. In ferrous materials, the permeability depends strongly on the applied field as well as on the past magnetic history of the material. Taking the volume of the ferrous material in the walls of the rocket casing to be 0.470 $\times 10^{-3}$ m³ (the payload portion is fabricated from aluminum) with a permeability of 100 in a field of 0.2×10^{-4} weber/m² results in a $K_I B_N^2$ of 150 dyne-cm. However, the value of 100 for μ_r which is generally considered the initial permeability of 410 stainless steel applies for a flux density of 200 gauss in the material. It is probably considerably less for an applied field of ± 0.2 oersted experienced in the geomagnetic field. There may well be a small component of induced torque, but this would act in the same general manner as the torque due to the permanent magnetism but with opposite sign. This would probably be compensated for by a slightly greater M_L than was assumed.

Since it appears that the gravitational and permanent magnetic torques are mainly responsible for the motion, an attempt was made

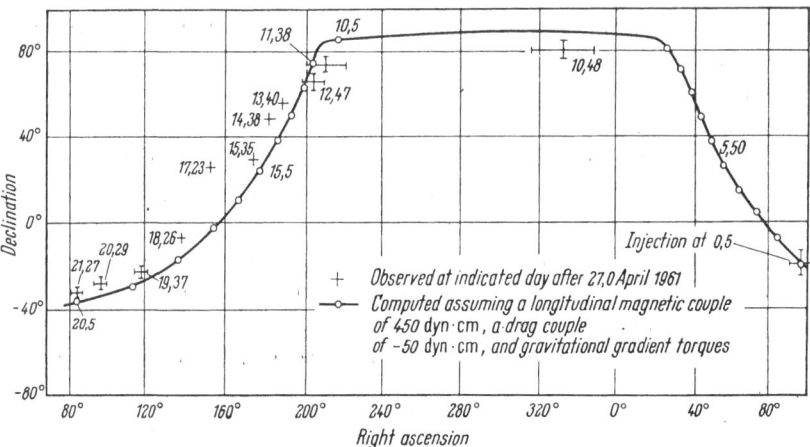

Fig. 4. Computed orientation of the angular momentum vector compared to observations for for the first 20 days of Explorer XI.

to supplement the few orientation data available between launch and day 20 before the satellite began tumbling. This was done by starting at the injection orientation and integrating the motion, assuming various

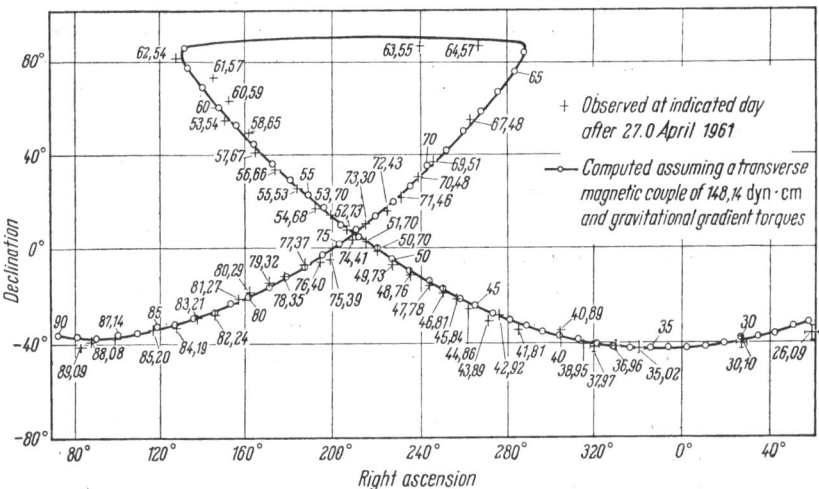

Fig. 5. Computed orientation of Explorer XI compared to observations for day 26 to 90.

values of M_3 to see if a motion could be made to result that was consistent with the observed data. It was found that a couple of $M_3 B_N$ = 450 dyne-cm gave a reasonable fit of the observed orientations. The addition of a drag couple of 50 dyne-cm further improved the fit. The results are shown in Figure 4.

Also, the motion was integrated from day 26 to day 90 and compared with observed data. As may be seen in Figure 5, there is excellent
agreement between computed and observed results.

IV. Explorer IV

The orientation of Explorer IV was deduced in much the same
manner as Explorer XI, except that in Explorer IV this analysis was
done as an afterthought. Therefore, the choice of tracking antenna and
recording techniques was far from optimum for such an analysis. As
a result, the amount and accuracy of the orientation data are much less
than for Explorer XI. The data were sufficient to explain the temperature measurements in terms of area presented to the sun [9]. Also,
the motion of a directional scintillation counter relative to the magnetic
field was deduced [10] which allowed a determination of the directional
flux of the trapped corpuscular radiation in the radiation belt [11].
The orientation measurements are shown in Figure 6. It appears that
the orientation changes at approximately the same rate as observed in
Explorer XI. However, the torques responsible for this motion re-
mained unexplained until quite recently. The possibility of induced
magnetic torques was suggested by COLOMBO [12], but starting at day
32.0 after 26.0 July 1958 and integrating the motion produced by gravitational and induced torques resulted in a motion that is not consistent
with the observed data shown by the short dashed line in Figure 6.

Integrating the torques produced by a permanent transverse magnetic couple of -100 dyne-cm, similar to that found in Explorer XI,
produced a motion that apparently matches the observed data for
several days before and after 32.0. A serious difficulty arises with this
assumption, however, since it was known that Explorer IV had a residual
ω_3 which would tend to cancel any effect due to a transverse magnetic
moment. Evidence for this roll is seen as modulations in the observed
count rate of a directional scintillation detector whose axis is directed
along a transverse axis [13]. Figure 7 shows the roll period history
deduced from these count rate modulations. Note that the roll period
increases rapidly after day 30. In Figure 6, the change in orientation
seems to be much less after this time. From the roll period data, it can
be determined that the half-cone angle α increased rapidly the first few
days, but approached $\pi/2$ very slowly. In fact, from Eq. (1) it may be
shown that α is about $83°$ on day 30. It then becomes clear that a longitudinal magnetic moment will give a constant projection on $\hat{L}$ that will
produce the same effect as a transverse magnetic moment. A rather
large value of -3500 dyne-cm must be assumed for this couple to give
a projected component sufficient to account for the motion between

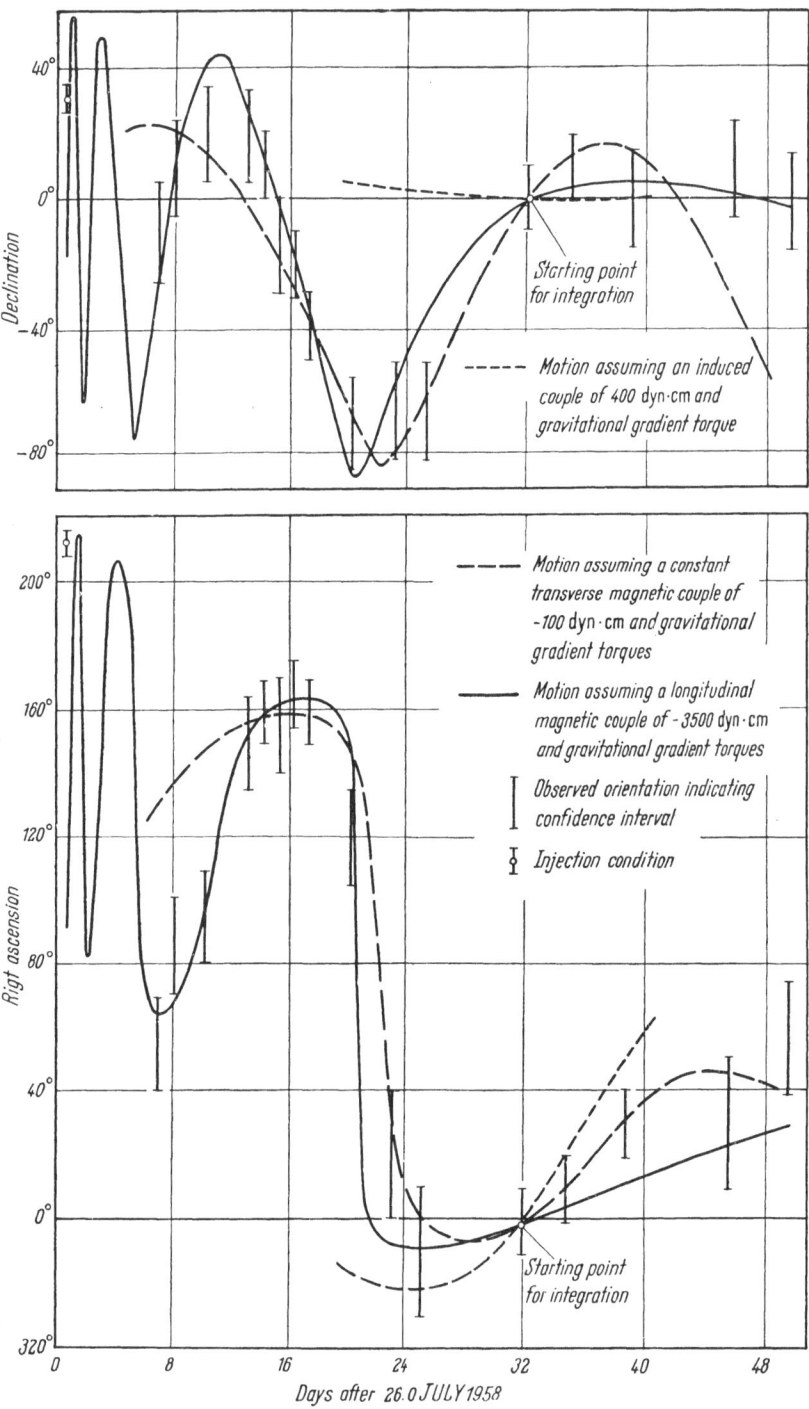

Fig. 6. Analysis of the computed orientation of Explorer IV due to various assumptions compared to observation.

day 25 and day 32, but this assumption appears to explain the observed data rather well. After day 32, the torques due to gravitational effects appear to be sufficient to explain the motion in a reasonable manner.

There may also be some significant second-order torques produced by the large longitudinal magnetic moment after day 32. These second-order torques arise from the fact that the geomagnetic field is changing due to geographical changes of the satellite during the 7-sec tumble cycle. Such a change prevents the longitudinal magnetic torques from averaging to zero due to rotational symmetry. Using the formulation

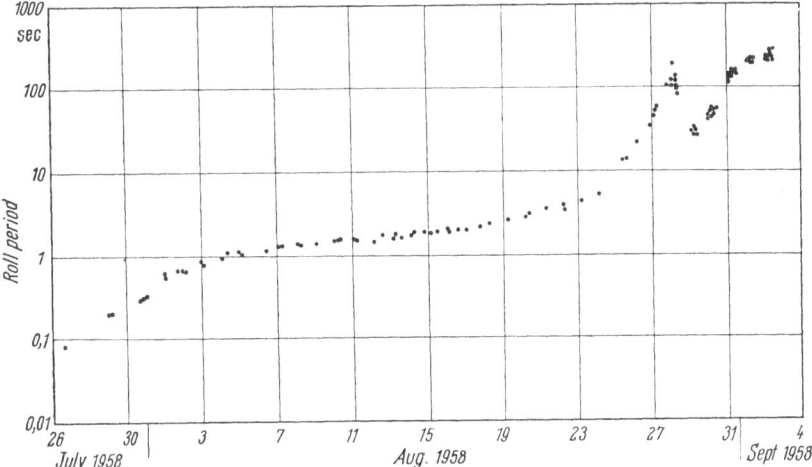

Fig. 7. Observed axial roll period for Explorer IV.

of the full six degree of freedom problem discussed in the introduction, it was found that the net second-order effect was approximately 10^{-3} of the maximum instantaneous magnetic torque. This effect would therefore become significant only if there exists a large magnetic moment having rotational symmetry and if the other torque-producing effects are small.

Some justification is necessary for the fact that a couple of -3500 dyne-cm corresponding to a longitudinal moment of 19 amp-m^2 is required to explain Explorer IV motion, whereas only 450 dyne-cm or 2.35 amp-m^2 is required for Explorer XI. This can be resolved, qualitatively at least, by the fact that the instrumented portion of Explorer IV is stainless steel rather than aluminum. Furthermore, this portion was placed on a magnetically driven shake table for vibrational testing prior to flight. This quite probably resulted in a rather large permanent magnetization. Taking the volume of ferrous material to be 450 cm^3, a flux density of approximately 500 gauss in the material is required to produce a magnetic moment of 19 amp-m^2. The saturation of 410 stainless steel is on the order of 16 kilogauss and the retentivity is approximately 8 kilogauss for 10 kilogauss applied. Hence, 500 gauss is well within the magnetic capabilities of the material.

It is curious to note that Explorer IV began tumbling very shortly after injection but required 30 or more days to approach $\pi/2$. The longitudinal roll ω_3 apparently never reached zero and, in fact, even appears to increase on day 33. This increase is possibly due to an interaction torque from a transverse magnetic moment. On the other hand, Explorer XI had a very small half-cone angle until about day 20, at which time it opened rapidly and lost all its ω_3 within a few days. These facts may be explained by the consideration that Explorer IV did not have a mercury damper and the dissipative forces became less effective in removing energy as ω_3 became small. It is difficult to understand why Explorer XI, with a mercury damper, took so long to begin tumbling. There is evidence that a tape recorder inside the satellite was accidentally commanded on and left running after launch. There is also evidence that this recorder failed on about the day the tumbling started. It has been suggested that the recorder, through bearing friction, actually put rotational energy into the satellite which overcame the energy dissipated by mechanical flexing [14]. This presumably allowed the satellite to remain in its maximum energy configuration despite dissipative losses.

V. Explorer VIII

Unlike the satellite discussed previously, Explorer VIII was of a stable configuration. No attempt was made to analyze the orientation

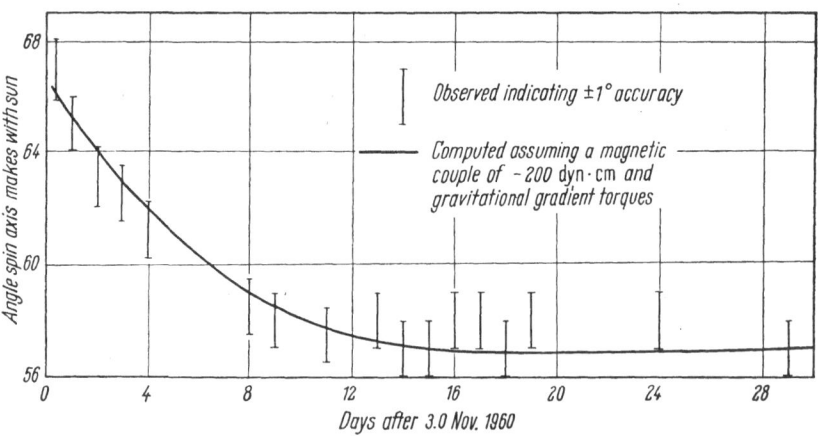

Fig. 8.
Computed solar aspect angle for Explorer VIII compared to observations of BOURDEAU et al [15].

by radio signal strength techniques, but the satellite contained a crossed-slit solar aspect angle measuring device along with a horizon sensor.

Presumably the data from both of these devices would determine the orientation of the spin axis. The solar aspect device yielded excellent data, accurate to about 1 degree. However, the horizon scan data are apparently difficult to interpret. The only published orientation data are the angles between the spin axis and the sun for the first 30 days [15].

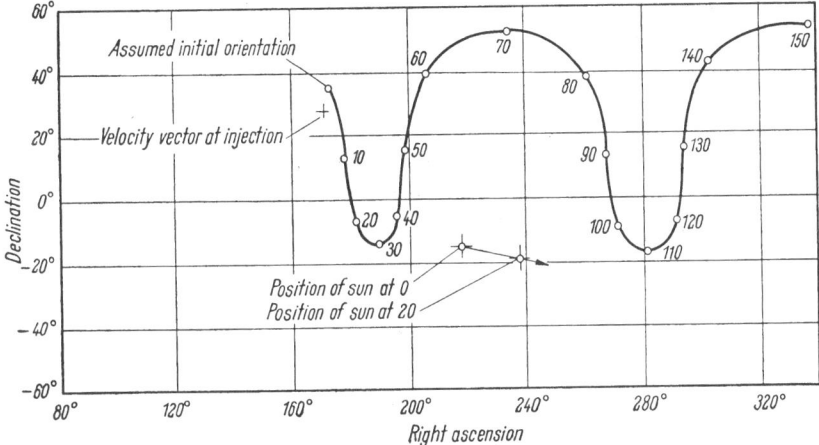

Fig. 9. Computed orientation of Explorer VIII.

From the previous analysis, it was felt that the torques acting on Explorer VIII must be predominately due to a permanent magnetic moment along the spin axis and gravitational torques. Therefore, an attempt was made to integrate the motion from the injection condition to produce the observed solar aspect angle.

The first measured solar aspect angle did not agree with an assumed initial spin axis coinciding with the injection velocity vector; hence, it was necessary to assume a spin axis injection condition a few degrees from the velocity vector at injection. This can be reconciled by a slight misalignment of the fourth stage velocity vector with the total injection velocity, or by assuming that the satellite was precessing slightly when the fourth stage was jettisoned.

Various values of $M_3 B_N$ were assumed until it was found that $M_3 B_N = -200$ dyne-cm would produce a computed solar aspect angle shown in Figure 8. This computed result compares with the measured solar aspect angle well within the limits of accuracy. The resulting trace of the spin axis shown in Figure 9.

VI. Explorer VII

A crude determination of Explorer VII orientation was accomplished by noting the times at which the fading due to Faraday rotation of the 20 mc/sec signal strength became the most severe [16]. The turnstile antenna emits omnidirectional radiation that is circularly polarized along the spin axis. The polarization becomes more elliptical as the angle from the spin axis is increased, becoming linearly polarized in the plane of the equator. Since the FARADAY fading becomes the most severe for a linearly polarized wave, the observance of this severe fading indicates the station to satellite vector is normal to the spin axis.

During the first two months, the orientation appeared to remain almost constant at the injection condition. A slight shift was indicated

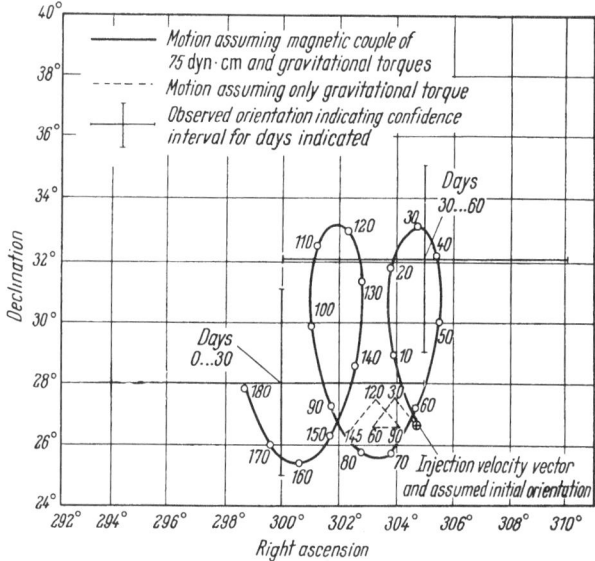

Fig. 10.
Computed orientation of Explorer VII with and without an assumed magnetic moment.

but the limits of accuracy of the measurement were not sufficient to ascertain this. Thermal measurements taken throughout the first year of orbiting were general agreement with the assumption of a space-fixed orientation, although again such measurements are not sensitive enough to detect a slight shift of the spin axis [16].

It is certain that gravitational torques act on Explorer VII and also it seems virtually impossible to construct a satellite without some magnetic moment unless particular care is taken to prevent such a moment. Explorer VII does have a very large angular momentum due to its fast spin rate which greatly reduces the forced precession of the spin axis.

The motion of the spin axis was integrated, assuming only gravitational torques, and resulted in the dashed curve in Figure 10. This is seen to result in a very slow net drift of the spin axis with time. The addition of a small magnetic moment with $M_3 B_N = 75$ dyne-cm produces a motion indicated by the solid line. Again, it is seen that although an oscillation of several degrees is present, the net drift is quite small. This motion also appears to fit the observed orientation data, although the accuracy is not really sufficient to confirm it.

VII. Conclusions

It now seems clear that permanent magnetic and gravitational torques are the dominating effects which produce changes in a satellite orientation. Much the same conclusion was reached by MANGER [17] for the Tiros satellites. In fact, the orientation of Tiros III was effectively controlled by applying currents in a loop to alter the magnetic moment. The orientation was predicted and the control currents were programmed prior to launch. According to HETCH and MANGER [18], this was quite successful. Other satellites have been passively oriented along the magnetic field lines by employing permanent magnets and hysteresis damping rods [19].

The magnetic effects in Explorer XI and Explorer IV were not as obvious as those in stable spinning satellites since it was not clear that a permanent magnetic moment projection on the axis of rotation existed. But now that the motions of these satellites have been successfully analyzed and understood, a heretofore disturbing problem has been resolved.

It is now felt that sufficient knowledge of the rotational behavior of satellites exists to make it worthwhile to formulate the equations of motion about the center of mass with those for the orbit and solve the full six degree of freedom equations of motion. Particular care should be taken to determine the moments of inertia and magnetic moments prior to flight.

It was also shown that simplifying approximations may be employed that give a fairly good representation of the motion without integrating the orbital equations. The orbital information necessary to use these approximate equations is the semi-major axis, the inclination, and right ascension of the ascending node. These equations are simple enough to be integrated on a small computer such as a ReComp II or LGP-30 or, if necessary, by hand.

Knowledge of the torque-producing effects is also of vital importance in active attitude-controlled satellites since such information can be used to determine the optimum saturation level for inertial devices and propellant storage requirements for mass expulsion devices.

References

[1] LUNDQUIST, C. A., and R. J. NAUMANN: Orbital and Rotational Motion of a Rigid Satellite, Seminar Proceedings; Tracking and Orbit Determination, Jet Propulsion Laboratory, Cal. Inst. of Techn., Pasadena, California (1960).

[2] WHITAKER, E. T.: A Treatise on the Analytical Dynamics of Particles and Rigid Bodies, Fourth Edition, Dover Publishing Company (1944).

[3] GOLDSTEIN, H.: Classical Mechanics, Cambridge, Mass.: Addison-Wesley Publishing Company, Inc. (1950).

[4] ROBERSON, R. E., and D. TATISTCHEFF: The Potential Energy of a Small Rigid Body in the Gravitational Field of an Oblate Spheroid, J. Franklin Inst., 262, 209—214 (1956).

[5] DOOLIN, B. F.: Gravity Torque on an Orbiting Vehicle, NASA TN D-10, Ames Research Center, Moffet Field, California (1960).

[6] COLOMBO, G.: On the Motion of Explorer XI Around its Center of Mass, American Astronautical Society, Goddard Memorial Symposium, Washington, D. C. March 16—17 (1962).

[7] FINCH, H. F., and B. R. LEATON: The Earth's Main Magnetic Field, Epoch 1955. 0., Mon. Notices of the Royal Astronomical Society, 7, 313—317 (1957).

[8] NAUMANN, R. J.: An Investigation of the Observed Torques Acting on Explorer XI, American Astronautical Society, Goddard Memorial Symposium, Washington, D. C., Mar. 16—17 (1962).

[9] NAUMANN, R. J.: Recent Information Gained from Satellite Orientation Measurement, Plan. and Space Sci., 7, 445—453 (1961).

[10] LUNDQUIST, C. A., R. J. NAUMANN, and S. A. FIELDS: Recovery of Further Data from 1958 Epsilon, Space Research, II (ed. by VAN DE HULST, JAGER, and MOORE), Amsterdam: N. Holland Pub. Co. (1961).

[11] LUNDQUIST, C. A., R. J. NAUMANN, and A. H. WEBER: Directional Flux Densities and Mirror-Point Distributions of Trapped Particles from Satellite 1958 Epsilon, Measurements, American Geophysical Union, First Western National Meeting, UCLA, Dec. 27—29, 1961. Also submitted for publication in J. Geophys. Research.

[12] COLOMBO, G.: The Motion of Satellite 1958 Epsilon Around Its Center of Mass, Research in Space Science, Special Report No. 70, Smithsonian Institution Astrophysical Observatory, Cambridge, Mass. July 18, 1961.

[13] FIELDS, S. A.: Body Motions of 1958 Epsilon, Army Ballistic Missile Agency Tech. Memo CL-TM-11-60 (1960).

[14] KUEBLER, M. E.: Spin-Tumble Transfer of Satellite S-15 (Explorer XI), Marshall Space Flight Center MTP-G & C-61-30 (1961).

[15] BOURDEAU, R. E., J. L. DONLEY, and E. C. WHIPPLE: The Ionosphere Direct Measurements Satellite Instrumentation (Explorer VIII), National IAS-ARS Joint Meeting, Los Angeles, Calif. June 13—16, 1961.

[16] HELLER, G. B., B. P. JONES, R. J. NAUMANN, and W. C. SNODDY: Correlation of the Thermal Behavior of Satellites and the Rotational Momentum Vector, From Peenemunde to Outer Space (edited by E. STUHLINGER, F. ORDWAY, III, J. MCCALL, and G. BUCHER) Marshall Space Flight Center, Huntsville, Alabama, 203—240, March 23, 1962. To be published by McGraw-Hill Book Co.

[17] BANDEEN, W. R., and W. P. MANGER: Angular Motion of the Spin Axis of Tiros I Meteorological Satellite Due to Magnetic and Gravitational Torque, J. Geophys. Res. 65, 2992—6 (1960).

[18] HETCH, E., and W. P. MANGER: Magnetic Attitude Control of the Tiros Satellites, American Astronautical Society, Goddard Memorial Symposium, Washington, D. C., March 16—17, 1962.

[19] FISCHELL, R. E.: Passive Magnetic Attitude Control for Earth Satellites, American Astronomical Society, Goddard Memorial Symposium, Washington, D. C., March 16—17, 1962.

The prediction of satellite orbits

By

Irwin I. Shapiro

Lincoln Laboratory[1], Massachusetts Institute of Technology
Lexington, Mass., U.S.A.

Abstract. In this paper we develop a perturbation method and apply it to the prediction of orbits of earth satellites that are disturbed by various external forces. The theory of resonances and stable orbit configurations, possible when sunlight pressure is important, are also analyzed. Finally, we describe experiments to determine the magnitude of perturbing forces such as charge drag.

I. Introduction

Two aspects of current work in satellite orbit theory which most differentiate it from that of the classical period in celestial mechanics are (1) the availability of high-speed digital computers, and (2) the necessity of considering orbital behavior over many thousands of revolutions. The orbit prediction method that we have developed during the past $3^1/_2$ years is based on both these factors, and is described in some detail in the following sections. After indicating the first-order effects on the orbit of several perturbing forces, we examine resonances and stable orbit configurations which can occur when these forces act in concert. Particular emphasis is given in these applications to satellites with large area-to-mass ratios, like the Echo balloon and the West Ford dipoles. (As was originally shown several years ago [1], orbits of these satellites are highly perturbed by sunlight pressure.) We also discuss experiments for improving our knowledge of various disturbing forces such as charge drag.

We assume throughout the validity of Newtonian mechanics, as the corrections required by the theory of relativity are, in general, far below present observational accuracy.

II. Method of orbit computation

All the disturbing accelerations considered in this paper are small compared to that of the central force field. We shall, therefore, describe all orbits in terms of osculating ellipse parameters. The equations

---

[1] Operated with support from the U.S. Air Force.

relating the time variation of these parameters to the perturbing accelerations are nonlinear first-order differential equations, and are derived in all standard texts on celestial mechanics [2]. However, we shall use the argument of latitude, u, as the independent variable, instead of time. In terms of u, the Gaussian form for the "variation of constants" is[1]

$$
\left.
\begin{aligned}
\frac{da}{du} &= \frac{2a^2 r^2 \gamma}{\mu p}\{e \sin\theta \cdot R + (1 + e\cos\theta) \cdot S\}, \\[2mm]
\frac{de}{du} &= \frac{r^2 \gamma}{\mu}\left\{\sin\theta \cdot R + \frac{r}{p}[2\cos\theta + e(1 + \cos^2\theta)] \cdot S\right\}, \\[2mm]
\frac{d\omega}{du} &= \frac{r^2 \gamma}{\mu e}\left\{-\cos\theta \cdot R + \left[1 + \frac{r}{p}\right]\sin\theta \cdot S\right\} - \cos i \cdot \frac{d\Omega}{du}, \\[2mm]
\frac{d\Omega}{du} &= \frac{r^3 \gamma}{\mu p}\frac{\sin u}{\sin i} \cdot W, \\[2mm]
\frac{di}{du} &= \frac{r^3 \gamma}{\mu p}\cos u \cdot W, \\[2mm]
\frac{dt}{du} &= \frac{r^2 \gamma}{(u p)^{1/2}},
\end{aligned}
\right\} \tag{1}
$$

where $\theta = u - \omega$ is the true anomaly, where

$$
\gamma \equiv \left\{1 - \frac{r^3}{\mu p}\cot i \cdot \sin u \cdot W\right\}^{-1}, \tag{2}
$$

and where R, S, and W represent the sums of the components of the perturbing accelerations in the $\hat{r}$, $\hat{s}$, and $\hat{w}$ directions, respectively. The unit vector $\hat{r}$ is directed radially from the center of the earth towards the instantaneous position of the satellite; $\hat{s}$ is perpendicular to $\hat{r}$, lies in the orbit plane, and makes an acute angle with the satellite's velocity vector; and $\hat{w} = \hat{r} \times \hat{s}$. The symbols for the osculating elements have their usual definitions (e.g., p denotes the *semilatus rectum*); while $\mu \approx 3.98 \times 10^{20}$ cm^3/sec^2 is the product of the gravitational constant and the mass of the earth, and r is the distance from the center of the earth to the satellite. It is sometimes convenient to use p as an independent element instead of a; the first of Eqs. (1) should then be replaced by

$$
\frac{dp}{du} = \frac{2r^3 \gamma}{\mu} S. \tag{3}
$$

Eqs. (1), together with the appropriate definitions, form a closed set: When the perturbing accelerations R, S, and W are given as functions of the elements and the independent variable, u, then the values of the elements for all values of u (and of t) will be uniquely determined by their initial values.

---

[1] If necessary when considering nearly circular and/or nearly equatorial orbits, we can modify Eqs. (1) in the standard manner; see, for example, ref. [2].

. When one perturbing force acts on the satellite, Eqs. (1) can be written symbolically in the form[1]

$$\frac{da^k}{du} = \lambda f^k(a^j; u); \qquad j, k = 1 \to 5, \tag{4}$$

where $a^1 = a$, $a^2 = e$, etc., and where $\lambda \ll 1$ is (unrigorously) defined as the ratio of the strength of the disturbing acceleration to that of the central force field.[2] The functions f^k depend, in general, on all the parameters a^j and on the independent variable, u.

Before describing our method of solving these equations, we examine briefly several other approaches. A simple first-order solution to Eqs. (4) is given by

$$a^k(u) = \lambda \int\limits^u f^k(a_0^j; u)\, du, \tag{5}$$

where the a^j on the right side of (5) retain their initial values, signified by the zero subscripts. The errors implicit in such an approximation usually build up rapidly as u increases.

Since the dependence of the f^k on u is almost always trigonometric, and since one is usually interested in long-term orbital behavior, Eqs. (4) are sometimes replaced by

$$\frac{d\bar{a}^k}{du} = \frac{\lambda}{2\pi} \int\limits_0^{2\pi} f^k(\bar{a}^j; u)\, du \equiv \lambda g^k(\bar{a}^j); \qquad j, k = 1 \to 5, \tag{6}$$

where the a^j on the right sides are held constant during integration. These equations are then solved by various techniques to determine the time behavior of the "mean" orbital elements, i.e., the elements obtained by "averaging out" the short-period perturbations. This averaging procedure is often referred to as the KRYLOFF-BOGOLIUBOFF method. Our approach is similar; thus to first order in λ, we write

$$a^k(u = 2\pi) - a^k(u = 0) \equiv a_1^k - a_0^k \equiv \Delta a_1^k = \lambda \int\limits_0^{2\pi} f^k(a_0^j; u)\, du, \tag{7}$$

and[3]

$$a_n^k = a^k(u = 2n\pi) = a_0^k + \sum_{i=1}^{n-1} \Delta a_i^k, \tag{8}$$

where

$$\Delta a_i^k = \lambda \int\limits_0^{2\pi} f^k(a_{i-1}^j; u)\, du. \tag{9}$$

---

[1] We assume $\gamma = 1$ while considering first-order methods, hence the right sides of Eqs. (4) are shown as linear in λ. [Compare Eqs. (12) where the nonlinearity introduced by γ is explicitly indicated.]. The last of Eqs. (1), ignored here as its right side is of lower order in λ, is similarly solvable.

[2] In Section III, an appropriate λ is defined for each perturbation considered.

[3] To obtain $a^k(u)$ when $2n\pi < u < 2(n+1)\pi$, Eq. (19) can be employed.

In general, the error increases much more slowly here than in Eqs. (5). By using TAYLOR's theorem with the remainder (provided that the appropriate functions are suitably differentiable in the regions of interest), one can obtain upper bounds on the accumulated errors for any value of u.

We can easily generalize our formulas to include simultaneously any number of (small) perturbations. We find

$$\Delta a_i^k = \sum_{q=1}^{n} \lambda_q \int_0^{2\pi} f_q^k(a_{i-1}^j; u)\, du, \tag{10}$$

where λ_q is the appropriate constant for the q th perturbing acceleration. Usually, one can integrate the functions $f_q^k(u)$ analytically, and the digital computer need be programmed only to perform the iterations necessary to determine the a_n^k.

This method can also be extended to higher orders in λ. Again, assuming that all functions have the necessary differentiability properties, we can write[1]

$$a^k(u) = a^{k(0)}(u) + \lambda\, a^{k(1)}(u) + \cdots. \tag{11}$$

With Eqs. (4) now of the form

$$\frac{da^k}{du} = \lambda\, f^k(a^j; \lambda; u); \quad j, k = 1 \to 5, \tag{12}$$

substituting the expansion for $a^k(u)$ and collecting terms proportional to the same power of λ, leads to the hierarchy of equations

$$\left. \begin{aligned} \frac{da^{k(0)}}{du} &= 0, \\ \frac{da^{k(1)}}{du} &= f^k(a^{j(0)}; 0; u). \\ &\vdots \end{aligned} \right\} \tag{13}$$

[These (in principle, at least) could be solved successively to obtain $a^k(u)$ accurate to an arbitrary order in λ.] To extend our method to the next order in λ, we solve the first two:

$$a^{k(0)}(u) = a^k(u = 0) \equiv a_0^k, \tag{14}$$

$$a^{k(1)}(u) = \int_0^u f^k(a_0^j; 0; u)\, du, \tag{15}$$

and replace Eq. (7) by

$$\Delta a_1^k = \lambda \int_0^{2\pi} f^k(a_0^j + \lambda\, a^{j(1)}(u); \lambda; u)\, du. \tag{16}$$

---

[1] In this development we do not approximate γ by unity and, for simplicity of notation, only one disturbing acceleration is considered.

After n periods we have, as in Eq. (8),

$$a_n^k \equiv a^k(u = 2n\pi) = a_0^k + \sum_{i=1}^{n-1} \Delta a_i^k, \qquad (17)$$

but now

$$\Delta a_i^k = \lambda \int_0^{2\pi} f^k\big(a_{i-1}^j + \lambda\, a_{i-1}^{j(1)}(u);\ \lambda;\ u\big)\,du, \qquad (18)$$

where

$$a_{i-1}^{j(1)}(u) = \int_0^u f^j(a_{i-1}^k;\ 0;\ u)\,du. \qquad (19)$$

Since this is only a second-order theory, the f^k in Eqs. (16) and (18) can be replaced by

$$f^k \approx f^k(a_{i-1}^j;\ 0;\ u) + \lambda \left\{ \sum_{j=1}^{6} \frac{\partial f^k}{\partial a^j} \frac{\partial a^j}{\partial \lambda} + \frac{\partial f^k}{\partial \lambda} \right\}, \qquad (20)$$

where the partial derivatives are evaluated for $\lambda = 0$. To include several perturbations simultaneously, we merely use multi-variable TAYLOR expansions. If the remainder is also included, upper bounds on the accumulated errors can be obtained. (Developing this method beyond second order is impractical at present.)

In the remainder of the paper, we confine ourselves to applications of the first-order iterative method described by Eqs. (8) to (10).

III. First-order results for various perturbing forces

In applying the first-order theory just discussed, we must first determine analytically the mean changes in the orbital elements per revolution. We summarize here the results for perturbations caused by[1]:

1. The higher harmonics of the earth's gravitational field;
2. Direct solar radiation pressure;
3. Solar radiation reflected from the earth;
4. The lunar and solar gravitational fields;
5. Atmospheric drag (both neutral and charge); and
6. The POYNTING-ROBERTSON drag.

Of these perturbations, all but the gravitational ones depend importantly on the shape and orientation of the satellite. Since the effects of sunlight pressure are relatively less well known, our treatment of these will be more detailed. Other forces (such as the LORENTZ force [3]) are usually quite small (but not necessarily smaller than some listed) and will be neglected in this discussion.

---

[1] Since our concern is mainly with the change in shape and orientation of the orbit, rather than with the location of the satellite along it, we omit the variations in (nodal) period implied by the last of Eqs. (1).

With these results it should be possible to predict accurately the long term orbital behavior of a spherical (or tumbling cylindrical) earth satellite, given the initial values of the orbital elements. We have therefore programmed an IBM-7090 computer to perform the iterations implied by Eqs. (8) and (10), using the analytical results obtained for the effects of the perturbations (1) through (5) listed above [4]. Unfortunately for one's theoretical calculations, however, air drag is usually an important perturbation: Since the density of the upper atmosphere (or exosphere) changes radically and, as yet, unpredictably with time, it is essentially impossible to make very precise long-term orbital predictions. Only for satellites whose perigees remain above about 1500 km and whose area-to-mass ratios are not too large can we obtain high accuracy.[1]

1. Gravitational field of the earth

When we neglect the axial asymmetries in the mass distribution, the gravitational potential energy of the earth can be expanded in spherical harmonics as follows:

$$\Phi(r, \delta) = -\frac{\mu}{r}\left\{1 - \sum_{n=2}^{\infty} J_n \frac{P_n(\cos\delta)}{r^n}\right\}, \tag{21}$$

where δ is the colatitude, P_n is the nth order LEGENDRE polynomial, and the J_n are constants, determined empirically. The values of the first few, divided by 10^{-6}, are [5]:

$$\frac{J_2}{R_0} = 1082.7 \pm .3; \; \frac{J_3}{R_0^3} = -2.3 \pm .2; \; \frac{J_4}{R_0^4} = -1.7 \pm .3; \; \frac{J_5}{R_0^5} = -0.2 \pm .2, \tag{22}$$

where $R_0 \approx 6,378.17$ km is the mean equatorial radius of the earth.

Given Φ, it is a simple matter to calculate the R, S, and W components of the perturbing acceleration for each of the higher harmonics:

$$R = -\frac{\partial\Phi}{\partial r}; \quad S = -\frac{\partial\Phi}{r\,\partial u}; \quad W = -\frac{\partial\Phi}{r\sin u\,\partial i}, \tag{23}$$

where $\cos\delta = \sin i \cdot \sin u$. From Eq. (7), we find that the nonvanishing mean changes in the orbital elements caused by these perturbations after one revolution, are[2]:

$$\left.\begin{array}{l} \Delta\omega_{2h} = \dfrac{3\pi}{2}\dfrac{J_2}{p^2}(5\cos^2 i - 1), \\[2ex] \Delta\Omega_{2h} = -3\pi\dfrac{J_2}{p^2}\cos i, \end{array}\right\} \tag{24}$$

---

[1] Charge drag must also be small; for example, see Section IV.
[2] Note that, in general, $\Delta a_{nh} = 0$ and $\Delta p_{nh} = 2p \tan i \cdot \Delta i_{nh}$.

$$\Delta e_{3h} = -\frac{3\pi}{4}\frac{J_3}{p^3}(1 - e^2)\cos\omega \sin i(5\cos^2 i - 1),$$

$$\Delta \omega_{3h} = \frac{3\pi}{4}\frac{J_3}{p^3}\frac{(1 + 4e^2)}{e}\sin\omega \sin i(5\cos^2 i - 1) - \cos i \cdot \Delta\Omega_{3h},$$

$$\Delta\Omega_{3h} = \frac{3\pi}{4}\frac{J_3}{p^3}e\sin\omega \cot i(15\cos^2 i - 11),$$

$$\Delta i_{3h} = \frac{3\pi}{4}\frac{J_3}{p^3}e\cos\omega \cos i(5\cos^2 i - 1),$$

$$(25)$$

$$\Delta e_{4h} = -\frac{15\pi}{16}\frac{J_4}{p^4}e(1 - e^2)\sin 2\omega \sin^2 i[7\cos^2 i - 1],$$

$$\Delta\omega_{4h} = -\frac{15\pi}{8}\frac{J_4}{p^4}\Big\{8 - 28\sin^2 i + 21\sin^4 i - $$

$$- \sin^2\omega \sin^2 i(7\cos^2 i - 1) + e^2\Big[6 - 14\sin^2 i + \frac{63}{8}\sin^4 i + $$

$$+ \sin^2\omega\Big(6 - 35\sin^2 i + \frac{63}{2}\sin^4 i\Big)\Big]\Big\},$$

$$(26)$$

$$\Delta\Omega_{4h} = \frac{15\pi}{16}\frac{J_4}{p^4}\cos i\{2(7\cos^2 i - 3) + $$

$$+ e^2[7\cos^2 i - 1 + 4\sin^2\omega(7\cos^2 i - 4)]\},$$

$$\Delta i_{4h} = \frac{15\pi}{32}\frac{J_4}{p^4}e^2 \sin 2\omega \sin 2i[7\cos^2 i - 1],$$

and[1]

$$\Delta e_{5h} = \frac{15\pi}{16}\frac{J_5}{p^5}\cos\omega \sin i\{8 - 28\sin^2 i + 21\sin^4 i + O(e^2)\},$$

$$\Delta\omega_{5h} = -\frac{15\pi}{16}\frac{J_5}{p^5}\frac{\sin\omega \sin i}{e}\{8 - 28\sin^2 i + 21\sin^4 i + O(e^2)\} - $$

$$- \cos i \cdot \Delta\Omega_{5h},$$

$$(27)$$

$$\Delta\Omega_{5h} = -\frac{15\pi}{16}\frac{J_5}{p^5}e\sin\omega \cot i\{8 - 84\sin^2 i + 105\sin^4 i + O(e^2)\},$$

$$\Delta i_{5h} = -\frac{15\pi}{16}\frac{J_5}{p^5}e\cos\omega \cos i\{8 - 28\sin^2 i + 21\sin^4 i + O(e^2)\}.$$

As is well known, the second harmonic causes a secular regression or progression of the ascending node, depending on the quadrant of the inclination angle, and a progression or retrogression of the argument of perigee, depending on whether the inclination angle satisfies the condition $63.5° \leq i \leq 116.5°$.

The variations caused by the third harmonic are all of long period. Most notable are the changes in e and in ω (when the orbit eccentricity is small). The characteristics of the perturbations caused by the fifth harmonic are similar. The fourth harmonic contributions to Δe and Δi

---

[1] The (usually small) higher order terms in e are quite lengthy, and therefore have not been included explicitly.

are of long period (half that of the argument of perigee), while the contributions to $\Delta\omega$ and $\Delta\Omega$ have secular parts.

In view of the relative sizes of the coefficients given in Eq. (22), it is clear that (except for $\Delta\omega_{3h}$ when e is small) terms of order J_2^2 will be as important as terms linear in J_3, J_4, and J_5; and should be included in a theory purporting to be accurate to first order in the higher harmonics. These terms are, in fact, being calculated and will be incorporated into our computer program [6].

2. Direct solar radiation pressure

The force exerted by sunlight pressure is proportional to the solar flux and to the area of the satellite as projected on a plane perpendicular to the direction of the flux. The magnitude of the acceleration is

$$Q_{\mathrm{sp}} = \left(\frac{K\,A}{M}\right)\left(\frac{I}{c}\right), \qquad (28)$$

where I is the solar flux in the vicinity of the earth, c is the speed of light, (A/M) is the appropriate cross-sectional area-to-mass ratio of the satellite, and K is a constant $(0 \leq K \leq 2)$ whose value depends on the reflection characteristics of the satellite.

The direction of the acceleration is not, in general, parallel to the sun's rays. Rather, this direction depends on the shape of the object, its scattering law, and its orientation with respect to the earth-sun line.[1] Thus K is usually a time-dependent tensor.

Since the earth's orbit is elliptical, the solar flux will vary during the year; we assume

$$I = I_0\left(\frac{a_E}{r_E}\right)^2, \qquad (29)$$

where a_E is the semimajor axis of the earth's orbit, and r_E is the earth-sun distance. (Since this distance is approximately 10^4 times the earth-satellite distance, we ignore the short-period variation in satellite-sun distance.) The solar constant, I_0, is approximately $2.00 \pm 2\%$ cal/cm² min and remains quite constant (within $\pm .1\%$), except for short periods of time when the fluctuations are as large as a few percent [7].[2]

I. Spherical satellites. For spherical satellites the direction of acceleration is parallel to the earth-sun line and its magnitude is constant (except when the satellite passes through the earth's shadow). Our results will therefore be valid for any satellite on which the force is constant and parallel to the impinging solar rays.

---

[1] For a specularly reflecting flat plate, the acceleration can be (almost) perpendicular to the direction of the sun's rays.

[2] Note that $I_0/c \approx 4.65 \times 10^{-5}$ dynes/cm².

If we let Ψ denote the angle between the ecliptic and the earth's equatorial plane ($\Psi \approx 23.5°$) and let β denote the angular position of the earth-sun line (measured from the vernal equinox), then the R, S, and W components of the disturbing acceleration caused by sunlight pressure are[1]

$$\begin{aligned}
R_{sp} &= -Q_{sp}\{\tilde{R}\cos(u-\omega) + \tilde{S}\sin(u-\omega)\}, \\
S_{sp} &= -Q_{sp}\{\tilde{S}\cos(u-\omega) - \tilde{R}\sin(u-\omega)\}, \\
W_{sp} &= -Q_{sp}\tilde{W},
\end{aligned} \tag{30}$$

where $\tilde{R}$, $\tilde{S}$, and $\tilde{W}$ are, respectively, the components of $\hat{x}$ (a unit vector in the direction of the sun) along the $\hat{r}$, $\hat{s}$, and $\hat{w}$ directions appropriate for perigee. Explicitly, we have

$$\begin{aligned}
\tilde{R} &= (\cos\omega\cos\Omega - \sin\omega\sin\Omega\cos i)\cos\beta + \\
&\quad + (\cos\omega\sin\Omega + \sin\omega\cos\Omega\cos i)\sin\beta\cos\Psi + \\
&\quad + \sin\omega\sin i\sin\beta\sin\Psi, \\
\tilde{S} &= -(\sin\omega\cos\Omega + \cos\omega\sin\Omega\cos i)\cos\beta + \\
&\quad + (-\sin\omega\sin\Omega + \cos\omega\cos\Omega\cos i)\sin\beta\cos\Psi + \\
&\quad + \cos\omega\sin i\sin\beta\sin\Psi, \\
\tilde{W} &= \sin\Omega\sin i\cos\beta - \cos\Omega\sin i\sin\beta\cos\Psi + \\
&\quad \cos i\sin\beta\sin\Psi.
\end{aligned} \tag{31}$$

Since β, ω, and Ω change with time, $\tilde{R}$, $\tilde{S}$, and $\tilde{W}$ exhibit certain periodicities. To show these clearly, we rewrite Eqs. (31) in the form [8]

$$\begin{aligned}
\tilde{R} &= \cos^2\frac{i}{2}\cos^2\frac{\Psi}{2}\cos(\omega+\Omega-\beta) + \sin^2\frac{i}{2}\cos^2\frac{\Psi}{2}\cos(\omega-\Omega+\beta) + \\
&\quad + \cos^2\frac{i}{2}\sin^2\frac{\Psi}{2}\cos(\omega+\Omega+\beta) + \sin^2\frac{i}{2}\sin^2\frac{\Psi}{2}\cos(\omega-\Omega-\beta) + \\
&\quad + \frac{1}{2}\sin i\sin\Psi[\cos(\omega-\beta) - \cos(\omega+\beta)], \\
\tilde{S} &= -\cos^2\frac{i}{2}\cos^2\frac{\Psi}{2}\sin(\omega+\Omega-\beta) - \sin^2\frac{i}{2}\cos^2\frac{\Psi}{2}\sin(\omega-\Omega+\beta) - \\
&\quad - \cos^2\frac{i}{2}\sin^2\frac{\Psi}{2}\sin(\omega+\Omega+\beta) - \sin^2\frac{i}{2}\sin^2\frac{\Psi}{2}\sin(\omega-\Omega-\beta) + \\
&\quad + \frac{1}{2}\sin i\sin\Psi\{\sin(\omega+\beta) - \sin(\omega-\beta)\}, \\
\tilde{W} &= \sin i\left\{\cos^2\frac{\Psi}{2}\sin(\Omega-\beta) + \sin^2\frac{\Psi}{2}\sin(\Omega+\beta)\right\} + \\
&\quad + \cos i\sin\Psi\sin\beta.
\end{aligned} \tag{32}$$

---

[1] Note that the direction of the earth-sun line is opposite to that of the solar rays.

Before giving the changes in the orbital elements caused by sunlight, we consider the shadow cast by the earth. We assume the shadow to be one-half of a right circular cylinder whose axis coincides with the continuation of the earth-sun line and whose radius is R_0.[1] The equation for the points on the orbit which lie on the shadow boundary is a quartic in either $\cos u$ or $\sin u$; the general solution can therefore be obtained in closed form in terms of algebraic operations.[2] This solution is quite unwieldy and other techniques are usually used to determine the shadow boundaries. In one method (used in our computer program), we evaluate the function

$$R_0 - r(u) \left\{ 1 - (R_{\mathrm{sp}}/Q_{\mathrm{sp}})^2 \right\}^{1/2}$$

systematically for a set of closely spaced values of u to determine the intervals during which it changes sign.[3] These intervals can be subdivided, and the procedure repeated, to determine the boundaries more precisely. (Of course, no matter how fine the net, the shadow region can still lie wholly between two successive trial values of u. The important point is to reduce the spacing sufficiently so that were the shadow region undetected, the error committed would be negligible.) To avoid a needless search, simple preliminary tests can be made to determine whether a shadow region exists. For example, if $a(1 - e) > a_s$, where a_s is the semimajor axis of the shadow ellipse, then the satellite is always in sunshine. On the other hand, if $a(1 + e) < a_s$, the satellite must pass through shadow. Once a shadow boundary has been determined, the new boundary applicable after the next orbital period can usually be found by successive approximations using the initial one as a first estimate.

There are two important special cases for which the quartic reduces to a biquadratic[4]:

1. The earth-sun line lies in the orbit plane;
2. The orbit is circular.

In 1., the shadow boundaries in the orbit plane are straight lines and the intersection of these with the orbit ellipse yields two second-

---

[1] The angular errors introduced in the shadow boundaries (because of our neglecting refraction by the atmosphere, diffraction by the earth, the oblateness of the earth, and the size and distance of the sun) are usually less than one degree.

[2] The equation is quartic because the shadow boundaries are determined from the intersection of two second-degree curves — the orbit ellipse and the half-ellipse formed by the intersection of the orbit plane with the shadow half-cylinder. (See Figure 1.) It can also be proven that these curves intersect at most twice.

[3] This function is just the difference between the radius of the earth and the projection of $\vec{r}$ on a plane perpendicular to the earth-sun line. If $\hat{r} \cdot \hat{x}$ is negative and the function positive, then the point in question lies in shadow.

[4] The quartic also simplifies when the major axes of the orbit and shadow ellipses coincide.

degree equations; in 2., the symmetry of the circle leads to the simplification. Both biquadratics are useful in determining the shadow boundaries for the general case. For example, the solution to 1. can be used to delimit the u-region to be explored.

In exhibiting the element changes, we use

$$\lambda_{\text{sp}} \equiv \frac{K A}{M}\left(\frac{I}{c}\right)\left(\frac{a^2}{\mu}\right), \tag{33}$$

which is the ratio of the magnitude of the acceleration caused by sunlight to that caused by the central force field at a distance a from the earth's center. We find[1]

$$\begin{aligned}
\Delta a_{\text{sp}} &= 2a\,\lambda_{\text{sp}}\left\{\tilde{R}\left(Q_1 + \frac{e}{1-e^2}\,Q_3\right) - \tilde{S}\left(Q_2 + \frac{e}{1-e^2}\,Q_4\right)\right\}, \\[4pt]
\Delta e_{\text{sp}} &= \lambda_{\text{sp}}\{\tilde{R}\cdot Q_3 - \tilde{S}\cdot Q_4\}, \\[4pt]
\Delta \omega_{\text{sp}} &= \frac{\lambda_{\text{sp}}}{e}\{\tilde{R}\cdot Q_5 - \tilde{S}\cdot Q_6\} - \cos i \cdot \Delta\Omega_{\text{sp}}, \\[4pt]
\Delta \Omega_{\text{sp}} &= \frac{\lambda_{\text{sp}}}{\sin i}\,\tilde{W}\{Q_1\cos\omega + Q_2\sin\omega\}, \\[4pt]
\Delta i_{\text{sp}} &= -\lambda_{\text{sp}}\,\tilde{W}\{Q_2\cos\omega - Q_1\sin\omega\},
\end{aligned} \tag{34}$$

where

$$\begin{aligned}
Q_1(u) &= -\cos E - \frac{e}{2}\sin^2 E, \\[4pt]
Q_2(u) &= -\frac{1}{2(1-e^2)^{1/2}}\{3e\,E - \sin E[2(1+e^2) - e\cos E]\}, \\[4pt]
Q_3(u) &= \frac{(1-e^2)}{2}\sin^2 E, \\[4pt]
Q_4(u) &= \frac{(1-e^2)^{1/2}}{2}\{3E - \sin E[4e - \cos E]\}, \\[4pt]
Q_5(u) &= \frac{(1-e^2)^{1/2}}{2}\{3E - \sin E[2e + \cos E]\}, \\[4pt]
Q_6(u) &= \frac{1}{2}\{\sin^2 E + 2e\cos E\}
\end{aligned} \tag{35}$$

and where the eccentric anomaly E is given in terms of u by[2]

$$\tan\left(\frac{E}{2}\right) = \left\{\frac{1+e}{1-e}\right\}^{1/2}\tan\left(\frac{u-\omega}{2}\right). \tag{36}$$

The Q_i integrals, of course, are not all independent; thus

$$Q_3 = e\,Q_1 + Q_6. \tag{37}$$

---

[1] We also note: $\Delta p_{\text{sp}} = 2p\,\lambda_{\text{sp}}\{\tilde{R}\cdot Q_1 - \tilde{S}\cdot Q_2\}$.

[2] The Q_i can also be expressed directly in terms of u [9]; however the resulting expressions are somewhat more complicated.

Limits of integration were not shown since they depend on the situation: If the satellite remains in sunlight during its entire orbital period, then the integration is from 0 to 2π, and we substitute $Q_i(2\pi) - Q_i(0)$ for Q_i in Eqs. (34). If the satellite enters the shadow at u_1, and leaves at u_2, then the integration will be either from 0 to u_1 and u_2 to 2π, or from u_2 to u_1, depending on whether the point $u = 0$ is in sunlight or shadow.[1] Hence we replace Q_i by

$$Q_i(u_1) - Q_i(u_2) + \sigma(u_2 - u_1)\,[Q_i(2\pi) - Q_i(0)], \qquad (38)$$

where

$$\sigma(x) = \begin{cases} 0, & x < 0 \\ 1, & x \geq 0. \end{cases} \qquad (39)$$

(When $u_2 - u_1 \geq 0$, $u = 0$ lies in sunlight; when $u_2 - u_1 < 0$, $u = 0$ lies in shadow.)

If the satellite remains in sunlight, only Q_2, Q_4, and Q_5 yield non-vanishing contributions:

$$\left.\begin{aligned} Q_2 &= -\frac{3\pi e}{(1 - e^2)^{1/2}}\,, \\ Q_4 &= 3\pi(1 - e^2)^{1/2} = Q_5, \end{aligned}\right\} \qquad (40)$$

and we find

$$\left.\begin{aligned} \Delta a_{\mathrm{sp}} &= 0, \\ \Delta e_{\mathrm{sp}} &= -3\pi\,\lambda_{\mathrm{sp}}(1 - e^2)^{1/2}\,\tilde{S}, \\ \Delta\omega_{\mathrm{sp}} &= \frac{3\pi\,\lambda_{\mathrm{sp}}}{e}(1 - e^2)^{1/2}\,\tilde{R} - \cos i \cdot \Delta\Omega_{\mathrm{sp}}, \\ \Delta\Omega_{\mathrm{sp}} &= \frac{3\pi\,\lambda_{\mathrm{sp}}\,e}{(1 - e^2)^{1/2}}\,\frac{\sin\omega}{\sin i}\,\tilde{W}, \\ \Delta i_{\mathrm{sp}} &= \frac{3\pi\,\lambda_{\mathrm{sp}}\,e}{(1 - e^2)^{1/2}}\cos\omega \cdot \tilde{W}. \end{aligned}\right\} \qquad (41)$$

Under these conditions, we expect no change in the orbital energy (i.e., in a) after a complete revolution since the radiation field (minus the shadow region) is conservative.[2]

The orbital energy after a complete revolution will be altered if the satellite enters and leaves the shadow at different distances from the sun (or, equivalently, from the center of the earth).[3] If the satellite leaves the shadow closer to the sun than when it entered, the satellite gains energy: It is not retarded during that distance. And, similarly,

---

[1] We assume $0 \leq u_1, u_2 \leq 2\pi$.

[2] We assume that both the sun's position and the orbital elements remain constant during an orbital period.

[3] This can occur only if the orbit is noncircular and the shadow region is asymmetric with respect to the orbit major axis.

the satellite loses energy when the relation between the distances is reversed.

The potential of a spherical satellite in this field is proportional to its separation from the sun:

$$\Phi_{\mathrm{sp}}(x) = \frac{KA}{M}\left(\frac{I}{c}\right)x, \tag{42}$$

where x is the satellite-sun distance. We find, therefore, that

$$
\begin{aligned}
\varDelta\varepsilon_{\mathrm{sp}} &= \frac{KA}{M}\left(\frac{I}{c}\right)[1 - \tilde{W}^2]^{1/2}\{r_1\cos(\alpha - \theta_1) - r_2\cos(\alpha - \theta_2)\} \\
&= \frac{\mu}{2a^2}\varDelta a_{\mathrm{sp}}, \tag{43}
\end{aligned}
$$

where $\varDelta\varepsilon_{\mathrm{sp}}$ is the change in energy caused by sunlight per orbital period, and where α is the angle between perigee and the extension of the projec-

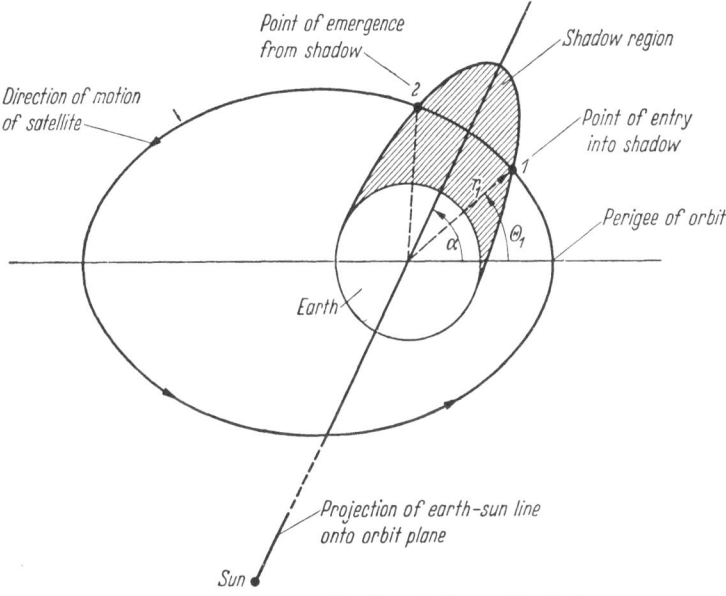

Fig. 1. Passage of a satellite through the earth's shadow.

tion of the earth-sun line onto the orbit plane (see Figure 1). The subscripts 1 and 2 indicate the points of entry into and emergence from the shadow region, respectively. We see that for $0 < \alpha < \pi$, point 2 will always be further from the sun than point 1. Similarly, for $0 < \alpha < \pi$, sunlight pressure will tend to decrease perigee height. It follows generally that whenever radiation pressure causes perigee height to decrease, it will also cause the orbital energy to decrease (if part of the orbit is in shadow), and vice-versa for perigee height increasing because of sunlight pressure.

For any (elliptical) orbit, and for any orientation of the earth-sun line, the change in the orbital energy of the satellite is always greater when the earth-sun line actually lies along its projection onto the orbit plane. We therefore restrict our analytical discussion to rectilinear shadow regions. In terms of the orbital elements and α, we find[1]

$$\Delta\varepsilon_{\mathrm{sp}} = -Q_{\mathrm{sp}}\, p \times$$

$$\times \left\{ \frac{-\nu\, e^2 \sin 2\alpha + [1 - \nu^2(1-e^2) + 2\nu\, e \sin\alpha]^{1/2} - [1 - \nu^2(1-e^2) - 2\nu\, e \sin\alpha]^{1/2}}{1 - e^2 \cos^2\alpha} \right\}$$

$$= -Q_{\mathrm{sp}}\, R_0\, e \sin\alpha \times$$

$$\times \left\{ \frac{2}{1 + e \cos\alpha} + \nu^2 + \frac{\nu^4}{4}[3(1 - e^2 \cos^2\alpha) + 4 e^2 \sin^2\alpha] + \cdots \right\}, \qquad (44)$$

where

$$\nu = \frac{R_0}{p} < 1. \qquad (45)$$

Each term in the complete power series in ν has the same sign. Therefore, for a given eccentricity and a given orientation of the orbit with respect to the sun, $\Delta\varepsilon_{\mathrm{sp}}$ will have maximum magnitude when ν is maximal:

$$\nu_{\max} = \frac{1}{1 + e}, \qquad (46)$$

corresponding to the minimum allowable value of a; i.e., $a_{\min}(1 - e) = R_0$.

For fixed e and $\nu \ll 1$ (i.e., $a \gg R_0$), we find that $|\Delta\varepsilon_{\mathrm{sp}}|$ is a maximum for $\cos\alpha = -e$:

$$|\Delta\varepsilon_{\mathrm{sp}}|_{\max} \approx \frac{2 e\, R_0}{(1 - e^2)^{1/2}}\, Q_{\mathrm{sp}}, \qquad (47)$$

independent of a. Hence, with $\nu \ll 1$, $|\Delta\varepsilon_{\mathrm{sp}}|$ is a maximum for α near $\pm\pi/2$ when $e \ll 1$ and for α near π when $e \lesssim 1$. (For α near π_+ or $-\pi/2$, $\Delta\varepsilon_{\mathrm{sp}}$ is near its maximum positive value; whereas for α near $\pi/2$ or π_-, $\Delta\varepsilon_{\mathrm{sp}}$ is near its maximum negative value.) We also have the asymptotic relation

$$|\Delta\varepsilon_{\mathrm{sp}}|_{\substack{\max \text{ over} \\ \alpha}} \underset{\substack{a \to \infty \\ e \to 1 \\ a(1-e) \gtrsim R_0}}{\sim} Q_{\mathrm{sp}}(2 a\, R_0)^{1/2}, \qquad (48)$$

and finallly, for the special case of resonance[2] (when $\alpha = \pm\pi/2$), we find

$$(\Delta\varepsilon_{\mathrm{sp}})_{\alpha = \pm\pi/2} = \mp Q_{\mathrm{sp}}\, R_0\, e \left\{ 2 + \nu^2 + \frac{\nu^4}{4}(3 + 4 e^2) + \cdots \right\}. \qquad (49)$$

(Of course, for $\alpha = 0$ or π, $\Delta\varepsilon_{\mathrm{sp}} = 0$.)

---

[1] See also S. P. Wyatt: Smith. Astrophys. Obs. Spec. Rep. No. 60 (1961).
[2] See Section IV.

II. Cylindrical satellites. The sunlight pressure on a specularly reflecting cylindrical satellite can be in a different direction from that of the incident rays. For convenience in studying such a case, we assume that the cylinder is tumbling end over end. Its angular momentum vector[1] is therefore perpendicular to the cylinder axis. Further, we assume that the tumbling period is small compared with the orbital period, and that the cylinder radius is negligible compared with its length.

Under these assumptions, the acceleration caused by sunlight pressure, averaged over a tumbling period, is given by

$$\langle \vec{Q}_{sr} \rangle = -\frac{4}{3} Q_{sr} \{ E_2 \cdot \hat{x} + E_3 \cos \xi \cdot \hat{c} \}, \tag{50}$$

where

$$Q_{sr} = \frac{A_{\max}}{M} \left(\frac{I}{c} \right), \tag{51}$$

$$E_2 (\sin^2 \xi) = \frac{2}{\pi} \int_0^{\pi/2} \sin^2 \phi [1 - \sin^2 \xi \cos^2 \phi]^{1/2} \, d\phi, \tag{52}$$

$$E_3 (\sin^2 \xi) = \frac{2}{\pi} \int_0^{\pi/2} \cos^2 \phi [1 - \sin^2 \xi \cos^2 \phi]^{1/2} \, d\phi, \tag{53}$$

and where $A_{\max}$ is the maximum geometric cross-sectional area of the cylinder. The unit vectors $\hat{x}$ and $\hat{c}$ are in the direction of the sun and in that of the angular momentum vector, respectively, while ξ is the angle between $\hat{x}$ and $\hat{c}$. Describing the direction of the satellite angular momentum vector by its colatitude, δ, and by its longitude, η[2], we find that

$$\cos \xi = \sin \delta \{ \cos \beta \cos \eta + \sin \beta \sin \eta \cos \psi \} + \cos \delta \sin \beta \sin \psi. \tag{54}$$

If $\xi = 0$ (cylinder axis perpendicular to incident radiation), then $\langle \vec{Q}_{sr} \rangle = -(4/3) Q_{sr} \hat{x}$ and the acceleration is in the direction of the impinging sunlight. This orientation, in fact, yields the maximum effective cross-section which is $1/3$ greater than the maximum geometric cross-section. If $\xi = \pi/2$, corresponding to the incident radiation being in the plane of tumble of the satellite, $\langle \vec{Q}_{sr} \rangle = -(16/9\pi) Q_{sr} \hat{x}$ and the acceleration is again in the direction of the sunlight.[3] The effective cross-section here is only $8/9$ of the average geometric cross-section which is itself $(2/\pi) A_{\max}$. When $\xi \approx 50°$, we obtain the largest

---

[1] "Angular momentum" here refers only to that of the satellite about its center of mass.

[2] We employ a nonrotating geocentric frame in which longitude is measured from the vernal equinox.

[3] This latter fact can, of course, be deduced from simple symmetry arguments.

possible angle between the acceleration vector and the direction of sunlight — approximately $17°$. (For a nontumbling cylinder, the maximum
angle is $90°$.)

To determine the changes in the orbital elements caused by $\langle \overrightarrow{Q_{sr}} \rangle$,
we first resolve $\hat{c}$ into its components in the $\hat{r}$, $\hat{s}$, and $\hat{w}$ directions appropriate for perigee. These components, denoted by $\tilde{R}_c$, $\tilde{S}_c$, $\tilde{W}_c$ are
given, respectively, by Eqs. (31) if we replace $\cos\beta$ by $\sin\delta\cos\eta$, $\sin\beta$
$\cos\psi$ by $\sin\delta\sin\eta$, and $\sin\beta\sin\psi$ by $\cos\delta$. Comparing Eq. (50) with
Eqs. (28) and (30), we find that the element changes after one revolution
are given by Eqs. (34), with the prescribed interpretations for the Q_i,
if we replace $\tilde{R}$, $\tilde{S}$, and $\tilde{W}$ by $\bar{R}$, $\bar{S}$, and $\bar{W}$, respectively, where

$$
\left.
\begin{aligned}
\bar{R} &= E_2\,\tilde{R} \;+\; E_3\,\tilde{R}_c\cos\xi, \\
\bar{S} &= E_2\,\tilde{S} \;+\; E_3\,\tilde{S}_c\cos\xi, \\
\bar{W} &= E_2\,\tilde{W} + E_3\,\tilde{W}_c\cos\xi,
\end{aligned}
\right\}
\tag{55}
$$

and if we replace λ_{sp} by λ_{sr} where

$$
\lambda_{sr} = \frac{4}{3}\left(\frac{A_{\max}}{M}\right)\left(\frac{I}{c}\right)\frac{a^2}{\mu}.
\tag{56}
$$

This analysis is easily generalized to allow an arbitrary fraction of
the incident radiation to be specularly reflected and the remainder to be
absorbed. If f_c is the fraction specularly reflected, we merely replace E_2
by $(3/4)\,(E_2 + E_3)\,(1 - f_c) + f_c\,E_2$ and E_3 by $f_c\,E_3$ in the expressions
for $\bar{R}$, $\bar{S}$, and $\bar{W}$.[1]

The sunlight pressure perturbations are more complicated here than
for a spherical satellite since ξ is a function of time—not only because
the earth revolves about the sun, but also because various torques act
on the cylinder and cause its angular momentum vector to precess. For
very small cylinders we have an additional complication: Collisions with
micrometeoroids change the angular momentum vector discontinuously
and randomly. To predict satellite orbits when these processes are
important, we have incorporated an appropriate Monte-Carlo method
into our computer program.

3. Solar radiation reflected from the earth

Of the total solar radiation striking the earth, part is reflected, the
remainder being absorbed and reradiated more or less isotropically. By
ignoring the variations in the effective area of a satellite along its orbit,
we conclude that the reradiation has no other effect than to change
the earth's gravitational attraction by an entirely negligible amount [10].

---

[1] Note that $E_2 + E_3$ is a complete elliptic integral of the second kind.

The fraction of sunlight reflected from a given portion of the earth depends on the surface characteristics and, in particular, on the cloud cover. We shall ignore these complicated dependencies and consider the fraction of reflected radiation to be a function only of the angle of incidence.

The law of reflection obeyed by the solar radiation is also difficult to determine. Hence, we make the simplifying assumption that at each point on the earth's surface $f_e (0 \leq f_e \leq 1)$ of the reflected light is specularly reflected and $(1 - f_e)$ is diffusely reflected. Since the resulting formulas that we have derived for diffusely reflected light (obeying LAMBERT's law) are quite complicated and not particularly edifying, we present explicitly only those for specularly reflected light.

The magnitude of the acceleration of the satellite caused by the specularly reflected radiation from the earth is

$$Q_{\text{re}} = f_e \, F(\phi) \, G(\phi, d) \, Q_{\text{sp}}, \tag{57}$$

where $F(\phi)$ is the fraction of the incident sunlight which is reflected by the earth, and $G(\phi, d)$ is the ratio of the reflected energy received at the position of the satellite [when $f_e = F(\phi) = 1$] to the direct radiation energy received.[1] This ratio clearly depends on both ϕ and d, where ϕ is the angle of incidence of sunlight at a point on the earth's surface, and d is the distance from that point to the satellite. From geometric considerations, we find

$$G(\phi, d) = \frac{R_0^2 \cos\phi}{(R_0 + 2d \cos\phi)(R_0 \cos\phi + 2d)}, \tag{58}$$

where

$$d^2 = R_0^2 + r^2 - 2r \, R_0 \cos(\tau - \phi) \tag{59}$$

and

$$r \sin(2\phi - \tau) = R_0 \sin\phi, \tag{60}$$

with τ being the angle between the radial position of the satellite and the direction of the earth-sun line:

$$\cos\tau = \tilde{R} \cos(u - \omega) + \tilde{S} \sin(u - \omega). \tag{61}$$

Given r and τ, the transcendental Eq. (60) can be solved quickly and accurately for ϕ, by using simple iterative methods [11]. For $F(\phi)$, we assume a functional form with two variable parameters:

$$F(\phi) = \begin{cases} F & 0 \leq \phi \leq \phi_0 \\ F + (1 - F) \dfrac{\phi - \phi_0}{(\pi/2 - \phi_0)} & \phi_0 < \phi \leq \pi/2 \end{cases} \tag{62}$$

where $0 \leq F \leq 1$, and $0 \leq \phi_0 \leq \pi/2$. This form is a reasonably good approximation to the experimental data for reflection from dielectrics

---

[1] We assume that the area of the satellite presented to the reflected sunlight is identical to that presented to the direct rays.

and conductors. Atmospheric attenuation, relatively more important for grazing incidence ($\phi \sim \pi/2$), can also be incorporated approximately in $F(\phi)$.

To obtain the element changes, we analyze the acceleration into two components—one along the radius vector to the satellite, and the other along the $-\hat{x}$ direction. The radial component is

$$\mathfrak{f}_e F(\phi) G(\phi, d) Q_{\text{sp}} \frac{\sin 2\phi}{\sin \tau},$$

whereas the one along the direction of direct sunlight is

$$\mathfrak{f}_e F(\phi) G(\phi, d) Q_{\text{sp}} \frac{\sin (2\phi - \tau)}{\sin \tau}.$$

Making use of the calculations for the direct sunlight pressure perturbations, we then find[1]

$$
\left.
\begin{aligned}
\Delta a_{\text{re}} &= \frac{2a\,\lambda_{\text{sp}}}{(1 - e^2)} \{e\,H_7 + \tilde{R}(H_1 + e\,H_4) - \tilde{S}(H_2 + e\,H_3 + e\,H_6)\}, \\[4pt]
\Delta e_{\text{re}} &= \lambda_{\text{sp}} \{H_7 + \tilde{R}(H_4 + e\,H_1) - \tilde{S}(H_3 + H_6 + e\,H_2)\}, \\[4pt]
\Delta \omega_{\text{re}} &= \frac{\lambda_{\text{sp}}}{e} \{-H_8 + \tilde{R}(H_3 + H_5) - \tilde{S} \cdot H_4\} - \cos i \cdot \Delta \Omega_{\text{re}}, \\[4pt]
\Delta \Omega_{\text{re}} &= -\lambda_{\text{sp}} \frac{\tilde{W}}{\sin i} \{H_2 \sin\omega + H_1 \cos\omega\}, \\[4pt]
\Delta i_{\text{re}} &= -\lambda_{\text{sp}} \tilde{W} \{H_2 \cos\omega - H_1 \sin\omega\},
\end{aligned}
\right\} \quad (63)
$$

where

$$
\left.
\begin{aligned}
H_i(u) &= \int du \, \frac{(1 - e^2)^2 \, h_i(u)}{[1 + e\cos(u - \omega)]^3} \, g(u); \quad i = 1 \to 6, \\[6pt]
H_i(u) &= \int du \, \frac{(1 - e^2)^2 \, h_{i-6}(u)}{[1 + e\cos(u - \omega)]^2} \, j(u); \quad i = 7, 8,
\end{aligned}
\right\} \quad (64)
$$

$$
\left.
\begin{aligned}
h_1 &= \sin(u - \omega); & h_2 &= \cos(u - \omega); & h_3 &= 1 + e\,h_2, \\
h_4 &= h_1 h_2; & h_5 &= h_1^2; & h_6 &= h_2^2
\end{aligned}
\right\} \quad (65)
$$

and

$$
\left.
\begin{aligned}
g(u) &= \mathfrak{f}_e F(\phi) G(\phi, d) \frac{\sin(2\phi - \tau)}{\sin \tau}, \\[6pt]
j(u) &= g(u) \frac{\sin 2\phi}{\sin(2\phi - \tau)}.
\end{aligned}
\right\} \quad (66)
$$

[The dependence of ϕ and τ on u is given in Eqs. (60) and (61).] The limits of integration in Eqs. (64) are identical with those in Eqs. (35). Unfortunately, here the integrals must be evaluated numerically.

Using our computer program (with an albedo for the earth of 0.4, with $\mathfrak{f}_e = 1$, and with various values for F and ϕ_0), we have investigated the

---

[1] Since Δp does not depend on the radial component of acceleration, we have $\Delta p_{\text{re}} = 2p\,\lambda_{\text{sp}} \{\tilde{R} \cdot H_1 - \tilde{S} \cdot H_2\}$, similar to the result for direct solar radiation.

long term (approximately half a year) effects of these complicated contributions to the changes in the orbital elements and find them to be seldom larger than 1% of the important changes due to direct sunlight. On the other hand, from a similar numerical analysis, we find that diffusely reflected sunlight can decrease the direct effects for close-in satellites by about 5%, because of the relative predominance of backscattered light in diffuse reflection.

It should be emphasized, however, that

1. Our model is a very crude approximation to the actual reflection properties of the earth, and hence cannot be used to obtain precise results but merely to indicate the order of magnitude of the effect;

2. The numerical investigations we carried out were quite limited; we by no means explored the entire spectrum of initial conditions.

4. Lunar and solar gravitational fields

We shall present here only those formulas necessary for the calculation of moon-induced perturbations. A description of the gravitational perturbations caused by the sun can be obtained by substituting the solar orbit for the lunar orbit.[1]

Since we are describing the satellite motion with respect to the earth, we must determine the *relative* acceleration of the satellite and the earth caused by the moon. Expanding in powers of the ratio of the earth-satellite distance to the earth-moon distance, we find

$$\vec{Q}_m = \frac{\mu_m}{r_m^3}\left\{\left[3\left(\frac{r}{r_m}\right)\cos\tau_m + \frac{3}{2}\left(\frac{r}{r_m}\right)^2 (5\cos^2\tau_m - 1) + \cdots\right]\vec{r}_m - \right.$$
$$\left. - \left[1 + 3\left(\frac{r}{r_m}\right)\cos\tau_m' + \cdots\right]\vec{r}\right\}, \tag{67}$$

where μ_m is the product of the gravitational constant and the mass of the moon, r_m is the earth-moon distance, and τ_m is the angle between the earth-satellite line and the earth-moon line.

The consequent changes in the orbit elements, to first order in r/r_m), are[2]

$$\left.\begin{aligned}
\Delta a_m &= 0, \\
\Delta e_m &= -5\lambda_m\, e\, \tilde{R}_m\, \tilde{S}_m, \\
\Delta \omega_m &= \lambda_m\{4\tilde{R}_m^2 - \tilde{S}_m^2 - 1\} - \cos i \cdot \Delta\Omega_m, \\
\Delta \Omega_m &= \lambda_m \frac{\tilde{W}_m}{\sin i}\left\{\left(\frac{1+4e^2}{1-e^2}\right)\tilde{R}_m \sin\omega + \tilde{S}_m \cos\omega\right\}, \\
\Delta i_m &= \lambda_m \cdot \tilde{W}_m\left\{\left(\frac{1+4e^2}{1-e^2}\right)\tilde{R}_m \cos\omega - \tilde{S}_m \sin\omega\right\},
\end{aligned}\right\} \tag{68}$$

---

[1] In our development all orbits are described geocentrically.
[2] We also find $\Delta p_m = -2a\, e\, \Delta e_m$.

where

$$\lambda_m = 3\pi (1 - e^2)^{1/2} \frac{\mu_m}{\mu} \left(\frac{a}{r_m}\right)^3. \tag{69}$$

$\tilde{R}_m$, $\tilde{S}_m$, and $\tilde{W}_m$ can be obtained by replacing Ω with $\Omega - \Omega_m$, ψ with i_m, and β with u_m in Eqs. (32). [The quantities Ω_m, i_m, and u_m refer to the lunar orbit which is defined with respect to the earth in the same manner as is the satellite orbit.]

Lunar perturbations are usually quite small for close-in earth satellites[1], but are about twice as large as corresponding solar perturbations.

5. Atmospheric drag

As mentioned earlier, an accurate theoretical prediction of the effects of the atmosphere on satellite orbits is not possible at present. Air densities are known to vary with time, longitude, and, of course, altitude, but in an as yet unpredictable manner. In addition, the mechanisms of neutral and charge drag are not completely understood. We therefore limit our discussion to simplified models that yield reasonable, approximate results in most cases. Our considerations on dust drag will be very brief.

I. Neutral drag. We assume that the neutral drag force is caused by the satellite's absorbing (and isotropically re-emitting) all molecules (and ions) in its path.[2] Since the speed of the molecules is usually negligible compared to that of the satellite, air is absorbed at the rate $\varrho \, A \, v$, where ϱ is the air density, $\vec{v}$ is the satellite velocity, and A its area as projected on the plane perpendicular to $\vec{v}$. The rate of momentum transfer is $\varrho \, A \, v \vec{v}$ which leads to a deceleration

$$Q_{nd} = -\varrho \left(\frac{A}{M}\right) v^2. \tag{70}[3]$$

This deceleration causes the following changes in the elements:

$$\left.\begin{aligned}
\Delta a_{nd} &= -2\left(\frac{A}{M}\right) a^2 \int_0^{2\pi} \varrho \, \frac{(1 + e^2 + 2e\cos\theta)^{3/2}}{(1 + e\cos\theta)^2} \, d\theta, \\[2ex]
\Delta e_{nd} &= -2\left(\frac{A}{M}\right) p \int_0^{2\pi} \varrho \, \frac{(1 + e^2 + 2e\cos\theta)^{1/2}}{(1 + e\cos\theta)^2} \, [e + \cos\theta] \, d\theta, \\[2ex]
\Delta \omega_{nd} &= -2\left(\frac{A}{M}\right) \frac{p}{e} \int_0^{2\pi} \varrho \, \frac{(1 + e^2 + 2e\cos\theta)^{1/2}}{(1 + e\cos\theta)^2} \sin\theta \cdot d\theta,
\end{aligned}\right\} \tag{71}$$

---

[1] For a satellite in a circular orbit at an altitude of 2400 miles, note that $\lambda_m \approx 10^{-6}$.

[2] We define charge drag as the difference between the total drag and that which would be found were there no coulomb interaction.

[3] Our model thus yields a drag coefficient of two.

where $\theta = u - \omega$. Since $\vec{v}$ lies in the orbit plane, the elements Ω and i are unaffected by drag in this approximation. Corrections to these formulas required by the rotation of the atmosphere and the flattening of the earth at the poles can be derived easily [12].

If ϱ is approximated by a function of altitude only (and, hence, a function of $\cos\theta$ only), then $\Delta\omega_{nd}$ vanishes. With ϱ an exponential function of altitude, the right sides of Eqs. (71) can be developed in quickly converging series of Bessel functions [12]. For a power law density model, the right sides can be developed in series involving Legendre polynomials [11]; however, a more realistic density model (such as one based on the work of Jacchia [13]) would generally require numerical integration. In our computer program we are currently using a power law density model with constants chosen to most closely match the ARDC 1959 Model Atmosphere and the densities deduced from the Echo I orbit [14].

II. Charge drag. The effect of charge drag is far more difficult to calculate[1] [16]. Fortunately, when predicting orbits for ordinary satellites this drag can usually be neglected. But for certain satellites which are small compared to the charge screening distance, it may be of considerable importance. We restrict ourselves mainly to such (metal) satellites and divide the remainder of our discussion into three parts:

1. Determination of the electrostatic potential (or, equivalently, of the charge) on the surface of a given satellite;

2. Determination of the perturbing force as a function of the plasma characteristics and the satellite potential;

3. Calculation of the orbital motion, given the dependence of the charge drag force on position and time.

1. As was recognized over 20 years ago [17], the potential on an object in space is contingent upon the photoelectric effect and the ambient charged particle fluxes. The potential depends as well on the shape, orientation, composition, and surface conditions of the satellite[2].

As we shall see, the photoelectron flux is exceedingly difficult to predict. However, almost all emitted electrons are of low energy, thus setting an upper limit on the positive potential of a few volts for conditions in which the satellite is in sunshine and in which the positive ion flux does not exceed the electron flux.

Were the ambient plasma in thermal equilibrium, the electron flux would greatly exceed the ion flux, tending to impart a negative potential

---

[1] An extensive bibliography of the recent literature on this subject can be found in a review article by Chopra [15].

[2] We assume that the variations in fluxes are slow compared with the time taken for the potential on the body to reach an equilibrium value. Hence, we can speak of an electrostatic potential (which may then vary slowly with time).

to the satellite. Its magnitude would depend on the temperature of the plasma, on the mass of the ions, and on the reflection and secondary emission coefficents for the incident charged particles. For "normal" values of these parameters, we would expect negative satellite potentials of a few tenths of a volt for temperatures of about $1500°$ K — unless the photoelectric effect is dominant, which is unlikely for altitudes below about 1500 km, where the charged particle density exceeds $\approx 10^4/\mathrm{cm}^3$. On the other hand, near the peak of the F region, one expects the electron distribution to have a high energy tail which could lead to considerably larger potentials.

The experimental situation is far from clear. Measurements of BOURDEAU et al [18] from Explorer VIII indicate potentials of a few tenths of a volt negative in the altitude region 800 to 1500 km in temperate latitudes. They also found conditions of thermal equilibrium to exist in the plasma. Others, using somewhat different LANGMUIR probe techniques, measured satellite potentials at several volts negative at altitudes both below and comparable to the low altitudes traversed by Explorer VIII [19].

Above 1500 km the plasma characteristics are probably even more complicated. Recent experiments indicate that the separation of the radiation belts into inner and outer ones may be artificial, caused by previous investigations of only selective particle energy ranges. It would seem now that the entire region contains relatively high charged particle fluxes, but that the constituents, both positive and negative, vary with altitude for a particular geomagnetic latitude in an as yet little understood manner [20]. Thus, the assumption that the plasma is in thermal equilibrium in this region may be a poor one. (The average time between particle collisions may be sufficiently long so that changes in the characteristics of the sources of and sinks for the fluxes prevent thermal equilibrium from being established.) In this situation, a meaningful calculation of potential is clearly impossible; we therefore present only a general method, applicable for small satellites, and a qualitative discussion of the values of the relevant parameters.

The potential reached on a satellite is determined from the balance of positive and negative currents:

where
$$I_e + I_i + I_{\mathrm{ph}} = 0, \tag{72}$$

$$
\left.
\begin{aligned}
I_e(\vec{r}, t) &= -e \int_{\vec{v}_e} v_e \, A_e(\Phi_s, \vec{v}_e)\, \{1 - \delta_e(v_e, \varepsilon_e)\}\, \{1 - R(v_e, \varepsilon_e)\} \times \\
&\qquad\qquad\qquad\qquad \times f_e(\vec{r}, \vec{v}_e, t)\, d^3 v_e, \\
I_i(\vec{r}, t) &= Z_i e \int_{\vec{v}_i} v_i \, A_i(\Phi_s, \vec{v}_i)\, \{1 - \delta_i(v_i, \varepsilon_i)\}\, \{1 - R(v_i, \varepsilon_i)\} \times \\
&\qquad\qquad\qquad\qquad \times f_i(\vec{r}, \vec{v}_i, t)\, d^3 v_i
\end{aligned}
\right\} \tag{73}
$$

and

$$I_{\text{ph}} = e\, A_{\text{ph}}(t) \int\limits_{\lambda \leq \lambda_0} \Gamma(\lambda,\, \varepsilon,\, s)\, N_\lambda\, d\lambda;$$

$$\frac{h\,c}{\lambda_0} = W_f + \frac{e}{2}(\Phi_s + |\Phi_s|); \qquad h = \text{PLANCK's constant}, \qquad (74)$$

and where $\vec{v}_e$ is the velocity of the electron with respect to the satellite, e is the magnitude of the electronic charge, and Φ_s is the potential on the satellite's surface[1], relative to that of the plasma "at infinity". The effective area of the satellite for electron impact, $A_e(\Phi_s,\, \vec{v}_e)$, can exceed the geometric cross-section if $\Phi_s > 0$, whereas for $\Phi_s < 0$, A_e is always lower and in fact vanishes for $v_e < v_{\min}(\Phi_s)$. $f_e(\vec{r},\, \vec{v}_e,\, t)\, d^3 v_e$ is the number of electrons per unit volume with velocities relative to the satellite between $\vec{v}_e$ and $\vec{v}_e + d\vec{v}_e$; hence

$$\int\limits_{\vec{v}_e} f_e(\vec{r},\, \vec{v}_e,\, t)\, d^3 v_e = N_e(\vec{r},\, t), \qquad (75)$$

where $N_e(\vec{r},\, t)$ is the density of the electrons at the position $\vec{r}$ at time t. (In this formulation, f is independent of the satellite's position; but, of course, the satellite will to some extent influence the plasma in its vicinity, as will the ejected photoelectrons.) The secondary emission coefficient, $\delta_e(v_e,\, \varepsilon_e)$, is not significant for energies below about 20 ev [22]. However, if the satellite is in sunlight, δ_e may increase noticeably, as even red light may provide sufficient energy to allow some electrons to escape [23].[2] For kilovolt electrons, this coefficient can exceed unity, resulting in a net positive current contribution to the satellite. (These secondary electrons are usually of very low energy, less than about 5 ev.) The reflection coefficient, $R(v_e,\, \varepsilon_e)$, is the fraction of electrons reflected from the surface. It is a function of the velocity and angle of incidence, ε_e, of the electron as well as of the satellite's surface properties. Little experimental evidence is available on these dependencies, but R seems to be governed strongly by surface conditions and to decrease with increasing energy. For most measured materials, it is below about 0.4 for energies below 5 ev.

---

[1] Φ_s is not necessarily constant over the surface; for large satellites one should also consider the effects of induced potentials caused by the satellite's moving with respect to the earth's magnetic field [21]. In these cases it is probably necessary to determine the electric field on the satellite's surface (instead of determining the satellite's charge).

[2] Experimental results of HINTEREGGER [24] suggest that at a 3500 km altitude, the electron population may consist largely of particles with energies of several ev; hence such an increase in soft secondaries might maintain the satellite's potential at a very low positive value during daylight.

The quantities with subscripts i all refer to ions; in particular $Z_i e$ denotes the ionic charge. The secondary emission coefficient for these particles is quite low for primary energies less than 100 ev. In the several hundred electron volt range, however, δ_i can reach values as high as ten. The reflection coefficient, R_i, is usually very small for low energies and is generally under 0.1 for energies up to 100 ev. Contributions to the current from each different type of ion should, of course, be considered separately.

In the expression for the photon current, $N_\lambda\, d\lambda$ is the number of incident photons (per unit area per sec.) with wavelengths between λ and $\lambda + d\lambda$. $\Gamma(\lambda,\,\varepsilon,\,s)$ is the photoelectron yield, i.e., the number of emitted electrons per incident photon, and must be determined experimentally. Near the threshold (≈ 4 ev for most metals) it depends critically on the material and on the condition of the satellite surface, and is of the order of 10^{-6} to 10^{-4} electrons per incident photon [25]. For higher energy photons ($\gtrsim 10$ ev, i.e., $\lambda \lesssim 1250\text{Å}$) the yield is about two to four orders of magnitude greater; it does not depend so critically on the composition and is almost independent of surface conditions; this is the so-called "volume effect". However, the photon flux near the threshold for most metals ($\approx 10^{13}$ ph/cm^2 sec-Å) is about four orders of magnitude greater than that near the peak yield point for the volume effect [26]. The total photoelectron yield will therefore still depend importantly on the satellite's surface condition. If the potential is zero or negative, the integration in Eq. (74) extends over photons with energies greater than the work function, W_f, of the material. (The effect on the threshold of negative Φ_s's is negligible for potentials of the magnitude under consideration here.) But if the potential is positive, only photoelectrons whose energy exceeds W_f by $e\,\Phi_s$ will escape. Therefore, neglecting small effects due to temperature, we integrate over the region indicated.

Eqs. (72) to (74) are not of much practical use, mainly because of the unknown behavior of so many of the important physical parameters; the uncertainty in Φ_s is therefore extremely large, potentials anywhere between 0.1 v and 10 v being perfectly conceivable over a wide range of altitudes in either sunshine or shadow. And if the high energy tail of the electron distribution contains fluxes considerably larger than the photoelectron and thermal proton fluxes (and if $\delta_e < 1$), then the satellite potential could become enormous.

2. Calculating the charge drag force, given the satellite potential, is also not a simple exercise, although the uncertainty is probably not so great as in 1. The basic force causing this drag is coulombic. It is sometimes convenient to view the interaction of the satellite with the plasma in terms of its separate interactions with the individual charged

particles (the consequent interactions of the latter being neglected), and sometimes in terms of the collective modes of interaction (e.g., the excitation of plasma oscillations [27]). Recent investigations of the latter type have centered around attempts to solve simultaneously the relevant BOLTZMANN equation and POISSON's equation, assuming the plasma to be thermal. Only linearized analytical solutions have so far been obtained, corresponding to an assumption of a weak interaction ($e \, \Phi_s / k \, T \ll 1$). The relation between these solutions and the real charge drag forces is not clear. Several attempts have also been made (using numerical techniques) to solve the nonlinear POISSON equation using more or less plausible velocity distributions for the positively and negatively charged particles [28]. The results showed screening distances of up to about 10 DEBYE lengths.[1]

The interactions of a charged particle with the individual constituents of a plasma have been discussed by SPITZER [29], and are based on formulas for dynamical friction in stellar systems that were developed by CHANDRASEKHAR [30]. Again, the explicit formulas given are applicable to thermal plasmas, and may therefore be far from an adequate description for the actual plasma at altitudes much above 1500 km. In addition, this development may not be applicable for a charged "particle" the size of a satellite.

A crude understanding of the charge drag force may be obtained by considering a very simple model[2]: Assume that the satellite is a point charge and that the ions[3] remain stationary during the interaction.[4] The momentum transfer to the ions (mass, m_i; speed, v_i) will be

$$\Delta \, (m_i \, v_i) = \int_{-\infty}^{\infty} F_{\perp} \, dt = \int_{-\infty}^{\infty} \frac{Z_s \, e^2}{r_{si}^2} \, \frac{b}{r_{si}} \, dt, \qquad (76)$$

where Z_s is the satellite charge (in units of the electron charge) and r_{si} is the distance between the satellite and the ion, whose charge is assumed to be unity. Since the satellite moves linearly during the time of effective interaction, we use

$$r_{si}^2 \approx b^2 + v^2 \, t^2, \qquad (77)$$

---

[1] The DEBYE length, λ_D, is the radius of a shell such that a stationary charged particle at its center in a thermal plasma would be effectively screened from the charged particles outside; $\lambda_D = (k \, T / 4\pi \, N e^2)^{1/2}$, where k is BOLTZMANN's constant, T is the temperature, N the charged particle density, and e the electronic charge. For $T = 1200 \, °K$ ($k \, T \approx .1$ ev) and $N = 10^4$ particle/cm³, we find $\lambda_D \approx 2.5$ cm.

[2] This model, due to BOHR, is also used in nuclear physics to determine the range energy relations for charged particles moving through matter [31].

[3] We ignore the interaction with the electrons since the total momentum transfer to them is generally far less than to the positive ions.

[4] Actually, at several thousand km altitudes and above, the ions, being mainly H^+, have mean thermal velocities comparable to that of the satellite.

where b is the impact parameter (which in this model corresponds to the distance of closest approach, reached at $t = 0$). Hence,

$$\Delta(m_i v_i) = \frac{2 Z_s e^2}{b v} . \qquad (78)$$

The momentum transfer to each ion is therefore proportional to Z_s. However, for this elastic collision, the energy transfer is proportional to the square of the momentum transfer. We find[1]

$$\Delta \mathcal{E}_i = \frac{2 Z_s^2 e^4}{m_i b^2 v^2} . \qquad (79)$$

This is the energy transfer per ion along the satellite path. If the ion density is N_i particle/cm³, then to obtain the transfer per unit distance along the satellite path, for ions with a given impact parameter, we multiply by $N_i 2\pi$ bdb. Integrating over all impact parameters, we find that the energy loss per unit path length (i.e., the drag force) is

$$F_{cd} = \frac{d\mathcal{E}}{ds} = -\frac{4\pi N_i Z_s^2 e^4}{m_i v^2} \ln\left(\frac{b_{max}}{b_{min}}\right) . \qquad (80)$$

This force is therefore proportional to the square of the satellite's charge. It also depends on the logarithm of the ratio of the largest contributing parameter to the smallest. (If there were no shielding, F_{cd} would be infinite, and similarly, if $b = 0$ were an allowable impact parameter.) For a very small satellite, we can approximate b_{max} by $n \lambda_D^2$ and b_{min} by the impact parameter corresponding to the maximum possible momentum transfer to an ion $-2 m_i v$. Thus,

$$\vec{Q}_{cd} = \frac{\vec{F}_{cd}}{M} = -\frac{4\pi N_i Z_s^2 e^4}{m_i M v^2} \ln\left(\frac{n \lambda_D m_i v^2}{Z_s e^2}\right) \hat{v} . \qquad (81)$$

It is assumed, of course, that $b_{max} > b_{min}$; otherwise Eqs. (80) and (81) make no sense. While b_{min} will increase linearly with Z_s, n will also increase [28], tending to preserve the inequality. As an example of the magnitudes involved, consider a West Ford dipole at a 3500 km altitude [3]. We find that with a dipole potential of 10 v, $Z_s \sim 10^7$ and $\lambda_D m_i v^2/Z_s e^2 \sim 1$. Hence, in order to use Eq. (81) for such an application, we must assume that the effective screening distance, $n \lambda_D$, does in fact increase sufficiently to ensure that $b_{max} \gg b_{min}$.

For a given effective screening distance, it is clear that we can obtain a (crude) upper bound on the charge drag deceleration by assuming that the satellite "sweeps up" all charged particles within this distance of

---

[1] The exact energy transfer in a two-body coulomb collision is somewhat more complicated and is not needed for our purposes.

[2] In view of the results described in ref. 28, we assume that $1 < n < 10$ for most cases. A more precise determination of n, for given conditions, is beyond the scope of our description.

its surface. Hence, for satellites with dimensions small compared to $n \lambda_D$, the effective cross-section would be increased relative to that for neutral drag by a factor of order $n^2 \lambda_D^2/A$, where A is the average geometric cross-section.

We now examine briefly the dependence of charge drag acceleration on satellite dimensions. For thin cylindrical satellites of radius r_c and length l_c (l_c, $r_c < n \lambda_D$), we adapt Eq. (80) by replacing b_{max} with $r_c + n \lambda_D$ and b_{min} with $r_c + (Z_s e^2/m_i v^2)$. We also replace $e Z_s$ by the product of the satellite's capacity, C, and its potential. Then, assuming Φ_s to be independent of the satellite's dimensions, we find

$$ Q_{cd} \sim \frac{C^2}{M} \sim \frac{l_c}{r_c^2}; \qquad r_c, l_c < n \lambda_D, \tag{82} $$

since the capacity of a thin cylinder is (apart from a logarithmic dependence) proportional to its length [3]. This contrasts sharply with the corresponding dependence for neutral drag which is proportional to r_c^{-1}.

For thin cylinders which are long compared to $n \lambda_D$, we can no longer use Eq. (80). A possibly better formula for such cases is [32]

$$ F_{cd} \approx - \frac{\pi (4 \pi e^2 N_i k T)^{1/2} Z_i^2 e^2}{m_i v^2 [1 - (k T/m_i v^2)]^{1/2} l_c} . \tag{83} $$

Hence, we find $Q_{cd} \sim \dfrac{C^2}{M l_c} \sim r_c^{-2}$ which again has a stronger dependence on r_c than does neutral drag.

For a sphere with radius r_s, we can replace b_{max} by $r_s + n \lambda_D$ and b_{min} by $r_s + (Z_s e^2/m_i v^2)$. With $r_s < n \lambda_D$,

$$ Q_{cd} \sim r_s^{-1}; \qquad r_s < n \lambda_D, \tag{84} $$

which has the same dependence on r_s as does neutral drag. Assuming Eq. (80) to be meaningful when $r_s \gg n \lambda_D$, we expand the logarithm and obtain

$$ Q_{cd} \approx - \frac{4 \pi N_i C^2 \Phi_s^2 e^2 n \lambda_D}{m_i M v^2 r_s} , \tag{85} $$

hence

$$ Q_{cd} \sim \frac{C^2}{M r_s} \sim r_s^{-2}; \qquad r_s \gg n \lambda_D. \tag{86} $$

We infer from these results that charge drag will be relatively most important for satellites with dimensions small compared to the shielding distance.

Unfortunately, almost every aspect of this treatment of charge drag involves gross oversimplifications of the actual physical situation. We therefore wish to emphasize that despite the length of our development (or, perhaps, because of it) extreme caution should be used in drawing from it any quantitative conclusions about the influence of charge drag on the behavior of real satellites.

3. Were the charge drag acceleration known, the calculation of the corresponding changes in the orbital elements would be straightforward. In the absence of any detailed understanding we are using in our computer program a somewhat flexible (but admittedly crude) parameterized model for the acceleration so as to study a wide range of conceivable physical conditions.

Since at low altitudes (where the percentage of ionization is very small) charge drag is almost always masked by neutral drag, we confine our study to altitudes greater than 500 km. We allow the satellite potential to depend on altitude and latitude, and on whether or not the satellite is in sunlight. But we ignore the dependence on solar activity and the possibility of nonnegligible components of the drag force being in the plane perpendicular to $\vec{v}$. In particular, we use the simplified formula

$$\vec{Q}_{cd} = - \frac{4\pi N_i C^2 \Phi_s^2 e^4}{M m_i v^2} \ln\left(\frac{b_{\max}}{b_{\min}}\right) \hat{v} \tag{87}$$

and substitute

$$\Phi_s^2 = \begin{cases} \Phi_{su}^2 \dfrac{[1 + \gamma \sin^{2n}\delta]}{1 + \left(\dfrac{r - r_1}{z_1}\right)^2} ; & \text{(sunlight)} \\[20pt] \Phi_{sh}^2 \dfrac{[1 + \gamma' \sin^{2n'}\delta]}{1 + \left(\dfrac{r - r_1'}{z_1'}\right)^2} ; & \text{(shadow)} \end{cases} \tag{88}$$

where Φ_{su}, Φ_{sh}, γ, γ', r_1, r_1', z_1, and z_1' are parameters. (δ denotes the satellite's colatitude.) Since N_i decreases slowly with altitude, as does the mean molecular weight of the ions, we assume that the ratio remains about constant. The logarithmic term is also slowly varying with respect to the satellite's position along the orbit and we write, finally,

$$\vec{Q}_{cd} = - P \frac{\Phi_s^2}{v^3} \vec{v}, \tag{89}$$

where P is a parameter:

$$P = \frac{4\pi N_i e^4}{m_i}\left(\frac{C^2}{M}\right) \ln\left(\frac{b_{\max}}{b_{\min}}\right). \tag{90}$$

This model allows us to investigate, for example, a peak in the potential which may occur during darkness at the center of the "inner" Van Allen belt. Φ_{sh} and z_1' determine the strength and width of the peak while r_1' determines the altitude of its center. We can easily generalize this form to include an arbitrary number of seperate peaks, and can also consider a potential which decreases monotonically with altitude (by choosing r_1 small enough). To simulate an altitude independent Φ_s, we let $z_1 \to \infty$. [When the photoelectric effect is dominant, the potential should be (within limits) almost independent of altitude and latitude.]

Various strong or gentle changes of potential with latitude can be approximated by varying γ and n. [At these altitudes the mean free path is quite long, and the charged particles spiral back and forth along magnetic field lines. As a result, the charged particle density is correlated with the dipolar pattern of these lines and, hence, depends on (geomagnetic) latitude.]

This model is currently being used in an investigation of the effects of charge drag on sunlight pressure resonances involving thin, cylindrical satellites (see Sections IV and VI).

III. Dust drag. The drag caused by dust particles depends on their trajectories. For example, suppose there exists a fairly uniform distribution of dust along the earth's path, moving in direct orbits about the sun at lower speeds than the earth because of the effects of sunlight pressure. Then, in the vicinity of the earth, dust particles will stream preferentially toward the earth in the direction opposite to that of the earth's orbital motion. A near-equatorial direct satellite orbit will therefore be displaced away from the sun; perigee will tend to form on the sunlit side of the earth. On the other hand, were dust to stream toward the earth uniformly from all directions, satellite orbits (if we neglect shielding by the earth) would be affected in essentially the same way as by neutral air drag. If there were substantial amounts of dust in unidirectional orbits about the earth, then, depending on the relative direction of motion of the satellite, the latter might experience a drag whose effects would be quite different from those of neutral air drag. However, after examining recent micrometeoroid data [33], we find that even the Echo balloon could not have been detectably perturbed by dust drag.

6. Poynting-Robertson drag

This drag is caused by a v/c correction to direct sunlight pressure [34]. To this order in v/c, the total acceleration of an absorbing (or specularly reflecting spherical) satellite due to sunlight is[1]

$$\vec{Q}_{\mathrm{sp}} = Q_{\mathrm{sp}} \left\{ \left(1 - \frac{v_n}{c}\right) \hat{n} - \frac{\vec{v}_s}{c} \right\}, \tag{91}$$

where $\hat{n}$ is a unit vector parallel to the direction of the incident rays $(\hat{n} = -\hat{x})$, $\vec{v}_s$ is the satellite's velocity with respect to the sun, and $v_n = \vec{v}_s \cdot \hat{n}$. The additional acceleration is, therefore,

$$\vec{Q}_{\mathrm{prd}} = -Q_{\mathrm{sp}} \left(\frac{v_n}{c} \hat{n} + \frac{\vec{v}_s}{c}\right). \tag{92}$$

---

[1] Transforming the acceleration from a heliocentric to a geocentric coordinate system, in conformity with the theory of relativity, introduces corrections of second order in v/c.

The first term on the right side is the DOPPLER modification: The sunlight pressure increases when the satellite moves towards the sun ($v_n < 0$) and decreases when it moves away ($v_n > 0$). The second term represents a tangential drag. To understand its origin, consider the satellite to be a perfect absorber in thermal equilibrium. It therefore absorbs and reradiates energy, both at the rate $Q_{\mathrm{sp}} M c (= I A)$. The reradiation is isotropic with respect to the satellite, which however is itself moving in the solar frame with velocity $\vec{v}_s$. The accompanying momentum loss [at a rate $Q_{\mathrm{sp}} M (\vec{v}_s/c)$] causes the tangential drag.

Defining $\tilde{R}(\beta)$, $\tilde{S}(\beta)$, and $\tilde{W}(\beta)$ as before [see Eqs. (32)], and writing

$$\vec{v}_s = \vec{v} + \vec{v}_E, \tag{93}$$

where $\vec{v}_E$ is the velocity of the earth relative to the sun, we easily obtain the changes in orbital elements

$$
\left.
\begin{aligned}
\Delta a_{\mathrm{prd}} &= -2\pi a\, \lambda_{\mathrm{sp}}\!\left(\frac{v}{c}\right)\{3 - \tilde{W}^2(\beta)\}, \\[4pt]
\Delta e_{\mathrm{prd}} &= 3\pi\, \lambda_{\mathrm{sp}}\!\left(\frac{v_E}{c}\right)\tilde{S}\!\left(\beta + \frac{\pi}{2}\right), \\[4pt]
\Delta \omega_{\mathrm{prd}} &= -3\pi\, \frac{\lambda_{\mathrm{sp}}}{e}\!\left(\frac{v_E}{c}\right)\tilde{R}\!\left(\beta + \frac{\pi}{2}\right) - \cos i \cdot \Delta \Omega_{\mathrm{prd}}, \\[4pt]
\Delta \Omega_{\mathrm{prd}} &= \pi\, \frac{\lambda_{\mathrm{sp}}}{\sin i}\!\left(\frac{v}{c}\right)\{\tilde{R}(\beta)\cos\omega - \tilde{S}(\beta)\sin\omega\}\,\tilde{W}(\beta), \\[4pt]
\Delta i_{\mathrm{prd}} &= -\pi\, \lambda_{\mathrm{sp}}\!\left(\frac{v}{c}\right)\{\tilde{R}(\beta)\sin\omega + \tilde{S}(\beta)\cos\omega\}\,\tilde{W}(\beta),
\end{aligned}
\right\} \tag{94}
$$

which are accurate to zeroth order in eccentricity[1] and to first order in (v_s/c) when we neglect the effects of the earth's shadow. We see that the actual drag (i.e., the cause of the decrease in a) is due mainly to $\vec{v}$. The orbit also undergoes displacement. For example, a near-equatorial direct orbit is displaced away from the sun[2]: Perigee tends to form on the sunlit side of the earth while the eccentricity is increased, proportional to v_E. These results could, of course, have been anticipated by inspection of Eqs. (92) and (93).

IV. Theory of resonances

In the last section, we discussed separately, for various forces, the mean first-order changes in the orbital elements. By considering these in combination, we find that, for certain initial conditions, the changes in some elements can accumulate over long periods of time to produce quite sizable effects on the orbit. Such "resonances", which sometimes

---

[1] The higher order terms have been calculated, but have not been included because they are cumbersome and not particularly informative.

[2] Note that simple sunlight pressure causes a perpendicular displacement.

drastically alter the satellite's lifetime, are usually obtained when the second harmonic of the earth's field causes motions of ω and Ω which preserve a "favorable" relative orientation between the orbit and the direction of the disturbing force. Offsetting the motion of the source of this force allows the perturbations produced by it to accumulate monotonically. Resonances caused by the solar and lunar gravitational fields have been treated by several authors [35]; we shall discuss only those resonances due to sunlight pressure [1, 8].

Qualitatively, the effect of sunlight pressure is to displace the orbit in a direction perpendicular to the projection of the earth-sun line onto the orbit plane. [We assume throughout that the force on the satellite is parallel to the incident light rays.] The amplitude and period of this displacement depend on the relative motions of the argument of perigee and the earth-sun line. The amplitude is, of course, also proportional to the satellite's area-to-mass ratio, which we shall tacitly assume to be of the order of 100 cm²/gm. For certain values of $\dot\omega$, $\dot\Omega$, and $\dot\beta$[1], the resonance requirements can be met; we investigate some below, but for simplicity consider only the perturbations of initially circular orbits caused by sunlight pressure and the second harmonic.[2] We also discuss briefly the possible effect of charge drag on the resonance phenomenon.

1. Equatorial orbits

From Eq. (41), we see that the change in orbit eccentricity is proportional to $\tilde{S}$. If $i_0 = 0$, then [see Eqs. (32)],

$$\tilde{S} = -\cos^2(\psi/2)\sin(\omega + \Omega - \beta) - \sin^2(\psi/2)\sin(\omega + \Omega + \beta). \quad (95)$$

Hence, Δe will increase (almost) monotonically if the equality $\dot\omega + \dot\Omega = \dot\beta$ is maintained.[3] As can be seen from Eqs. (32) and (41), $\dot\omega_{sp} + \dot\Omega_{sp}$ will then become quite small; since $\cos^2(\psi/2) \approx .96 \gg \sin^2(\psi/2) \approx .04$, we obtain a strong resonance when

$$\dot\omega_{2h} + \dot\Omega_{2h} = \dot\beta \quad (96)$$

i.e., when

$$5\frac{5\cos^2 i - 1 - 2\cos i}{a^{7/2}(1 - e^2)^2}\ \text{deg/day} \approx 1\ \text{deg/day}, \quad (97)$$

where a is expressed in earth radii. With $e_0 = 0$, this equality is satisfied for a mean altitude of about 0.93 e.r. ≈ 5950 km. A weaker resonance would result for the rate equality $\dot\omega_{2h} + \dot\Omega_{2h} = -\dot\beta$, but this condition can never be satisfied.

---

[1] A dot over a symbol denotes its mean rate of change with respect to time. In this analysis, $\dot\beta$ is assumed constant, but in our computer program $\dot\beta$ varies in accordance with the actual motion of the earth about the sun.

[2] We neglect the effects of the shadow as well.

[3] Almost immediately, we find $\omega + \Omega - \beta \approx \pi/2$.

For retrograde equatorial orbits ($i_0 = 180°$), we find

$$\tilde{S} = -\cos^2(\psi/2)\sin(\omega - \Omega + \beta) - \sin^2(\psi/2)\sin(\omega - \Omega - \beta), \quad (98)$$

and we would have a strong resonance when $\dot{\omega}_{2h} - \dot{\Omega}_{2h} = -\dot{\beta}$. But this condition can also never be satisfied. We can only obtain a very weak resonance when $\dot{\omega}_{2h} - \dot{\Omega}_{2h} = \dot{\beta}$, which for $e_0 = 0$ occurs again at a mean orbit altitude of about 0.93 e.r.

Returning to Eq. (97) we see that this equality cannot be strictly maintained: As e increases because of the action of sunlight, the left side increases. In addition, a decreases due to the effects of the earth's shadow, causing a further increase of the left side. Resonant behavior (i.e., a monotonic increase in eccentricity) will therefore continue only until the relative angular position of perigee and the earth-sun line changes by 90°.[1] For satellites with sufficiently high KA/M values, this behavior will be maintained until the orbit intersects the earth's surface. But, for KA/M values much below 100 cm²/gm, the behavior throughout the satellite's lifetime will not be resonant: Perigee height will undergo long period oscillations with an amplitude proportional to KA/M. In order words, a rather long time is required for the (small) inequality in angular rates to produce a change in angular positions of 90°. Only for satellites with relatively low values of KA/M is sufficient time available.

Resonant behavior throughout the lifetime of a satellite will be obtained for a range of initial values of a (given $e_0 = i_0 = 0$) whose extent depends on the area-to-mass ratio of the satellite. The center of this range (when it exists) lies above $a_0 = 1.93$ e.r., since the resonance condition can be more nearly satisfied, on average, by choosing an initial mean altitude somewhat greater than that which satisfies Eq. (97).

Using our computer program, we have investigated the orbits of satellites with different area-to-mass ratios and for a wide range of resonant and near-resonant initial conditions. These (unpublished) results illustrate the qualitative features described above.

2. Polar orbits

For polar and near-polar orbits the description of resonances is considerably more complicated, partly because of the importance of the change in inclination angle.[2]

---

[1] As the relative angular position changes, $\omega + \Omega - \beta$ becomes larger than 90° and we see from Eq. (32) and (41) that $\dot{\omega}_{sp} + \dot{\Omega}_{sp}$ then tends to restore the resonance orientation.

[2] This change has a negligible effect for equatorial (and near-equatorial) orbits because in these latter cases the dependence of the resonance condition on inclination (insofar as the contribution of the second harmonic is concerned) is only of second order.

With $i = \pi/2$, we find

$$2\tilde{S} = -\cos^2(\psi/2)\{\sin(\omega + \Omega - \beta) + \sin(\omega - \Omega + \beta)\} -$$
$$-\sin^2(\psi/2)\{\sin(\omega + \Omega + \beta) + \sin(\omega - \Omega - \beta)\} +$$
$$+ \sin\psi\{\sin(\omega + \beta) - \sin(\omega - \beta)\}, \tag{99}$$

and the only realizable resonance condition is

i.e.,
$$\dot{\Omega}_{2h} - \dot{\omega}_{2h} = \dot{\beta}, \tag{100}$$

$$a^{7/2}(1 - e^2)^2 \approx 5(1 - 2\cos i - 5\cos^2 i), \tag{101}$$

where a is again expressed in earth radii. By defining $\delta i = i - \pi/2 \ll 1$, we have
$$a^{7/2}(1 - e^2)^2 \approx 5(1 + 2\delta i) \tag{102}$$

as an approximate resonance condition for polar and near-polar orbits. For $e_0 = 0$ and $i_0 = \pi/2$, this condition is satisfied for $a \approx 1.59$ e.r.

The only important change in i is caused by sunlight pressure. From the last of Eqs. (41), we see that the magnitude of Δi_{sp} increases approximately linearly with e, but that the sign of the change depends on the relative signs of $\tilde{W}$ and $\cos\omega$. To illustrate the effect of Δi_{sp} on the resonance condition, we explain this relationship for an initially circular polar orbit. When the sun passes the longitude of the ascending node, ω lies within $\pm\psi$ of the value $\pi/2$.[1] $\tilde{W}$ remains positive from that point until the sun passes the longitude of the descending node. During this time, ω decreases (through zero) from within $\pm\psi$ of $\omega = \pi/2$ to within approximately the same angular distance from $\omega = 3\pi/2$. For this half-year, then, $\tilde{W}\cos\omega$ is predominantly positive. The same result also obtains for the other half-year when both $\tilde{W}$ and $\cos\omega$ change sign.[2] Consequently while e increases from zero, i and δi decrease secularly, tending to maintain the equality in Eq. (102). With $KA/M = 50$ cm²/gm, we find that when $e = .25$ [and hence $(1 - e^2)^2 \approx .88$], the value of $1 + 2\delta i$ is $\approx .90$; and Eq. (102) seems to remain nearly satisfied. However, sunlight also causes a decrease in a which brings the left side about 5% below the right side by the time $e = .25$. (In addition, there is the smaller effect of $\dot{\Omega}_{\mathrm{sp}} - \omega_{\mathrm{sp}}$ on the resonance orientation.)

With Eq. (102) satisfied there are still two qualitatively different types of resonances possible for polar and near-polar orbits. In the

---

[1] With the resonance condition satisfied, perigee remains 90° from the in-plane component of the earth-sun line at the position in which the satellite is moving "away" from the sun. (Note, however, the difference for a polar resonance of the second type, which is described below.)

[2] The sign relationship becomes more complicated after i has changed appreciably.

first, the apsidal line rotates at the same rate and in the same direction as the in-plane component of the earth-sun line. In the second, these two equal rotations are in opposite directions, and we find small amplitude, seasonal oscillations of perigee height superposed on the secular decrease. This latter type of resonance can clearly be obtained from the former by a reversal in the direction of motion of the satellite, i.e., by a change of π in Ω_0.

We discuss analytically these types for the initial conditions $e_0 = \beta_0 = 0$ and $i_0 = \pi/2$. The most pronounced differences between the two

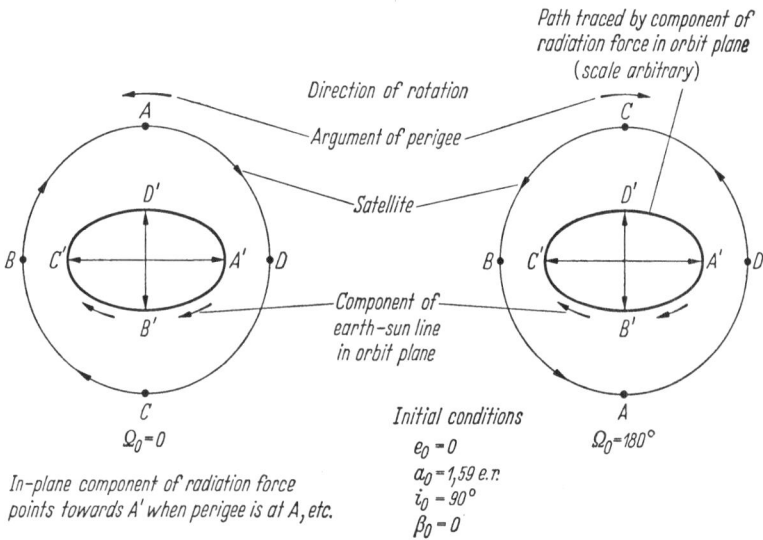

Fig. 2. Relative motion of perigee and in-plane component of earth-sun line.

then occur for the pair of initial values $\Omega_0 = 0$ and $\Omega_0 = \pi$. The corresponding relative motions of perigee and the earth-sun line are shown in Figure 2. For $\Omega_0 = 0$, we find

$$2\tilde{S} \approx -[1 - \sin\psi]\cos(\dot\omega + \dot\beta)\,t - [1 + \sin\psi]\cos(\dot\omega - \dot\beta)\,t, \quad (103)$$

because $\dot\Omega \approx 0$. $[\dot\Omega_{2h} = 0$ and $\dot\Omega_{\mathrm{sp}} \approx 0$ in these cases.] With the resonance condition satisfied,

$$2\tilde{S} \approx -[1 - \sin\psi] - [1 + \sin\psi]\cos 2\dot\beta\,t. \quad (104)$$

Initially, $\tilde{S}$ is negative, leading to an increase in e, but for

$$\beta_1 \leqq 2\dot\beta\,t \leqq \beta_2, \quad (105)[1]$$

---

[1] This equation must, of course, be modified in the obvious manner for $t > 6$ mos. ($t = 0$ at launch.)

where

$$\cos 2\beta_i = -\frac{1 - \sin\psi}{1 + \sin\psi} ; \qquad i = 1,2; \qquad \pi/2 < \beta_1 < \pi < \beta_2 < 3\pi/2, \quad (106)$$

we find $\widetilde{S}$ is positive, leading to a decrease in e during this part of the year. Since $\sin\psi \approx .4$, we obtain $\beta_2 - \beta_1 \approx 65°$. Hence, for a little more than a month both before and after each solstice, perigee height increases, while during the remainder of the year it decreases. The more accurate computer calculations shown in Figure 3 closely follow this prescription.

If $\Omega_0 = \pi$, then $\omega_{0+} = 3\pi/2$ and with the resonance condition satisfied

$$2\widetilde{S} = -[1 + \sin\psi] -$$
$$- [1 - \sin\psi]\cos 2\dot{\beta}\,t. \quad (107)$$

In this case, $\widetilde{S}$ remains negative throughout the year, and e increases monotonically but with a varying rate. The ratio of the minimum to the maximum rate is $\sin\psi$ with the former occurring (in this example) at the solstices $(2\dot{\beta}\,t = \pi$

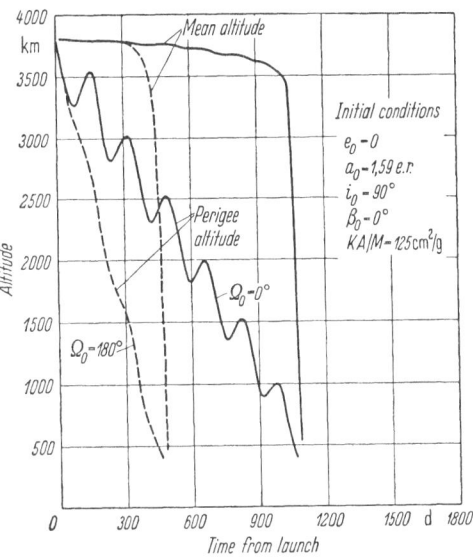

Fig. 3. Examples of resonant behavior in polar orbits.

or $3\pi)$ and the latter at the equinoxes $(2\dot{\beta}\,t = 0$ or $2\pi)$. These same results hold for $\beta_0 = \pi$; i.e., for launch at the autumnal instead of the vernal equinox.

For the intermediate cases $(\Omega_0 = \pm\pi/2)$, we find $\omega_{0+} = \pm\pi/2 \pm \psi$ and[1]

$$2\widetilde{S} = -|\sin 2\dot{\beta}\,t|. \quad (108)$$

The distinction between the two types thus disappears. Perigee height decreases monotonically but the rate vanishes every three months. The general case can be similarly analyzed; but the resulting expressions are, of course, not as simple as these.

On the basis of the above analysis alone, it is difficult to make accurate general statements about the relative lifetimes of satellites in

---

[1] Straightforward substitution into Eq. (99) leads to $2\widetilde{S} = -\sin 2\dot{\beta}\,t$. However, every half year the direction of the in-plane component of the earth-sun line changes discontinuously by $180°$; hence we have an absolute value sign in Eq. (108).

these resonant orbits. While we would expect the initial condition
$\Omega_0 = \pi$ to lead to the shortest lifetime, comparisons are complicated
by the effects of the changes in i.[1] For example, because of these changes,
we find that in the longer lifetime cases, there is a continuous drift
from one type of resonance to the other.

In addition to studying the lifetime as a function of the time of day
of launch, given the time of year (Ω_0 varying; β_0 fixed), we can also
inquire about the dependence on the season of launch, for insertion
into orbit at a given time
of day (β_0 varying, $\Omega_0 - \beta_0$
fixed). These and other varia-
tions have been studied
with our computer program.

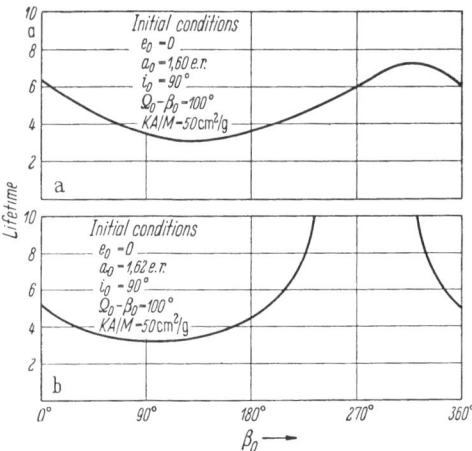

As an example, we show
in Figure 4 the dependence
of lifetime on the season of
launch for satellites with
$KA/M = 50$ cm²/gm and
with $\Omega_0 - \beta_0 \approx \pi/2$. In Figure 4a, where $i_0 = \pi/2$ and
$a_0 = 1.6$ e.r., the lifetime
varies almost sinusoidally,
the maximum and minimum
differing by about four years.

Fig. 4. Dependence of lifetime on season of launch
for initially polar orbits.

For different initial values
of a and i, the contrast can be much more extreme. For example with $i_0 = \pi/2$ but $a_0 = 1.62$ e.r., about 200 km larger than the
nominal resonance value for $e_0 = 0$, we find (see Figure 4b) that
resonance behavior is obtained only for $-40° \lesssim \beta_0 \lesssim 230°$. For
other values of β_0, the lifetimes are far greater: Perigee height
undergoes oscillations of very long period, increasing the lifetime
by a possibly large factor whose value depends critically on the air
density.

The results of a complementary calculation of lifetimes for different
values of a and i (with Ω_0 and β_0 fixed) have been given elsewhere [36].
There we showed that the size of the region of the $a - i$ plane in which
resonant behavior is obtained throughout a satellite's lifetime, depends
on the value of KA/M. For $i \approx \pi/2$ the width is roughly 300 km, when
$KA/M = 50$ cm²/gm, and increases very sharply, reaching approximately 6000 km for $KA/M = 250$ cm²/gm. For KA/M about 25 cm²/gm
or below, this resonance region vanishes.

---

[1] Note that $\dot{\Omega}_{2h}$ no longer vanishes when $i \neq \pi/2$.

3. Influences of charge drag

While for satellites with sufficiently high values of KA/M resonance behavior is virtually independent of neutral drag, it may be influenced by charge drag. This possibility seems to exist only for satellites small compared to the shielding distance in the plasma (see Sections III and VI). To investigate this influence in meaningful detail is impossible at present. We therefore describe it qualitatively and then discuss a few numerical results obtained for a simple model.

If charge drag causes a small decrease in the mean altitude of the satellite, we would expect its lifetime to be shortened if it has an initial mean altitude near the upper edge of the altitude range in which resonant behavior is obtained in the absence of drag. On the other hand, if the initial mean altitude is near the lower edge, we expect an increase in lifetime. In either case, for not too large a value of KA/M, we could anticipate that as Δa_{cd} increases from a negligible to a large magnitude, the lifetime will eventually increase and then decrease, when charge drag itself becomes a limiting factor. The net effect of moderate values of Δa_{cd} is, therefore, to raise a resonance region (in altitude) and to change its size.[1]

The precise position of such a region (in the $a-i$ plane) would clearly depend on the values of Δa_{cd} which are, unfortunately, unknown. Ideally, then, we would want the resonance regions for *all* values of Δa_{cd} to overlap so that some (smaller) region would be common to all. In that case, we could *a priori* be sure to obtain resonant behavior. A preliminary numerical investigation—based on a very simple model in which the charge drag is kept constant—indicates that, for KA/M $= 50$ cm²/gm, a narrow, overlap resonance region does exist, extending in altitude from about 3950 km to 4000 km at $i = \pi/2$. At the center of the original resonance region, (see Figure 4 of Ref. [36]) we find that only for 50 km/year $< \Delta a_{cd} < 250$ km/year is the lifetime increased substantially: In these cases the sunlight pressure resonance is sufficiently "spoiled" so that perigee height undergoes several long-period oscillations before the satellite ceases to orbit.

For equatorial orbits, with $KA/M = 125$ cm²/gm, the overlap resonance region appears to extend over several hundred km.

The sensitivity of these conclusions to the model used for charge drag is still under investigation, as are the effects of charge drag on the tumbling motion of very thin cylindrical satellites. (Since the acceleration caused by sunlight depends on the cross-sectional area of the satellite as projected on a plane perpendicular to the earth-sun line, it

---

[1] Of course, if charge drag is large, the difference in the lifetime resulting from resonant and from nonresonant behavior will be considerably muted.

is necessary to know the motion of such satellites about their centers of mass.) One might think that if charge drag were important enough to seriously alter the tumbling pattern, then the sunlight pressure resonance would be of little significance; however, the intermediate region bears a closer investigation.

4. Simplified theory

An approximate analytical solution for the long-term variations of the orbital elements can be derived easily after certain simplifications have been made. While not applicable for quantitative predictions, such solutions indicate the main features of the orbital behavior. For example, suppose that the eccentricity is very small (so that terms in e^2 can be ignored), that the ecliptic coincides with the equator[1], and that the earth is transparent. From Section III we find that for direct equatorial orbits the mean rates of change of e and $\omega + \Omega$ are given by

$$\left.\begin{array}{l} \dfrac{2\pi a^{3/2}}{\mu^{1/2}}\dfrac{de}{dt} = 3\pi\,\lambda_{\mathrm{sp}}\sin(\omega + \Omega - \beta), \\[3mm] \dfrac{2\pi a^{3/2}}{\mu^{1/2}}\dfrac{d(\omega + \Omega)}{dt} = \dfrac{3\pi\,\lambda_{\mathrm{sp}}}{e}\cos(\omega + \Omega - \beta) + \dfrac{3\pi J_2}{a^2}, \end{array}\right\} \quad (109)$$

when we consider only the effects of sunlight pressure and the second harmonic. If we denote the relative orientation of the orbit major axis and the earth-sun line by ϕ, we find

$$\left.\begin{array}{l} \dfrac{de}{dt} = \Lambda\sin\phi, \\[3mm] \dfrac{d\phi}{dt} = \dfrac{\Lambda\cos\phi}{e} + (\dot{\Pi}_{2h} - \dot{\beta}_s), \end{array}\right\} \quad (110)$$

where

$$\left.\begin{array}{l} \Lambda = \dfrac{3}{2}\left(\dfrac{K\,A}{M}\right)\left(\dfrac{I}{c}\right)\left(\dfrac{a}{\mu}\right)^{1/2}\text{ rad/sec}, \\[3mm] \dot{\Pi}_{2h} = \dfrac{3}{2}\dfrac{J_2}{a^{7/2}}\mu^{1/2}\text{ rad/sec}, \\[3mm] \dot{\beta}_s \approx 2\times 10^{-7}\text{ rad/sec}. \end{array}\right\} \quad (111)$$

To solve Eqs. (110), we introduce

$$\left.\begin{array}{l} l = e\sin\phi, \\[2mm] h = e\cos\phi, \end{array}\right\} \quad (112)$$

and obtain

$$\left.\begin{array}{l} \dfrac{dl}{dt} = [\dot{\Pi}_{2h} - \dot{\beta}_s]\,h + \Lambda, \\[3mm] \dfrac{dh}{dt} = -[\dot{\Pi}_{2h} - \dot{\beta}_s]\,l, \end{array}\right\} \quad (113)$$

---

[1] We make this assumption only to ensure that the general behavior of the solution is not obscured by algebraic complications.

which leads to
$$\frac{d^2 l}{d t^2} + [\dot{\Pi}_{2h} - \dot{\beta}_s]^2 l = 0, \tag{114}$$

and
$$\left.\begin{aligned} l(t) &= \frac{\Lambda}{(\dot{\Pi}_{2h} - \dot{\beta}_s)} \sin(\dot{\Pi}_{2h} - \dot{\beta}_s)\, t, \\ h(t) &= \frac{\Lambda}{(\dot{\Pi}_{2h} - \dot{\beta}_s)} \{\cos(\dot{\Pi}_{2h} - \dot{\beta}_s)\, t - 1\}, \end{aligned}\right| \tag{115}$$

for the initial condition $e_0 = 0$. Returning to our original variables, we have
$$\left.\begin{aligned} e(t) &= \left| \frac{2\Lambda}{\dot{\Pi}_{2h} - \dot{\beta}_s} \sin[(\dot{\Pi}_{2h} - \dot{\beta}_s)\,(t/2)] \right|, \\ \phi(t) &= \frac{1}{2}\{\pi + (\dot{\pi}_{2h} - \dot{\beta}_s)\, t\}. \end{aligned}\right| \tag{116}$$

In this case, therefore, the relative motion of perigee and the earth-sun line is periodic, with period $4\pi/(\dot{\Pi}_{2h} - \dot{\beta}_s)$. [We also note that $\phi(t = 0^+) = \pi/2$.] The period of oscillation of eccentricity is half as long, while the maximum value reached by e is $2\Lambda/|\dot{\Pi}_{2h} - \dot{\beta}_s|$.

Resonance occurs when $\dot{\Pi}_{2h} = \dot{\beta}_s$. The solution to Eqs. (113) is then
$$\left.\begin{aligned} l(t) &= \Lambda t, \\ h(t) &= 0, \end{aligned}\right\} \tag{117}$$

for the initial condition $e_0 = 0$. Hence,
$$\left.\begin{aligned} e(t) &= \Lambda t, \\ \phi(t) &= \pi/2, \end{aligned}\right\} \tag{118}$$

i.e., the eccentricity increases linearly with time while perigee remains perpendicular to the earth-sun line. (Some of the features complicating this simple description are given in Section IV.1.)

The general solution to Eqs. (113) is
$$\left.\begin{aligned} l(t) &= e' \sin[\phi' - (\dot{\Pi}_{2h} - \dot{\beta}_s)\, t], \\ h(t) &= -e' \cos[\phi' - (\dot{\Pi}_{2h} - \dot{\beta}_s)\, t] - \frac{\Lambda}{(\dot{\Pi}_{2h} - \dot{\beta}_s)}, \end{aligned}\right\} \tag{119}$$

where e' and ϕ' are constants, determined by the initial conditions. If we choose conditions such that $l_0 = 0$ and $h_0 = -\Lambda/(\dot{\Pi}_{2h} - \dot{\beta}_s)$, when $\dot{\Pi}_{2h} > \dot{\beta}_s$, then $e' = 0$ and
$$\left.\begin{aligned} l(t) &= 0, \\ h(t) &= -\frac{\Lambda}{(\dot{\Pi}_{2h} - \dot{\beta}_s)}, \end{aligned}\right\} \tag{120}$$

i.e.,
$$\left.\begin{aligned} e(t) &= \frac{\Lambda}{|\dot{\Pi}_{2h} - \dot{\beta}_s|}, \\ \phi(t) &= \pi, \end{aligned}\right\} \tag{121}$$

which corresponds to a stable orbit configuration: The eccentricity remains constant, and the apsidal line remains aligned with the earth-sun
line (see Section V). If $\dot{\Pi}_{2h} < \dot{\beta}_s$, then $\phi(t) = 0$, i.e., perigee is on the
sunlit side of the earth in this case.

More realistic conditions, at any inclination angle, can be treated
analytically in this same manner; but to obtain accurate predictions
for the long-term variations of the orbital elements, we resort to our
computer program.

V. Theory of stable orbits

We will now discuss a situation complementary to that described
in the previous section: If the in-plane component of the earth-sun line
remains aligned with the orbit major axis, then sunlight pressure will
have no effect on eccentricity, and, hence no effect on the orbit's shape.
The alignment can be maintained by choosing orbits in which the combined effect of the second harmonic and sunlight pressure on the motion
of perigee offsets that of the earth-sun line. We call orbits with this
characteristic stable. Actually, even neglecting air drag, there are no
orbits that are strictly stable in the above sense. For example, consider
equatorial orbits. While the sun moves in the ecliptic at a nearly constant
angular rate, its projection on the orbit plane moves at a varying rate.[1]
But, the corresponding variations in the motion of the perigee (caused
by the varying in-plane component of sunlight pressure) cannot compen-
sate for the varying motion of the earth-sun line. The resulting semiannual oscillations in eccentricity are, however, of small amplitude—
proportional to $\sin^2(\psi/2)$. In our discussion, we will therefore still refer
to these orbits as stable.

We shall again consider only the perturbations caused by sunlight
pressure and by the second harmonic. (The higher harmonics could be
included with only a slight modification in our development.) Neglecting
air drag means, in effect, that our analysis is meaningful only for orbits
with perigee heights that lie at sufficiently high altitudes.

1. Equatorial orbits

We examine direct equatorial orbits first. For these, $i = 0$ and Ω
becomes undefined. Therefore we introduce the element Π[2]:

$$\Pi = \omega \cdot \cos i + \Omega. \tag{122}$$

---

[1] The conclusion to our argument is unaffected by the nonzero ellipticity of
the earth's orbit (which we ignore in this section).

[2] This definition, rather than the more orthodox one, is introduced so that Π
and β will increase in the same direction, both for direct and for retrograde orbits.

The combined first-order effects per revolution of the two perturbations being considered are:

$$
\begin{aligned}
\Delta e_{2h} + \Delta e_{\mathrm{sp}} &= \lambda_{\mathrm{sp}}\{[\cos^2(\psi/2)\cos(\Pi - \beta) + \sin^2(\psi/2) \times \\
&\quad \times \cos(\Pi + \beta)]Q_3 + [\cos^2(\psi/2)\sin(\Pi - \beta) + \\
&\quad + \sin^2(\psi/2)\sin(\Pi + \beta)]Q_4\}, \\
\Delta\Pi_{2h} + \Delta\Pi_{\mathrm{sp}} &= \frac{\lambda_{\mathrm{sp}}}{e}\{[\cos^2(\psi/2)\cos(\Pi - \beta) + \sin^2(\psi/2) \times \\
&\quad \times \cos(\Pi + \beta)]Q_5 + [\cos^2(\psi/2)\sin(\Pi - \beta) + \\
&\quad + \sin^2(\psi/2)\sin(\Pi + \beta)]Q_6\} + \frac{3\pi J_2}{p^2}.
\end{aligned} \qquad (123)
$$

If $\Delta\beta$ represents the angular motion of the earth-sun line per revolution of the satellite, then we wish to find initial conditions such that

$$
\begin{aligned}
\Delta e &= 0, \\
\Delta\Pi - \Delta\beta &= 0.
\end{aligned} \qquad (124)
$$

The analysis is complicated by the presence of the earth's shadow. All direct (and almost all retrograde) stable orbits with $a \leq 1.93$ e.r. continuously intersect the shadow. For $a \geq 2.5$ e.r., most stable orbits are, at times, not in shadow (e.g., when the sun-line is maximally inclined to the orbit plane). When the orbit is always in shadow, the magnitude of $\cos E$ at the boundary increases monotonically from

$$
\{1 - [R_0^2/a^2(1 - e^2)]\}^{1/2}
$$

to[1]

$$
\left| \frac{\{(1 - e^2)^2\cos^2\psi - (R_0^2/a^2)(\cos^2\psi - e^2)\}^{1/2} \pm e\sin^2\psi}{\cos^2\psi - e^2} \right|, \qquad e \neq \cos\psi,
$$

as the inclination of the earth-sun line to the equatorial plane increases from 0 to ψ. To simplify our development we assume the shadow to have a constant, mean angular extent, 2η.[2] While making no essential use of it, nevertheless we sometimes retain η in our formulas to indicate their general dependence on the extent of the shadow region.

We consider first direct equatorial stable orbits with $a \leq 1.93$ e.r. Substituting Eqs. (35) in (123), we find the total changes in e and Π:

$$
\begin{aligned}
\Delta e &= \lambda_{\mathrm{sp}}(1 - e^2)^{1/2}\{\cos^2(\psi/2)\sin(\Pi - \beta) + \sin^2(\psi/2)\sin(\Pi + \beta)\} \times \\
&\quad \times \{3(\pi - \eta) + \sin\eta(4e - \cos\eta)\}, \\
\Delta\Pi &= \frac{3\pi J_2}{p^2} + \frac{\lambda_{\mathrm{sp}}(1 - e^2)^{1/2}}{e}\{\cos^2(\psi/2)\cos(\Pi - \beta) + \sin^2(\psi/2) \times \\
&\quad \times \cos(\Pi + \beta)\}\{3(\pi - \eta) + \sin\eta(2e + \cos\eta)\}.
\end{aligned} \qquad (125)
$$

---

[1] Note that the upper sign applies if perigee is in sunlight and the lower sign if it is in shadow.

[2] $\cos\eta = \langle\cos E\rangle$, where $\langle\cos E\rangle$ is a yearly average of the magnitude of $\cos E$ at the shadow boundary.

We wish to choose e_0 and Π_0 so that $\Delta \dot{\Pi} - \Delta \dot{\beta} \equiv \dot{\Pi} - \dot{\beta} = 0$[1] (i.e., we want $\dot{\Pi}$ to equal the average rate of motion of the in-plane component of the earth-sun line[2]). Since $\dot{\Pi}_{2h} > \dot{\beta}$ when $a < 1.93$ e.r., we must select $\Pi_0 = \beta_0 + \pi$ and

$$e_s = \lambda_{\mathrm{sp}}(1 - e_s^2)^{1/2}\{\cos^2(\psi/2) + \sin^2(\psi/2)\cos 2\beta\} \times$$
$$\times \left\{\frac{3(\pi - \eta) + \sin \eta\,(2e_s + \cos \eta)}{\dot{\Pi}_{2h} - \dot{\beta}}\right\} \quad (126)$$

in order to obtain $\dot{\Pi} = \dot{\beta}$. Differentiating, we have

$$\dot{e}_s = -\lambda_{\mathrm{sp}}(1 - e_s^2)^{1/2}\sin^2(\psi/2)\sin 2\beta\{3(\pi - \eta) + \sin \eta\,(2e_s + \cos \eta)\}\frac{2\dot{\beta}}{\dot{\Pi}_{2h} - \dot{\beta}}$$
$$(127)$$

if we neglect $\dot{e}_s$ on the right side since $\dot{e}_s \ll \dot{\beta}$. But from the first of Eqs. (125) we also have

$$\dot{e}_s = -\lambda_{\mathrm{sp}}(1 - e_s^2)^{1/2}\sin^2(\psi/2)\sin 2\beta\{3(\pi - \eta) + \sin \eta\,(4e - \cos \eta)\}. \quad (128)$$

For Eqs. (127) and (128) to agree, we must choose $\dot{\Pi}_{2h}(=\dot{\Pi}_{2hs}$, since a subscript s denotes the value of a quantity appropriate for a stable orbit) such that

$$\dot{\Pi}_{2hs} = \dot{\beta}\left\{1 + 2\frac{3(\pi - \eta) + \sin \eta\,(2e + \cos \eta)}{3(\pi - \eta) + \sin \eta\,(4e - \cos \eta)}\right\} \approx 3\dot{\beta}. \quad (129)^3$$

For $e_s \ll 1$, the value of a which satisfies Eq. (129) is 1.41 e.r.[4] Hence, we find that our orbit is (almost) stable: The eccentricity oscillates with a half-year period and a very small amplitude, proportional to $\sin^2(\psi/2) \approx .04$. This oscillation cannot be eliminated because of the varying rate of motion of the in-plane component of the earth-sun line.

For different values of a and e there are other (almost) stable orbits; but the corresponding oscillations in e are not so simple, despite amplitudes

---

[1] Here all time derivatives are expressed in radians per satellite period; thus $\dot{\Pi}_{2h} = \dfrac{3\pi J_2}{p^2}$. (Compare Section IV. 4.)

[2] The maximum difference between the angular distance moved by the earth-sun line in the ecliptic and the corresponding angular distance moved by its projection on the orbit plane is about 2.5° and occurs for $\beta \approx 46°$. Of course, such a maximum is reached every three months. For comparison, we note that the maximum difference between the true anomaly and the mean anomaly for the earth's orbit is almost 2° and is reached every half-year. In a more accurate analysis this variation would also be included.

[3] We can neglect e on the right side since the error is no larger than that incurred by always using the mean extent, 2η, for the shadow region. Also, $\sin \eta \cos \eta$ is usually small compared to $3(\pi - \eta)$.

[4] Given KA/M, the pair of values (e_s, a_s) satisfying Eqs. (126) and (129) is unique. Considering all possible values of KA/M, we find $0 \le e_s \le .33$ and 1.41 e.r. $\lesssim a_s \lesssim 1.50$ e.r.

proportional to $\sin^2(\psi/2)$. If we completely ignore terms proportional to it [and replace $\cos^2(\psi/2)$ by unity], we find from Eqs. (125) that

and
$$\dot e \approx \lambda_{\rm sp}(1 - e^2)^{1/2} \sin(\Pi - \beta)\{3(\pi - \eta) - \sin\eta\cos\eta\}, \quad (130)$$

$$\dot\Pi - \dot\beta \approx (\dot\Pi_{2h} - \dot\beta) + \frac{\lambda_{\rm sp}}{e}(1 - e^2)^{1/2}\cos(\Pi - \beta)\{3(\pi - \eta) + \sin\eta\cos\eta\}. \quad (131)$$

Therefore, we obtain (almost) stable orbits with

$$e_s \approx \lambda_{\rm sp}(1 - e_s^2)^{1/2}\left\{\frac{3(\pi - \eta) + \sin\eta\cos\eta}{|\dot\Pi_{2hs} - \dot\beta|}\right\}, \quad (132)$$

if $\Pi_0 = \beta_0 + \pi$ when $\dot\Pi_{2hs} > \dot\beta$ and if $\Pi_0 = \beta_0$ when $\dot\Pi_{2hs} < \dot\beta$ (when $\dot\Pi_{2h} = \dot\beta$ we have the condition of resonance).

Extracting the explicit dependence on eccentricity, and neglecting the (small) η dependence, we find

$$\frac{e_s\left|3\pi\mu^{1/2}J_2 - 2\pi a_s^{7/2}(1 - e_s^2)^2\dot\beta_s\right|}{\mu^{1/2}a_s^2(1 - e_s^2)^{5/2}} \approx 3\pi\lambda_{\rm sp}, \quad (133)$$

where $\dot\beta_s$ is defined in Eq. (111). Given $a_s < 1.93$ e.r., this equation has at most one solution for e_s for each value of KA/M. But for $a_s > 1.93$ e.r. we can find as many as three stable orbits for the same value of KA/M. Presenting the analytical proof for these statements is somewhat tedious; instead we summarize the results graphically. In Figure 5, we show for several values of a the eccentricity of stable orbits as a function of KA/M. The solid portions of the curves signify that the perigee distance, q, exceeds 1.25 e.r., whereas for the dotted portions 1.0 e.r. $\leqq q \leqq 1.25$ e.r. In other words, the solid segments refer to altitude regions where air drag has a relatively small effect on the stability, at least for several years. In Figure 6, we show, as a function of

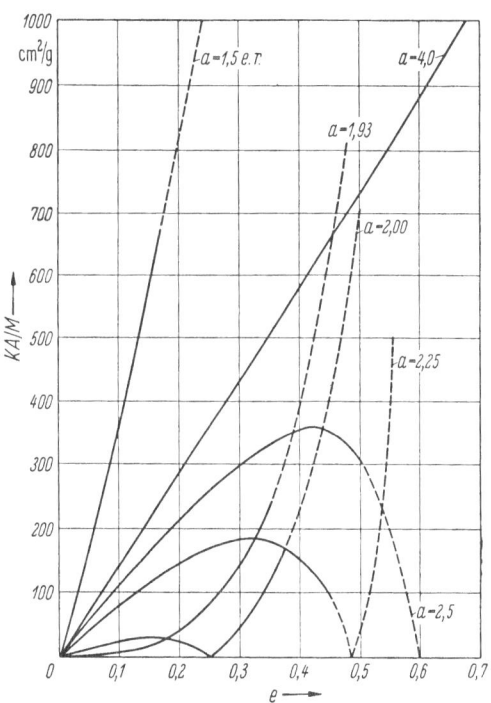

Fig. 5. Relationship between eccentricity and area-to-mass ratio for direct stable equatorial orbits.

a, the maximum value of KA/M for which at least one direct stable orbit exists. (We have again considered two cases: $q_{min} \geq 1.25$ e.r. and $q_{min} \geq 1.0$ e.r.)

By confining ourselves to stable orbits with perigee heights greater than 1.25 e.r., we find that there are at most three direct stable orbits for each a satisfying 1.93 e. r. $\leq a \leq 2.20$ e. r., at most two for each a

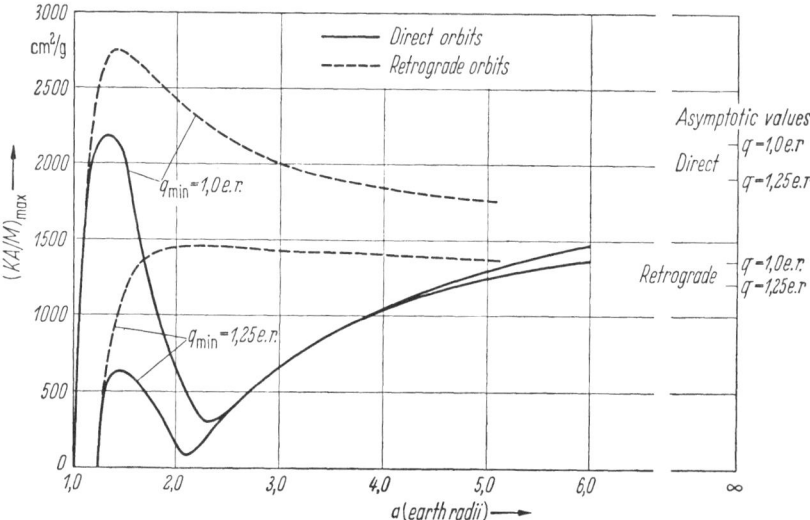

Fig. 6. Criteria for existence of stable equatorial orbits.

satisfying 2.20 e.r. $\leq a \leq 3.5$ e.r., and at most one for $a > 3.5$ e.r. There are no direct stable orbits, for any value of a, if KA/M exceeds 2200 cm²/gm.

For direct stable orbits, perigee is always in shadow if $a_s < 1.93$ e.r., while for $a_s > 1.93$ e.r., perigee is on the sunlit side of the earth except for those stable orbits corresponding to the "second branch" of the curves in Figure 5, i.e., the "branch" with the larger eccentricities (when two branches exist).

For retrograde equatorial stable orbits ($i = \pi$), a similar analysis can be performed. Thus,

$$\dot{e} = -\lambda_{sp}(1 - e^2)^{1/2}\{\cos^2(\psi/2)\sin(\Pi - \beta) + \\ + \sin^2(\psi/2)\sin(\Pi + \beta)\}\{3(\pi - \eta) - \sin\eta\cos\eta\},$$

$$\Pi - \dot{\beta} = -(|\dot{\Pi}_{2h}| + \dot{\beta}) + \frac{\lambda_{sp}(1 - e^2)^{1/2}}{e}\{\cos^2(\psi/2)\cos(\Pi - \beta) + \\ + \sin^2(\psi/2)\cos(\Pi + \beta)\}\{3(\pi - \eta) + \sin\eta\cos\eta\}. \qquad (134)^1$$

---

[1] $|\dot{\Pi}_{2h}| = -\dot{\Pi}_{2h} = \dfrac{3\pi J_2}{p^2}$ for retrograde equatorial orbits.

For $\dot\Pi - \dot\beta$ to vanish, we must choose $\Pi_0 = \beta_0$ (i.e., for a retrograde stable orbit, perigee will always be on the sunlit side of the earth) and

$$e_s = \lambda_{\rm sp}(1 - e_s^2)^{1/2}\{\cos^2(\psi/2) + \sin^2(\psi/2)\cos 2\beta\}\left\{\frac{3(\pi - \eta) + \sin\eta\,\cos\eta}{|\dot\Pi_{2hs}| + \dot\beta}\right\}. \tag{135}$$

To be consistent with the first of Eqs. (134), we must have

$$|\dot\Pi_{2hs}| \approx \dot\beta, \tag{136}$$

which is the analog of Eq. (129) and is satisfied for $a_s \approx 1.93$ e.r. if e_s is small. Again, if we neglect terms proportional to $\sin^2(\psi/2)$, we find

$$e_s \approx \lambda_{\rm sp}(1 - e_s^2)^{1/2}\left\{\frac{3(\pi - \eta) + \sin\eta\,\cos\eta}{|\dot\Pi_{2hs}| + \dot\beta}\right\} \tag{137}$$

and $\Pi_0 = \beta_0$ as the conditions for stable orbits. Here, given a_s, there is at most one solution, e_s, for each value of KA/M. The eccentricity

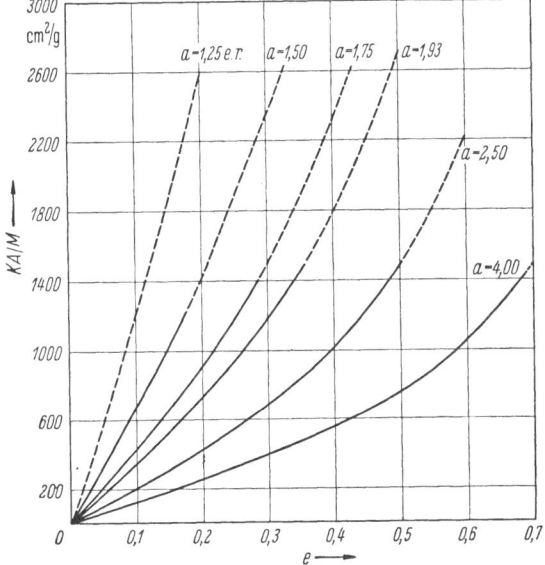

Fig. 7.
Relationship between eccentricity and area-to-mass ratio for retrograde stable equatorial orbits.

as a function of KA/M, for several different values of a is shown in Figure 7. Again, the solid portions of the curves signify $q \geq 1.25$ e.r., whereas the dotted portions refer to values of q satisfying 1.0 e.r. $\leq q \leq 1.25$ e.r. In Figure 6, we also give, as a function of a, the maximum value of KA/M for which a retrograde stable orbit exists.

For both direct and retrograde equatorial orbits, care must be exercised in drawing conclusions about long-term stability, especially for the "stable" orbits with high eccentricity. While $\dot{\Pi}_{2h}$ depends quadratically on inclination (see Section IV), $\dot{\Pi}_{sp}$ contains terms linear in i [see Eqs. (32) and (34)]; hence the changes in inclination caused by sunlight pressure (which are essentially linear in eccentricity) will in general result in a drift of the apsidal line away from the in-plane component of the earth-sun line. In addition, for high eccentricities, lunar and solar gravitational perturbations become important.

A preliminary analysis using our computer program indicates that the stability of many high eccentricity orbits can be expected to be maintained for at least two or three years.

2. Near-stable equatorial orbits

Because of the difficulty in achieving precisely an orbit with a predetermined eccentricity and perigee position, we investigate the behavior of orbits near those that are stable. To simplify the analysis, we neglect terms proportional to $\sin^2(\psi/2)$, and also the effects of the earth's shadow.

Consider a direct orbit whose initial center lies a distance ϱ from the center of the corresponding stable orbit (i.e., a stable orbit with

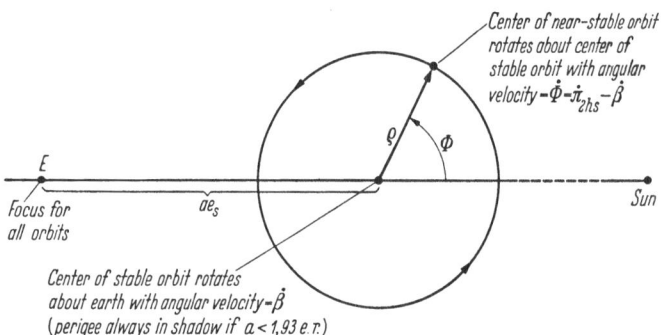

Fig. 8. Motion of the centers of direct near-stable equatorial orbits.

the same value of a for a satellite with the same value of KA/M). Let the line joining the centers make an angle Φ with the earth-sun line, as in Figure 8. For convenience, we use a coordinate system rotating in the orbit plane with angular velocity $\dot{\beta}$, since the stable orbit remains (almost) at rest in this system. We have also assumed that $a < 1.93$ e.r.; hence the center of the stable orbit lies on the sunlit side of the earth.

From the figure, we see that

$$\varrho^2 \approx a_s^2\{e^2 + e_s^2 + 2\,e\,e_s\cos(\Pi - \beta)\}, \tag{138}$$

since $\Pi_s \approx \beta + \pi$, and that

$$\varrho\,\dot{\varrho} \approx a_s^2\{\dot{e}[e + e_s\cos(\Pi - \beta)] - e\,e_s(\dot{\Pi} - \dot{\beta})\sin(\Pi - \beta)\}, \tag{139}$$

since $\dot{e}_s \approx 0$. Using Eqs. (125) and (132), we find

$$
\left.
\begin{aligned}
\dot{e} &\approx 3\pi \lambda_{\mathrm{sp}}(1 - e^2)^{1/2} \sin(\Pi - \beta), \\
\dot{\Pi} - \dot{\beta} &\approx \dot{\Pi}_{2hs} - \dot{\beta} + \frac{3\pi \lambda_{\mathrm{sp}}(1 - e^2)^{1/2}}{e} \cos(\Pi - \beta),
\end{aligned}
\right\}
\tag{140}
$$

and

$$
e_s \approx \frac{3\pi \lambda_{\mathrm{sp}}(1 - e_s^2)^{1/2}}{\dot{\Pi}_{2hs} - \dot{\beta}},
\tag{141}
$$

since $\dfrac{3\pi J_2}{p^2} \approx \dot{\Pi}_{2hs}$. Substituting in Eq. (139) leads to $\dot{\varrho} \approx 0$.

Similarly we have

$$
a_s^2 e^2 = a_s^2 e_s^2 + \varrho^2 + 2\varrho\, a_s\, e_s \cos \Phi
\tag{142}
$$

and (since $\dot{e}_s \approx \dot{\varrho} \approx 0$),

$$
\dot{\Phi} \approx \frac{a_s e}{\sin \Phi} \frac{\dot{e}}{\varrho\, e_s} \approx \frac{\varrho}{\sin(\Pi - \beta)} \frac{\dot{e}}{\varrho\, e_s} \approx \dot{\Pi}_{2hs} - \dot{\beta}.
\tag{143}
$$

In this approximation, therefore, we find that the center of the near-stable orbit moves, with a constant angular velocity $\dot{\Pi}_{2hs} - \dot{\beta}$, in a circular path about the center of the corresponding stable orbit. The angular velocity of the center depends only on the characteristics of the stable orbit. The initial conditions e_0 and ω_0 of the near-stable orbit determine only the radius of the circular motion of its center in the rotating frame.

For stable orbits (both direct and retrograde) with their centers on the dark side of the earth-sun line, we find for the motion of the corresponding near-stable orbit that $\dot{\varrho} \approx 0$ and that

$$
\dot{\Phi} \approx \dot{\beta} + |\dot{\Pi}_{2hs}|,
\tag{144}
$$

where Φ is again measured from the earth-sun line—counter-clockwise when looking down from the North Pole. With a proper choice of initial conditions, near-stable orbits can be obtained whose centers trace out (in a nonrotating geocentric frame) different well-known (closed) two-dimensional curves, such as cardioids, ellipses, etc. [37].

While the perigee height of a stable orbit is by definition constant, the perigee heights of near-stable orbits vary. With $q = a(1 - e)$ being the perigee height, we find the amplitude of its oscillation for near-stable orbits to be given by

$$
q_{\max} - q_{\min} = |\varrho + a_s\, e_s| - |\varrho - a_s\, e_s| =
\begin{cases}
2 a_s e_s; & \varrho \gtrless a_s e_s \\
2 \varrho; & \varrho \lessgtr a_s e_s
\end{cases}
\tag{145}
$$

In general, the variation of perigee height with time will be smooth. However, if the center of the near-stable orbit passes through the center of the earth ($\varrho = a_s\, e_s$), then there will be a discontinuity in the first derivative of perigee height with respect to time when $e = 0$. To show this, we note that

$$e^2 = 2 e_s^2 [1 + \cos(\dot{\Pi}_{2hs} - \dot{\beta})\, t], \qquad (146)$$

when $\varrho = a_s\, e_s$, and hence

$$\dot{q} = -a_s\, \dot{e} = \frac{(\dot{\Pi}_{2hs} - \dot{\beta}) \sin(\dot{\Pi}_{2hs} - \dot{\beta})\, t}{2^{1/2} [1 + \cos(\dot{\Pi}_{2hs} - \dot{\beta})\, t]^{1/2}}$$

$$= \begin{cases} \dot{\Pi}_{2hs} - \dot{\beta}; & \text{if } (\dot{\Pi}_{2hs} - \dot{\beta})\, t = \pi_- \\ -(\dot{\Pi}_{2hs} - \dot{\beta}); & \text{if } (\dot{\Pi}_{2hs} - \dot{\beta})\, t = \pi_+ \end{cases} \qquad (147)$$

For all $\varrho \neq a_s\, e_s$, we find $\dot{q} = 0$ when $(\dot{\Pi}_{2hs} - \dot{\beta})\, t = \pi$.

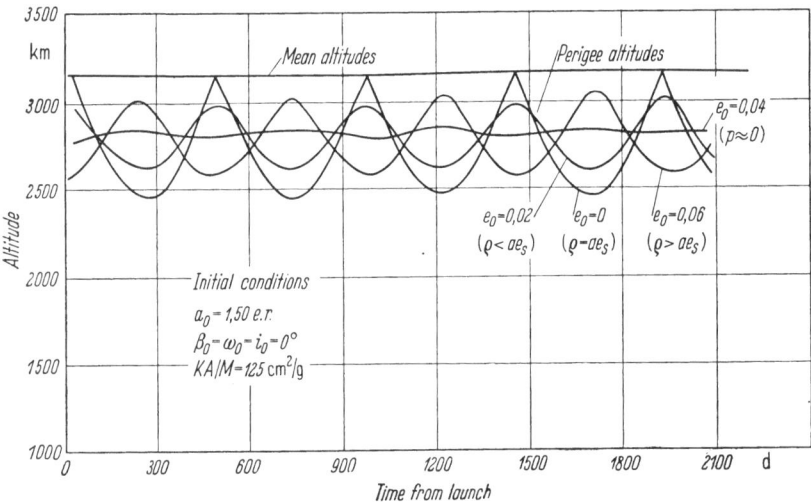

Fig. 9. Variations in perigee height for stable and near-stable equatorial orbits.

In Figure 9, we show the variations in perigee height obtained with our computer program for a stable equatorial orbit and some corresponding near stable orbits.

3. Dispersion of cylinders launched in a stable equatorial orbit

As an application of these concepts, we consider the dispersion of a cloud of thin cylinders (e.g., West Ford dipoles) launched into a direct stable equatorial orbit with $a < 1.93$ e.r. These cylinder satellites we assume to tumble rapidly compared with an orbital period. The effective area for acceleration by sunlight pressure would then vary from $A_{\max} = l\, d$ to $A_{\min} = (2/\pi)\, l\, d$, where l is the length and d the diameter of the cylinders [36]. We denote the corresponding values of

λ_{sp} by λ_{max} and λ_{min}, and assume that the cylinders are initially in the stable orbit corresponding to $\bar{\lambda} = (1/2)\,(\lambda_{min} + \lambda_{max})$, and that the relative velocities at launch were sufficiently small so that we can neglect the consequent differences in five orbital elements. The center of the orbits of those cylinders with values of $\lambda_{sp} = \lambda_{max}$ start rotating (with angular velocity $\dot{\Pi}_{2hs} - \dot{\beta}$) about the point $C_{\lambda_{max}}$ —the center of the stable orbit corresponding to λ_{max} and the given initial semimajor axis. The center of the orbits of those cylinders with $\lambda = \lambda_{min}$ start rotating with the same angular velocity about the point $C_{\lambda_{min}}$. (See Figure 10.) Were the effective area of each cylinder to remain constant, then after $2\pi/(\dot{\Pi}_{2hs} - \dot{\beta})$

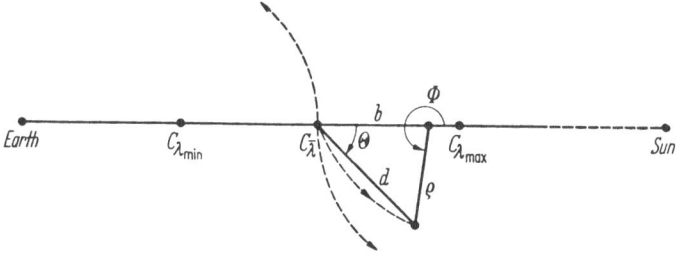

Fig. 10. Dispersion of cylinders initially in a stable orbit.

satellite periods the centers of the orbits of all the cylinders would again coincide. In such an event, the dispersion in orbit centers would be periodic, reaching a maximum of twice the distance between $C_{\lambda_{max}}$ and $C_{\lambda_{min}}$ after every $\pi/(\dot{\Pi}_{2hs} - \dot{\beta})$ satellite periods:

$$2(C_{\lambda_{max}} - C_{\lambda_{min}}) \approx (\lambda_{max} - \lambda_{min}) \frac{6\pi a_s}{\dot{\Pi}_{2hs} - \dot{\beta}}. \qquad (148)$$

The maximum possible dispersion (or, equivalently, the maximum separation of orbit centers) would be attained if those cylinders with $A_{eff} = A_{max}(A_{min})$ would periodically change their orientation instantaneously so that A_{eff} becomes $A_{min}(A_{max})$ and vice-versa. These changes should occur only after each half-revolution of an orbit center about the appropriate center of the corresponding stable orbit. [By symmetry, we see that the maximum separation in perigee heights increases after each half-revolution by $2(C_{\lambda_{max}} - C_{\lambda_{min}})$.]

To prove that this mechanism yields the maximum dispersion, we note that (see Figure 10)

$$d^2 = b^2 + \varrho^2 + 2\varrho\,b\cos\Phi, \qquad (149)$$

where b is the distance from $C_{\bar{\lambda}}$ to the center of the stable orbit about which the center of the near-stable orbit is instantaneously rotating.

Thus

$$\dot{d} \approx b \sin\Theta \cdot \dot{\Phi}. \tag{150}$$

For all pairs d, Θ, $(0 < \Theta < \pi)$, $\dot{d}$ is clearly a maximum if $b = (C_{\lambda_{\max}} - C_{\bar{\lambda}})$, since $\dot{\Phi}$ is the same for all centers of rotation.[1] When Θ passes through zero and $\dot{d} \neq 0$, $\dot{d}$ remains a maximum only if the center of rotation instantaneously moves from $C_{\lambda_{\max}}$ to $C_{\lambda_{\min}}$.

We easily see that the mean maximum rate of dispersion, $\bar{D}_s$, for the perigee heights of cylinders launched in this stable orbit is

$$\bar{D}_s \approx 6a(\lambda_{\max} - \lambda_{\min}), \tag{151}$$

(where by "rate" we mean per orbital period of the satellite).

For a resonant orbit, the corresponding mean maximum rate of dispersion, $\bar{D}_r$, would be[2]

$$\bar{D}_r \approx 3\pi a(\lambda_{\max} - \lambda_{\min}), \tag{152}$$

which is only a factor of $\pi/2$ greater. However, in this latter case, the maximum dispersion occurs when cylinders with $A_{\mathrm{eff}} = A_{\max}(A_{\min})$ maintain their relative average orientation to the earth-sun line. In the former case, the values of $A_{\mathrm{eff}} = A_{\max}(A_{\min})$ are required to change in a very special manner. A more realistic description of the dispersion rate can be obtained by considering all the torques acting on the (tumbling) cylinder and how they change the average relative orientation of it with respect to the sun. For very small cylinders (like the West Ford dipoles), the appropriate random walk problem must be solved by a Monte Carlo technique, since collisions with micrometeoroids cause the tumbling pattern of the dipoles to change discontinuously and randomly (see Section III. 2. II).

4. Polar orbits

Consider now the possible existence of polar stable orbits. One can conceive of various circumtances which might lead to a stable (or almost stable) polar (or near-polar) orbit. For example, suppose

1. The orbit plane remains perpendicular to the plane formed by the earth-sun line and the polar axis. The in-plane component of the earth-sun line (when nonzero) would then always lie along the polar

---

[1] We ignore the dependence on eccentricity of terms like $(1 - e^2)$ since for reasonably small λ_{sp}, $e_s \ll 1$ (see Fig. 5).

[2] Of course, if there is a stable orbit with a given a_s and $e_s \ll 1$, then there is no resonant orbit for the same value of a, unless $a_s \geqslant 1.93$ e.r. and then only with e satisfying $a^{7/2}(1 - e^2)^2 \approx 10$.

axis, and, if perigee were to remain over the pole, the alignment necessary for a stable orbit would be maintained; or

2. The orbit plane remains fixed, and

(I) ω rotates with varying rate to match the varying rotation of the in-plane component of the earth-sun line; or

(II) The apsidal line remains inclined by $\psi°$ to the equator, since the in-plane component of the earth-sun line always remains along this line if $\Omega_0 = \pi/2$ or $3\pi/2$ and $\beta_0 = 0$ or π.

It is easy to demonstrate that none of these conditions can be maintained: Even for a near-polar orbit with $\dot{\Omega}_{2h} \approx 1$ deg/day, perigee could not remain over the pole because the varying in-plane component of the earth-sun line would cause a variation in $\dot{\omega}_{sp}$ and hence $\dot{\omega} \approx \dot{\omega}_{sp} + \dot{\omega}_{2h}$ could not always vanish. Similar analyses show the other possibilities also to be untenable. While one can select initial conditions leading to relatively small changes in orbit shape over relatively long intervals of time, the possible reduction for polar and near-polar orbits does not approach that for equatorial and near-equatorial stable orbits.

VI. Proposed experiments

Finally, we mention several experiments which might be performed with satellites to improve our knowledge of some of the forces that perturb them. Only two (sunlight pressure and charge drag) have been analyzed in any detail.

1. Gravitational field of the earth

To refine the determination of the coefficients in the expansion of the earth's potential, we should

1. Orbit satellites with different inclinations and eccentricities to separate more easily the effects of the different coefficients. (Devising the "optimum" sets of parameters for a given number of satellites would require a careful analysis.)

2. Minimize the effects of air drag[1] (e.g., by orbiting satellites with sufficiently low area-to-mass ratios in high altitude orbits[2];

3. Improve the precision of orbit determination. One suggestion which has been made is to have a satellite emit very short bursts of light at accurately known times to aid precision photographic measurements. If the intensity of the light could be adjusted for prevailing

---

[1] Air drag affects all the orbital elements. Not only are a, e, and, to a lesser extend, ω, Ω, and i affected directly, but all are changed indirectly as well. Thus an imprecise knowledge of air drag causes severe difficulties in determining precisely the coefficients of other perturbing forces.

[2] The higher the altitude, of course, the smaller will be the orbital effects of the higher harmonics.

conditions, it would then be possible to increase the accuracy of locating the satellite relative to the star background.

2. Sunlight pressure

Putting a high area-to-mass ratio rigid spherical satellite in a high altitude orbit (to minimize the effects of air drag and of solar radiation reflected from the earth), should enable us to improve the accuracy of the measurement of the solar constant, perhaps by almost an order of magnitude.[1] Crucial to this improvement are, of course, the precision of orbit determination and the *a priori* knowledge of the reflection characteristics of the satellite's surface.[2] For such accuracy, K should be independent of the satellite's orientation and should be known to within about 0.2% for the (visible) radiation which contributes the bulk of the incident energy. Measurements of K in the ultra-violet need not be made so accurately.

It is also necessary to improve the determination of the shadow boundaries (by considering in detail the influences mentioned in Section III. 2. I) so that the uncertainty is less than half a degree [*38*]. This is easily done. However, to avoid any problem with the shadow boundaries for short-term determinations, the satellite can be placed in a near-polar orbit which remains completely in sunshine for several months.

From the orbital behavior, estimates of the solar constant can then be made once every few days (or less often to obtain an improved average determination).[3] The observed variations can then be compared with ground based measurements or with direct measurements obtained from orbiting instruments.

3. Atmospheric densities

The determination of atmospheric densities from observations of satellites has several weaknesses, one being the low resolving power of the measurements. The behavior of a satellite must be averaged over

---

[1] If a stable orbit is selected, then air drag is more easily minimized since, for a given a, the minimum perigee height will be higher. In addition, for a direct stable orbit with $a < 1.93$ e.r., perigee remains in darkness where the air density is relatively low.

[2] However, K can change in orbit due to impacts of micrometeoroids, and this effect must be estimated. (It might be desirable to have the satellite surface approximate a perfect absorber so that such collisions would have less effect on K; a disadvantage would be the necessity of attaching a transmitter and antenna to aid in tracking.)

[3] Conversely, if the solar constant is known very accurately, a similar experiment can be performed to study characteristics of the radiation reflected from the earth. However, here the interpretation of the orbital behavior will be more difficult.

at least a one-day interval, resulting in a comparable temporal uncertainty. The resolution in position along the orbit is usually no better than about $30°$. To improve this situation, it may be possible to orbit two satellites simultaneously, only one being sensitive to air drag. The differences in their orbital behavior could be determined in several ways. For example, the insensitive one might have very accurate distance measuring equipment (e.g., a millimeter radar or, looking ahead, an optical maser) which could continually monitor the separation between the two. The measurements, unfortunately, could only be taken periodically, when the distance between the two is sufficiently small to allow detection.[1] As an illustration of the accuracy required, we note that with $KA/M \approx 100$ cm^2/gm, densities of the order of 10^{-15} gm/cm^3 could be determined with a 30% accuracy after an interval of 30 sec only if the differences in distances between the centers of mass can be measured within 10 cm. (To insure reasonable lifetime, e should be large.)

One might also place an object inside a balloon satellite and attach it suitably to the balloon surface with tension-measuring cables. From an analysis of the time variations in tensions, one may then be able to deduce almost instantaneous density values [39]. However, not only are extremely sensitive instruments required, but the effects of small torques may severely limit the applicability of such methods.

4. Charge drag

As seen from Section III. 5. II, charge drag should be most easily detected at high altitudes by using very thin cylinders. A worthwhile experiment can therefore be performed by orbiting a UHF dipole[2] at an altitude of, say, 3500 km and with a polar inclination. The resulting changes in period caused by charge drag can be measured by observing the dipole with radar as it transits a given latitude.[3] After several months, changes of the order of -10^{-5} sec/rev (corresponding to a change in a of about $-.02$ km/yr, and a drag acceleration of -2×10^{-8} cm/sec^2) may be just barely detectable. For a dipole constructed from 250μ copper wire ($A_{max}/M \approx 5.7$ cm^2/gm), this acceleration corresponds to a potential of about 0.1 v if we assume $N_i = 3 \times 10^3$ particle/cm^3 at this altitude. However, assuming an ambient density of 8×10^{-21} gm/cm^3 (corresponding to about 60% ionization in a hydrogen dominated environment), we find that the neutral drag acceleration will also be about -2×10^{-8} cm/sec^2. The acceleration caused by sunlight pressure,

---

[1] Only for $i_0 = 0$ or $\pi/2$ will their orbital planes continue to coincide.

[2] Dipoles resonant at a UHF frequency of 400 mc will be approximately 35 cm long.

[3] The difference between the time of latitude crossing of the dipole and of the parent satellite can also be measured with high precision.

being approximately 3×10^{-4} cm/sec², is an even more important limitation on the sensitivity of the experiment. Specifically, this limitation is due to the uncertainty in the change in energy caused by the shadow. To partly circumvent this difficulty, the dipole should be launched in a near-polar orbit such that it will remain continuously in sunlight for as long as possible, at least for several months. The remaining uncertainty in the effect of sunlight pressure on the orbital period is caused by possibly rapid changes of the dipole's tumbling axis or by a tumbling period not very short compared to the orbital one. To guard against these possibilities, the surface of the dipole can be prepared so as to insure that sunlight pressure produces a high rate of tumble. Even with all these conditions satisfied, for a deceleration to be unambiguously attributable to charge drag it should be at least 2×10^{-6} cm/sec², corresponding to a potential of about 1 v.

The sensitivity of this experiment could, of course, be improved by using a thinner wire since the effect of charge drag increases as r_c^{-2} but that of sunlight pressure and neutral drag only as r_c^{-1}. For very thin wire, material with a high strength-to-weight ratio (like beryllium) should be employed.

While there are several radars (e.g., Millstone Hill) which could easily detect one such UHF dipole, it is desirable to orbit several. One can then make measurements showing the dispersive characteristics of the force, and can also guard against the (unlikely) possibility that a micrometeoroid will sever the dipole and drastically reduce the radar cross-section. Contributions to the dispersion can come from natural fluctuations in the plasma, and from different surface characteristics and tumbling patterns of the dipoles. The last is partially subject to direct measurement: With a reasonable radar sensitivity, the tumbling period of the dipole should be obtainable from the scintillations in the received signals.

Had the original West Ford experiment been successful, it would have provided important information on charge drag. Since these dipoles were made from 25μ copper wire, a potential of 1 volt would have produced a charge drag deceleration of about 2×10^{-4} cm/sec². Sunlight pressure produces a maximum acceleration of these dipoles of magnitude about 3×10^{-3} cm/sec², and separation of the two effects would have been easier here. However, in the West Ford experiment, the relatively large dispersion of the myriad dipole orbits would have lowered the sensitivity, perhaps to a comparable level.

In any case, it should be understood that such an experiment at best serves only to delimit the average magnitude of the charge drag deceleration. It will not yield any direct results on the dipoles' potential or on the details of their interaction with the plasma.

5. Poynting-Robertson drag

As shown in Eqs. (94), the decrease in a is so small that were the air density (at great distances from the earth) to decrease to $\approx 3 \times 10^{-21}$ gm/cm³ the effect of the POYNTING-ROBERTSON drag would be hard to distinguish unambiguously. The changes in e, Ω, and i are likewise quite small. However, if an Echo balloon were launched in a nearly-circular resonant orbit ($e_0 \leq .01$), then the change in ω, caused by this force, would be at the rate of about 6×10^{-4} deg/day during the first few months and might possibly be identifiable after this time. (In a resonant orbit, the apsidal line remains perpendicular to the earth-sun line, and the v/c corrections to sunlight pressure exert their maximum effect on ω.) Of course, a second-order theory is absolutely necessary for a theoretical analysis to establish the existence of this motion. In addition, in calculating these v/c corrections, consideration should be given to the effects of the satellite's spin and the temperature distribution over its surface. As pointed out in Section III, dust drag may also have effects similar to these v/c corrections, and therefore may make the identification even more difficult.

6. Corpuscular radiation

We investigated the possibility of observing the perturbations produced by corpuscular radiation on close-in, high area-to-mass ratio satellites. Even were fluxes of the order of 10^{10} charged particle/cm² — sec to impinge on the earth in the polar regions, it is doubtful that the direct momentum transfer to such satellites could be clearly identified from their orbital behavior.

The author wishes to express his gratitude to Dr. HARRISON M. JONES with whom he collaborated on almost all of the theoretical developments discussed in this paper.

References

[1] PARKINSON, R. W., H. M. JONES, and I. I. SHAPIRO: Science **131**, 920 (1960).
[2] See, for example, F. R. MOULTON: Celestial Mechanics (New York: Macmillan Co., 1914).
[3] SHAPIRO, I. I., and H. M. JONES: J. Geophys. Res. **66**, 4123 (1961).
[4] The major architect of this program was H. M. JONES. Other contributors include C. W. PERKINS, P. WILLMANN, M. ASH, and C. BERGER. G. STRANG, with the help of P. WILLMANN, also constructed a triple-precision RUNGE-KUTTA integration program which includes the effects of the earth's gravitational potential, direct sunlight pressure, and air drag, simulated by a simple model. This latter has been very useful for checking purposes.
[5] KING-HELE, D. G.: Geophys. J. **4**, 3 (1961).
[6] This work is being carried out by C. W. PERKINS.
[7] ABBOT, C. G.: Smith. Cont. Astrophys. **3**, 13 (1958).
[8] See also P. MUSEN: J. Geophys. Res. **65**, 1391 (1960), and Y. KOZAI: Smith. Astrophys. Obs. Spec. Rep. No. 56, 25 (1961).

[9] JONES, H. M.: Unpublished notes.

[10] JONES, H. M., I. I. SHAPIRO, and P. E. ZADUNAISKY in Space Research II: Proceedings of the Second International Space Science Symposium (North Holland, Amsterdam, 1961), p. 342.

[11] SHAPIRO, I. I.: Unpublished notes.

[12] STERNE, T. E.: J. Amer. Rocket Soc. 29, 777 (1959).

[13] JACCHIA, L. G.: Nature 183, 1662 (1959).

[14] ZADUNAISKY, P. E., I. I. SHAPIRO, and H. M. JONES: Smith. Astrophys. Obs. Spec. Rep. No. 61 (1961). See also ref. 10 and I. I. SHAPIRO and H. M. JONES: Science 132, 1484 (1960); 133, 579 (1961).

[15] CHOPRA, K. P.: Revs. Modern Phys. 33, 153 (1961).

[16] Our interest in this problem was restimulated by communications with S. F. SINGER.

[17] SPITZER, L.: Ap. J. 93, 369 (1941).

[18] BOURDEAU, R. E., et al: NASA Tech. Note D-164 (1961).

[19] See, for example, V. I. KRASSOVSKY: Proc. IRE, 47, 289 (1959).

[20] O'BRIEN, B. J.: private communication.

[21] BEARD, D. B., and F. S. JOHNSON: J. Geophys. Res. 65, 1 (1960).

[22] MASSEY, H. S. W., and E. H. S. BURHOP: Electronic and Ionic Impact Phenomena (Oxford: Oxford University Press, 1952).

[23] HUGHES, A. L., and L. A. DuBRIDGE: Photoelectric Phenomena (New York: McGraw Hill, 1932).

[24] HINTEREGGER, H. E.: Private communication. We also wish to thank Dr. HINTEREGGER for several stimulating discussions.

[25] WEISSLER, G. L.: Handbuch der Physik 21, 304 (1956).

[26] HINTEREGGER, H. E.: J. Geophys. Res. 66, 2367 (1961).

[27] BOHM, D., and D. PINES: Phys. Rev. 82, 625 (1951).

[28] DAVIS, A. H., and I. HARRIS: NASA Tech. Note D-704 (1961).

[29] SPITZER, L.: Physics of Fully Ionized Gases (Interscience, New York, 1956).

[30] CHANDRASEKHAR, S.: Ap. J. 97, 255 (1943).

[31] FERMI, E.: Nuclear Physics (University of Chicago Press, Chicago, 1951), p. 27.

[32] RAND, S.: Phys. Fluids 3, 265 (1960).

[33] MCCRACKEN, C. W., W. M. ALEXANDER, and M. DUBIN: Nature 192, 441 (1961).

[34] ROBERTSON, H. P.: M. N. 97, 423 (1937).

[35] See, for example, Y. KOZAI: Smith. Astrophys. Obs. Spec. Rep. 22, 7 (1959); UPTON, E., A. BAILIE, and P. MUSEN: Science 130, 1710 (1959).

[36] SHAPIRO, I. I., and H. M. JONES: Science 134, 973 (1961).

[37] SHAPIRO, I. I.: Paper presented at VIII Annual Meeting of Amer. Astronaut. Soc. (1962).

[38] We wish to thank L. G. JACCHIA for a discussion on this point.

[39] COLOMBO, G.: Smith. Astrophys. Obs., is investigating the possibility of using such techniques.

Discussion

During the discussion of the above paper, Dr. W. M. IRVINE did express the following remarks:

For sufficiently small orbiting particles high field emission may be very important in determining the electric charge, and consequently the charge drag, on the particle (H. C. VAN DER HULST: Rech. Astron. Obs. Utrecht XI, part 2). However, because of the dependence of this current on (const.) exp. (const. $E^{1/2}$), this would be a negligible effect even on a satellite as small as a WEST FORD needle.

Sur certains problèmes mathématiques du mouvement relatif des satellites

Par

N. N. Moisseyev

Computation Centre of the USSR Academy of Sciences, Moscow

Le présent travail est consacré à certaines questions nouvelles engendrées par le problème du départ d'un appareil cosmique de l'orbite des satellites artificiels. Afin de procéder au départ de l'orbite, l'appareil en cause doit inévitablement comporter à l'intérieur une certaine quantité de liquide. Ce fait impose des particularités importantes au caractère du mouvement du satellite par rapport au centre d'inertie et, par conséquent, l'orientation de celui-ci se trouve influencée par la circonstance en question. C'est ainsi qu'on arrive à la nécessité d'étudier le mouvement relatif du satellite dans le cadre de la théorie du mouvement du corps solide comportant des masses liquides.

Cette théorie fut l'objet de plusieurs publications des hommes de science aux Etats-Unis, au Japon et en Union Soviétique. Cependant, il est à noter que toutes ces recherches traitaient la dynamique du corps solide avec liquide dans les conditions où le mouvement du liquide dépendait essentiellement des forces de masse. Quant aux problèmes du mouvement relatif du satellite, ceux-ci nécessitent une étude du mouvement du liquide dans les conditions d'impondérabilité ou des champs très faibles des forces de masse. En raison des circonstances citées, le système mécanique corps liquide prend des particularités nouvelles qui ne peuvent pas être étudiées dans le cadre des théories courantes. Il en est de même de la nature mathématique des problèmes ainsi posés qui possèdent des particularités importantes par rapport aux problèmes de la dynamique d'un corps solide comportant un liquide lourd. D'où la nécessité directe de créer une théorie mathématique sur le mouvement d'un satellite contenant à l'intérieur un liquide subissant l'influence des forces de la tension de surface et de la viscosité.

Le présent mémoire traite les énoncés des problèmes nouveaux, les problèmes d'existence des solutions ainsi que quelques exemples sur le mouvement du corps solide avec un liquide dans les conditions d'impondérabilité.

1. Sur la forme d'équilibre d'une surface libre

1. Considérons une enveloppe dure Σ à l'intérieur de laquelle se trouve un liquide idéal et incompressible (voir Figure 1). Le liquide ne

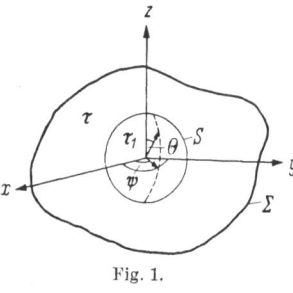

Fig. 1.

remplissant pas complètement le contenu de l'enveloppe, il se trouve à l'intérier du liquide une bulle; admettons que cette dernière est remplie de gaz dont la pression constante est P_0. La première question qui surgit dans la théorie du mouvement du corps solide comportant une cavité c'est ʻ celle qui concerne les formes possibles de l'équilibre d'une surface libre.

Soit l'enveloppe Σ au repos et les forces de masse absentes. Il s'ensuit qu'à la frontière liquide-gaz doit être réalisée une condition (voir 1).

$$\overline{K} = \text{const} \tag{1.1}$$

où $\overline{K}$ est la courbure moyenne de la surface limite

$$\overline{K} = \frac{1}{R_1} + \frac{1}{R_2}, \tag{1.2}$$

R_1 et R_2 — rayons principaux de la courbure.

Ainsi, la frontière S constitue une surface sphérique et $\overline{K} = 2/R$ où R est le rayon de la spèhre (nous n'allons pas considérer le cas des surfaces de la courbure constante négative). Lorsque la surface S n'a

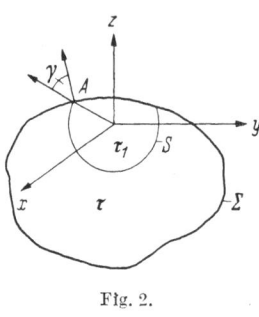

Fig. 2.

pas de points communs avec la surface Σ, cela signifie qu'à l'intérieur du liquide remplissant la cavité se trouve une bulle sphérique (voir Figure 1).

2. Soit les surfaces Σ et S croisées (voir Figure 2). Aux points de la ligne d'intersection les normales aux surfaces Σ et S forment un angle constant que nous convenons de nommer angle extrême. Cet angle (γ) ne dépend que des propriétés du matériel constituant l'enveloppe Σ ainsi que de celles du liquide remplissant la cavité τ, et du gaz. C'est cet angle qui constitue la constante du problème.

La forme de la surface Σ étant connue, aussi bien que l'angle γ et le volume du liquide, il n'est pas difficile de déterminer le rayon de la sphère R et la position de la ligne d'intersection.

Ceci est surtout simple à faire dans le cas où la surface Σ constitue de même une sphère. Le problème peut être résolu à l'aide d'un système d'équations trigonométriques, comme suit (voir Figure 3 où se trouvent représentées toutes les désignations).

$$\frac{1}{3} \pi R^3 [4 - (1 - \cos\alpha)^2 (2 + \cos\alpha)] -$$

$$- \frac{1}{3} \pi R^2 (1 - \cos\beta)^{*\,2} (2 + \cos\beta) = V \qquad (1.3)$$

$$\alpha = \beta + \gamma \qquad (1.4)$$

$$R \sin\alpha = R^* \sin\beta \qquad (1.5)$$

où R et R^* les rayons des sphères S et Σ, le V volume de la bulle.

3. Faisons coïncider le système sphérique des coordonnées et celui de DESCARTES comme il est indiqué à la Figure 1.

$$\left. \begin{array}{l} x = z \operatorname{Sin}\theta \operatorname{Cos}\psi, \\ y = r \operatorname{Sin}\theta \operatorname{Sin}\psi, \\ z = r \operatorname{Cos}\theta. \end{array} \right\} \qquad (1.6)$$

Examinons une certaine surface S_ζ définie par l'équation:

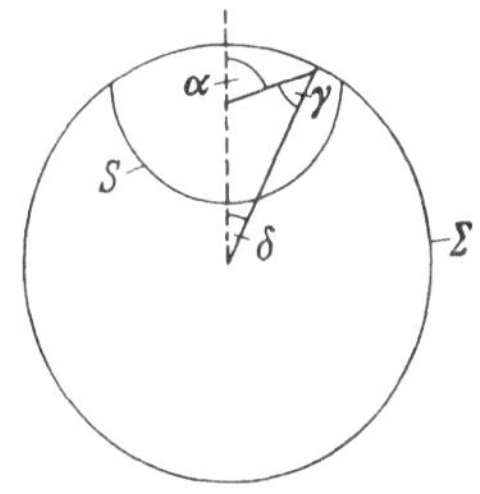

$$\zeta = \zeta(\theta, \psi) = r(\theta, \psi) - R. \qquad (1.7)$$

Fig. 3.

La surface S_ζ se distinguant peu de la sphère S, la corbure moyenne de ladite surface $\overline{K}_\zeta$ est déterminée par la formule ci-dessous, aux valeurs petites du premier ordre près:

$$\overline{K} = \frac{2}{R} - \frac{1}{R^2} L \zeta, \qquad (1.8)$$

où

$$L = \frac{1}{\sin^2\theta} \frac{\partial^2}{\partial \psi^2} + \frac{1}{\sin\theta} \frac{\partial}{\partial \theta} \left(\sin\theta \frac{\partial}{\partial \theta} \right) + 2.$$

Relevons quelques propriétés de l'opérateur dont on aura besoin dans la suite.

Considérons l'équation:

$$L \zeta = 0, \qquad (1.9)$$

et étudions ces solutions représentant des fonctions uniformes et bornées des points de la sphère S.

Il est aisé de se convaincre en recourant à une simple vérification que l'équation (1.9) a des solutions non nulles suivantes:

$$\left. \begin{array}{l} \zeta_1 = P_1(\cos\theta) = \cos\theta, \\ \zeta_2 = P_1^1(\cos\theta) \cos\psi = \sin\theta \cos\psi, \\ \zeta_3 = P_1^1(\cos\theta) \sin\psi = \sin\theta \sin\psi, \end{array} \right\} \qquad (1.10)$$

où $P_i(x)$ désigne les polynômes de LEGENDRE et $P_i^1(x)$ les fonctions adjointes de LEGENDRE de premier ordre.

Il résulte de la théorie des polynômes de Legendre que l'équation (1.9) n'a pas d'autres solutions uniformes bornées excepté celles qui ont été énoncées. Il s'ensuit que l'équation:

$$L \zeta = f \tag{1.11}$$

ne peut avior une solution que dans le cas où f appartient à un espace orthogonal aux fonctions:
$$\zeta_1, \zeta_2 \text{ et } \zeta_3.$$

Les solutions (1.10) ont un simple sens physique; elles décrivent des déplacements infinitésimaux de la sphère S prise dans son ensemble, le long des axes ox, oy et oz. L'existence de ces solutions reflète le fait évident suivant; si la bulle se trouve entièrement dans un liquide et ne touche pas son enveloppe, sa position est instable c'est-à-dire qu'elle peut être changée par des forces infinitésimales qui s'y trouvent appliquées.

Tâchons de changer le problème pour l'équation (1.9). Dès lors on va chercher des solutions uniformes et bornées lorsque $\theta = \pi$ et satisfaisantes à la condition:

$$\left. \frac{d\zeta}{d\theta} \right\} = 0 \tag{1.12}$$

quand

$$\theta = \theta_0 > 0.$$

Les solutions de ce problème doivent s'écrire sous la forme:

$$\zeta = f(\theta) \begin{Bmatrix} \cos \\ \sin \end{Bmatrix} m \, \psi$$

où m est un nombre entier.

Posons $m = 0$, il s'ensuit alors que $f(\theta)$ satisfait à l'équation:

$$\frac{1}{\sin\theta} \frac{d}{d\theta} \left(\sin\theta \frac{df}{d\theta} \right) + 2f = 0. \tag{1.13}$$

Sa solution générale s'écrit sous la forme:

$$f = C_1 \, P_1(\cos\theta) + C_2 \, Q_1(\cos\theta)$$

où Q_1 est la fonction de Legendre du deuxième ordre. La solution étant bornée, $C_2 = 0$ mais:

$$\frac{dP_1}{dx} \neq 0, \qquad x \in (0, \pi).$$

Par suite, l'équation (1.13) n'a qu'une solution nulle. De la même façon, il est aisé de démontrer que si $\theta_0 \in [0, \pi/2]$ les équations (1.9) n'admettent aucune solution excepté la solution nulle.

Ce qui a été énoncé permet d'affirmer la possibilité de résoudre l'équation (1.11).

4. Après avoir relevé certaines propriétés de l'opérateur L nous pouvons procéder à l'étude de la forme de l'équation libre d'une surface libre dans le cas où le liquide est soumis à l'influence d'un champ uniforme des forces de masse d'intensité $\vec{g}$. Désignons par U une fonction telle que:

$$V U = -\vec{g}.$$

La condition d'équilibre thermodynamique d'une surface libre s'écrit sous la forme:

$$p + \alpha \overline{K} = p_0, \qquad (1.14)$$

où α est le coefficient de tension de surface, p-pression dans le liquide, p_0-pression dans la bulle. La courbure moyenne $\overline{K}$ étant admise positive à condition que la concavité soit tournée du côté de la normale positive, vers le volume occupé par le liquide, la pression dans la bulle est supérieure à celle dans le liquide.

Lorsque dans le liquide règne la position d'équilibre, la pression est déterminée par le champ U:

$$p + \varrho\, U = \text{const.}$$

La condition d'équilibre va donc s'écrire ainsi:

$$U - \gamma\, \overline{K}_\zeta = \text{const} \qquad (1.15)$$

avec

$$\gamma = \alpha/\varrho.$$

Faisons coïncider l'axe OZ avec la direction du vecteur $-\vec{g}$. En considérant l'intensité assez petite et en linéarisant l'équation (1.15) près de la sphère, nous arriverons à l'équation suivante:

$$g \cos\theta (R + \zeta) - \frac{\gamma}{R^2} L \zeta = 0. \qquad (1.16)$$

Introduisons des variables sans-dimensionnelles $\zeta' = \zeta/R$, l'équation (1.16) prendra alors la forme:

$$\delta \cos\theta (1 + \zeta') + L \zeta' = 0 \qquad (1.17)$$

avec

$$\delta = \frac{g\, R^2}{\gamma}; \qquad \zeta' = \frac{\zeta}{R}.$$

La forme de la surface libre sera donc definie uniformément par le paramètre sans-dimensionnel δ.

Nous sommes arrivés à équation (1.17) à force de l'avoir linéarisée en ζ. De ce fait, l'équation (1.17) n'a de sens que dans le cas où δ est une variable du même ordre que ζ'. Ainsi, on peut omettre le terme de type $\delta\zeta'$ figurant dans l'équation de la surface libre. Par la suite, l'équation d'équilibre de la surface libre s'écrira sous la forme:

$$\delta\cos\theta + L\zeta' = 0 \qquad (1.18)$$

ou bien sous une forme dimensionnelle:

$$g\cos\theta\, R^3 + \gamma\, L\,\zeta = 0. \qquad (1.19)$$

La surface de la bulle étant croisée par la surface Σ, l'équation (1.19) n'a qu'une seule solution.

Lorsque la bulle se trouve entièrement à l'intérieur du liquide, l'équation (1.19) n'a pas de solutions bornées. Cela signifie que, si réduit que soit le champ des forces de masses, la bulle à l'intérieur de ce champ ne peut pas se trouver en équilibre: elle va se déplacer. Le problème de la définition du mouvement de la bulle ne présente pas de grandes difficultés (sans compter le calcul). Il est aisé par exemple de démontrer que dans le cas où le rayon de la sphère extérieure est assez grand, la bulle va se déplacer dans le sens opposé à celui du champ avec une accélération faisant le double de son intensité. Au fur et à mesure qu'elle s'approche de la frontière Σ, l'accélération de la bulle va diminuer et s'annulera au moment où la bulle va atteindre la surface Σ. Dans la suite, la bulle ne va pas se déplacer comme un ensemble et le mouvement de sa surface portera un caractère oscillatoire autour d'une certaine position ζ^* constituant la solution de l'équation (1.19) à condition (1.12):

$$\left(\frac{\partial\zeta}{\partial\theta}\right)_{\theta\,=\,\theta_0} = 0 \qquad (1.12)$$

du fait que la condition (1,12), comme il a été démontré, garantit la résolubilité de l'équation (1.19).

Remarque. La surface Σ étant une sphère, le problème du déplacement de la bulle peut être résolu exactement. En effet, ce problème se réduit à la solution du problème de Neumann pour le domaine limité par les surfaces des sphères excentriques, et en recourant aux coordonnées bipolaires d'espace, on peut obtenir la solution de ce problème en fonctions sphériques.

5. Or, le problème de définition de la surface d'équilibre est ramené à la recherche d'une solution de l'équation (1.19) qui doit être bornée au point $\theta = \pi$ et soumise à la condition (1.12). D'après ce qui précède la solution de cette équation existe et est unique. Il est aisé de démontrer que cette solution n'est qu'une fonction de θ et aisé de démontrer que cette solution n'est qu'une fonction de θ et ne dépend pas de ψ. L'équation (1.19) est donc une équation différentielle ordinaire dont la solution est surtout facile à rechercher en recourant à la méthode de Riesz.

D'après la théorie générale, le problème. de recherche de la solution de l'équation (1.19) se réduit au problème du minimum pour le fonctionnel.

$$A u = \frac{\gamma}{R^2} \int\limits_0^\pi L u u \, d\theta + 2g \int\limits_{\theta_0}^\pi u \cos\theta \, d\theta .$$

2. Problème sur l'oscillation du liquide

1. Le paragraphe précédent a été consacré au problème d'équilibre du liquide se trouvant à l'intérieur d'un satellite lorsque ce dernier est en mouvement de translation et possède une accélération peu considérable dont la valeur et le sens sont constants. Ces conditions n'étant pas observées, le liquide ne gardera pas l'équilibre par rapport au corps. Le mouvement de ce corps va provoquer celui du liquide et ce mouvement, à son tour, va influencer celui du corps. Plus précisément, il se produira des houles à la surface du liquide; celles-ci vont influencer la position relative de centre d'inertie du système corps-liquide. Le mouvement du liquide portera donc un caractère oscillatoire près d'une position quelconque d'équilibre. Ici, la question suivante qu'on doit analyser avant de procéder à l'étude de la dynamique d'un corps comportant un liquide, consiste à étudier les oscillations du liquide.

2. Considérons le problème des oscillations linéaires d'un liquide idéal et impondérable renfermé dans une enveloppe immobile Σ. On va considérer le mouvement du liquide comme potentiel, c'est-à-dire:

$$\vec{v} = V \varphi ,$$

où $\vec{v} = \vec{v}(P, t)$ est la vitesse d'une particule liquide, $P(r, \theta, \psi)$ le point du volume τ (voir Figure 1 ou 2), V l'opérateur de HAMILTON sur les variables spatiales $\varphi(P, t)$ la fonction harmonique en P. A la surface Σ la fonction φ doit satisfaire à la condition:

$$\frac{\partial \varphi}{\partial n} = 0. \tag{2.1}$$

Sur la sphère S doit être réalisée la relation cinématique:

$$\frac{\partial \zeta}{\partial t} = \frac{\partial \varphi}{\partial r} , \tag{2.2}$$

ainsi que la condition d'équilibre thermodynamique d'une surface libre (1.14):

$$p + \alpha \overline{K}_\zeta = p_0 . \tag{1.14}$$

Prenant en considération l'intégrale de LAGRANGE:

$$\varrho \frac{\partial \varphi}{\partial t} + p = \text{const}$$

ainsi qu'en appliquant l'expression linéarisée de l'opérateur [voir (1.18)] et rejetant la constante inexistante, nous ramenons la condition (1.14) à la forme:

$$\frac{\partial \varphi}{\partial t} + \frac{\gamma}{R^2} L \zeta = 0. \tag{2.3}$$

Si l'on introduit les variables non-dimensionnelles:

$$\zeta' = \frac{\zeta}{R}; \quad t' = \sqrt{\frac{\gamma}{R^3}} t; \quad \varphi' = R^{-\frac{1}{2}} \gamma^{-\frac{1}{2}} \varphi$$

la condition (2.3) s'écrira sous la forme:

$$\frac{\partial \varphi'}{\partial t'} + L \zeta' = 0. \tag{2.4}$$

Convenons d'appliquer le terme d'oscillations libres en propre aux solutions du problème étudié qui se présentent sous la forme:

$$\varphi' = \cos \sigma t \, \Phi(P),$$

$$\zeta' = \frac{1}{\sigma} \sin \sigma t \, f(P). \tag{2.5}$$

Le nombre σ est appelé fréquence propre. En substituant l'expression (2.5) dans l'équation du problème, nous aboutirons au résultat suivant:

le problème des propres oscillations se réduit à la recherche des fonctions Φ et f et du nombre σ d'après les conditions ci-dessous:

1. $\Phi(P)$ fonction harmonique en

$$\Delta \Phi = 0. \tag{2.6}$$

2. $\frac{\partial \Phi}{\partial n} = 0$

$$P \in \Sigma. \tag{2.7}$$

3. $\frac{\partial \Phi}{\partial r} = f$

$$P \in S. \tag{2.8}$$

4. $\sigma^2 \Phi = L f$

$$P \in S. \tag{2.9}$$

Elucidons la question d'existence de solution du problème posé et celle de la structure du spectre. Introduisons la notion de l'opérateur de Neumann. Celui-ci définit la conformité de la fonction Φ à la fonction f, la fonction Φ étant harmonique et telle que sur Σ elle satisfait à la condition (2.7) et sur S à la condition (2.8). Il en résulte que:

$$\Phi(P) = H f(Q)$$

$$P \in T, \quad Q \in S. \tag{2.10}$$

On sait que l'opérateur H est complètement continu et symétrique. En appliquant cet opérateur, on peut ramener le problème à l'équation:

$$(\sigma^2 H - L) f = 0. \tag{2.11}$$

Il est évident que l'opérateur L est symétrique. La bulle atteignant la surface Σ, l'opérateur est positivement déterminé du fait que, dans la position d'équilibre, l'énergie potentielle des forces de la tension de surface est minimale. C'est ainsi que nous arrivons au problème standard des valeurs propres:

$$\lambda A u = B u \tag{2.12}$$

où A et B sont des opérateurs symétriques, A étant complètement continu et B positivement déterminé. De ce fait, on peut appliquer à ce problème les résultats principaux de la théorie des opérateurs linéaires, ce qui conduit à la conclusion suivante:

(a) il existe dans le liquide des oscillations propres, c'est-à-dire des solutions de la forme (2.5);

(b) les nombres σ_n forment une suite discrète ascendante indéfiniment $\lim\limits_{n \to \infty} \sigma_n = \varphi$;

(c) à chaque nombre σ_n correspond un nombre fini de fonctions propres f_{nm} décrivant la forme d'une surface libre;

(d) la définition des nombres σ_n et des fonctions f_{nm} peut être effectuée à l'aide de la méthode de RIESZ.

La bulle étant entièrement renfermée dans le liquide, tous les énoncés précédents sont valables à l'espace des fonctions orthogonales aux fonctions ζ_i $(i = 1, 2, 3)$ constituant des solutions non nulles de l'équation:

$$L \zeta = 0.$$

Lorsque le liquide est soumis à l'influence des faibles forces de masse, le problème des propres oscillations peut encore être ramené au problème (2.12) avec:

$$A = H, \quad B = L^* - \delta I,$$

L étant la dérivée de FRÉCHET de l'opérateur $\overline{K}$ au point $\zeta = \zeta^*$ où ζ^* est une fonction décrivant la position d'équilibre d'une surface libre dans le champ de force de masse. Pour pouvoir appliquer dans ce cas l'affirmation que nous venons de formuler, il faut que l'opérateur $L^* - \delta I$ soit positivement déterminé. Cette exigence est équivalente à celle de la stabilité de la solution du problème d'équilibre.

Le calcul réel des fréquences propres aussi bien que la définition des fonctions f_{nm} peut être effectué au moyen de la méthode de RIESZ;

21

celle-ci est bien commode lorsqu'on a recours aux calculatrices électroniques numériques. Lorsque la cavité Σ est une cavité sphérique et le champ des forces de masses petit, on peut effectivement faire les calculs en utilisant les fonctions sphériques.

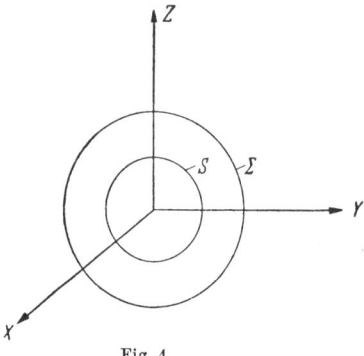

Fig. 4.

3. Analysons en tant qu'exemple le problème des oscillations libres d'une couche sphérique comprise entre deux sphères concentriques Σ et S (voir Figure 4). On va considérer le problème avec des variables sans-dimensionnelles prenant comme échelle linéaire le rayon R_1, de la sphère antérieure. Posons $q = R_2/R_1$.

A l'intérieur de la couche sphérique τ, la fonction Φ satisfait à l'équation (2.6) qui, dans le cas des coordonnées sphériques, se présente sous laforme:

$$\frac{\partial}{\partial r}\left(r^2\frac{\partial\Phi}{\partial r}\right) + \frac{1}{\sin\theta}\frac{\partial}{\partial\theta}\left(\sin\theta\frac{\partial\Phi}{\partial\theta}\right) + \frac{1}{\sin^2\theta}\frac{\partial^2\Phi}{\partial\psi^2} = 0. \quad (2.13)$$

Les solutions particulières des équations (2.13) constituant des fonctions limites et uniformes des variables r, θ et ψ et satisfaisant à la condition (2.7):

$$\frac{\partial\Phi}{\partial r} = 0 \quad \text{avec} \quad r = q,$$

s'écrivent sous la forme:

$$\Phi_{n,m}^{(1),(2)} = c\left[\frac{n+1}{n}q^{-(2n+1)}r^n + r^{-(n+1)}\right]p_n^m(\cos\theta)\frac{\sin}{\cos}m\,\psi, \quad (2.14)$$

c étant une constante arbitraire et n un nombre entier arbitraire, $m = 0, 1, 2\ldots n$, et $P_n^m(x)$ étant une fonction sphérique.

Relevons la surface libre-fonction $f(\theta, \psi)$ en recourant à la condition (2.8)

$$\left(\frac{\partial\Phi}{\partial r}\right)_{r=1} = f. \quad (2.8)$$

Substituant les grandeurs obtenues dans la condition (2.9):

$$(\sigma^2\Phi = L f)_{r=1}, \quad (2.9)$$

nous arriverons à l'expression ci-dessous déterminant les fréquences propres.

$$\sigma_n^2 = (n-1)(n+2)\frac{1 - q^{-(2n+1)}}{1 + \dfrac{n+1}{n}q^{-(2n+1)}}. \quad (2.15)$$

La formule (2.15) montre que la fréquence des oscillations ne dépend pas du nombre m. Il s'ensuit que toutes les oscillations principales dont les formes sont décrites par les fonctions $f_{n,m}^{(1),(2)}(m = 0, 1, 2 \ldots n)$ ont la même fréquence. Ainsi, notre problème possède un spectre multiple, et le nombre propre σ_n a une divisibilité $2n + 1$.

Soit $\sigma_0 = \sigma_1 = 0$. D'après ce qui précède, on voit qu'à ces valeurs du nombre n correspondent les solutions non nulles de l'équation $L \zeta = 0$ c'est-à-dire qu'elles ont à correspondre un mouvement séculaire de la bulle et non pas oscillatoire.

La moindre fréquence propre correspond au nombre $n = 2$.

Relevons les valeurs des premiers nombres propres:

$$\sigma_2^2 = 12 \,\frac{1 - q^{-5}}{1 + \dfrac{3}{2} q^{-5}}\,; \qquad \sigma_3^2 = 40 \,\frac{1 - q^{-7}}{1 + \dfrac{4}{3} q^{-7}}\,; \qquad \sigma_4^2 = 90 \,\frac{1 - q^{-9}}{1 + \dfrac{5}{4} q^{-9}}\,.$$

Les fréquences σ_n ne dépendent des profondeurs que dans les cas de petites profondeurs. Même avec $q = 2$ (l'épaisseur de la couche est égale au rayon de la bulle antérieure), l'on a:

$$\frac{\sigma_2^2}{\sigma_{2\infty}^2} = 0{,}961\,; \qquad \frac{\sigma_3^2}{\sigma_{3\infty}^2} = 0{,}9909\,,$$

et ainsi de suite.

On connaît bien la solution du problème des ondes capillaires à la surface d'une goutte sphérique [voir, par exemple, (1)]:

$$\sigma_n^{*2} = n(n - 1)(n + 2). \tag{2.16}$$

En comparant les formules (2.15) et (2.16), nous avons:

$$\sigma_n^2 = \sigma_n^{*2} \,\frac{n + 1}{n}\, \frac{1 - q^{-(2n+1)}}{1 + \dfrac{n + 1}{n} q^{-(2n+1)}}\,.$$

Citons en conclusion l'expression de la formule (2.15) sous forme dimensionelle:

$$\omega_n^2 = \frac{\gamma}{R^3}\,\sigma_n^2. \tag{2.116}$$

3. Sur certains problèmes de la propulsion relative d'un satellite

Le problème général du mouvement d'un satellite renfermant des masses liquides, par rapport au centre d'inertie, constitue un problème non-linéaire bien compliqué. Afin de se faire une idée sur le problème posé ainsi que sur les difficultés mathématiques qu'on a à surmonter en le résolvant, considérons deux exemples simples dont tous les calculs peuvent être accomplis effectivement. Supposons dans les deux cas que la surface Σ renfermant le liquide soit d'une forme sphérique.

a) **Changement de l'orientation du satellite lorsque l'angle de bra-
quage est petit et que la bulle se trouve au centre de la sphère.** Analysons
la rotation du satellite autor de l'axe $\vec{\omega}$. Cet axe est parallèle à l'axe
$ox (\vec{\omega} = \omega \vec{x}^0)$ et rencontre l'axe oy au point o_1 (voir Figure 5).

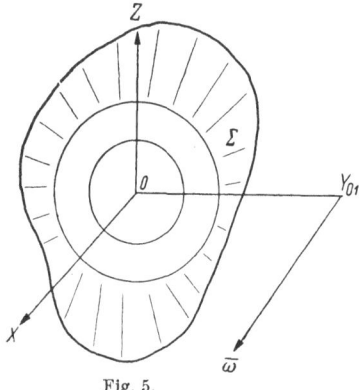

On va décrire le mouvement du
système corps-liquide au moyen de
l'équation des moments. Calculons
le moment de la quantité de mouvement de ce système par rapport au
point o_1;
$$\vec{G} = \vec{G}_1 + \vec{G}_2$$
où $\vec{G}_1$ est le moment de la quantité
de mouvement du solide:

$$\vec{G}_1 = J \vec{\omega}. \qquad (3.1)$$

Fig. 5.

J est le moment d'inertie du solide par rapport à l'axe $\vec{\omega}$.

$\vec{G}_2$ le moment de la quantité de mouvement du liquide:

$$\vec{G}_2 = \varrho \int_\tau [(-d\vec{y}^0 + \vec{r}) \times \nabla\varphi] \, d\tau \qquad (3.2)$$

où φ est le potentiel de la vitesse et $d = o \, o_1$.

Comme on considère des petits mouvements, ce problème admet
une linéarisation. En particulier, lorsqu'on calcule $\vec{G}_2$ le volume τ est
la couche sphérique comprise entre Σ et S. Le potentiel de la vitesse
sur Σ doit satisfaire à la condition suivante:

$$\frac{\partial \varphi}{\partial r} = -\omega d (\vec{x}^0 \times \vec{y}^0) \vec{n}^0 = -\omega d \cos\theta, \qquad (3.3)$$

où $\vec{\eta}^0$. est la normale extérieure de la surface Σ.

Substituons à la fonction φ une somme $\varphi = \omega \varphi^* + \tilde{\varphi}$, où le premier
terme est une fonction harmonique en τ et satisfait à la condition (3.3)
sur Σ et à la condition $\partial\varphi/\partial r = 0$ sur S. Il est aisé de calculer la fonction φ^* sous la forme explicite:

$$\varphi^* = d \cos\theta (A \, r + B \, r^{-2})$$

avec
$$A = \frac{R_1^3}{R_2^3 - R_1^3}; \qquad B = \frac{R_1^3 \, R_2^3}{2 \, (R_2^3 - R_1^3)},$$

où R_1 et R_2 sont respectivement les rayons des sphères S et Σ.

On a donc:

$$\vec{G}_2 = \varrho \int_\tau [(-d\vec{y}^0 + \vec{r}) \times \nabla\varphi^* \, \omega] \, d\tau + \varrho \int_\tau [(-d\vec{y}^0 + \vec{r}) \times \nabla\varphi^*] \, d\tau.$$
$$(3.4)$$

Faisons le calcul:

$$-d\omega \, \varrho \int_\tau (\vec{y}^0 \times \nabla \varphi^*) \, d\tau = m_1 \, \omega \, \vec{x}^0,$$

$$m_1 = \varrho \, d \int_\tau (\nabla \varphi^*)^2 \, d\tau,$$

$$\int_\tau (\vec{r} \times \nabla \varphi^*) \, d\tau = 0,$$

$$-d\vec{y}^0 \times \varrho \int_\tau \nabla \widetilde{\varphi} \, d\tau = -\varrho \, d \int_\tau \varphi^* \frac{\partial \widetilde{\varphi}}{\partial r} \, ds \, \vec{x}^0 + \vec{z}^0 (\cdots).$$

Comme on n'aura pas besoin dans la suite de la projection sur l'axe $\vec{z}^0$, nous ne la relevons pas.

D'une manière tout à fait similaire, on calcule le dernier terme dans l'expression (3.4):

$$\int_\tau (\vec{r} \times \nabla \widetilde{\varphi}) \, d\tau = \varrho \int_S \widetilde{\varphi} (y \cos(n\,z) - z \cos(n\,y)) \, ds \, \vec{x}^0 + \cdots$$

Cependant, il est évident que, sur S,

$$y \cos(n\,z) - z \cos(n\,y) = 0.$$

C'est pourquoi la projection du vecteur $\int_\tau (\vec{r} \times \nabla \widetilde{\varphi}) \, d\tau$ sur l'axe $o\,z$ est nulle.

On a donc:

$$\vec{G}_2 = m_1 \, \omega \, \vec{x}^0 - \varrho \, d \int_S \varphi^* \frac{\partial \widetilde{\varphi}}{\partial r} \, ds \, \vec{x}^0 + \vec{y}^0 (\cdots) + \vec{z}^0 (\cdots).$$

Introduisons la fonction décrivant une surface libre:

$$\frac{\partial \zeta}{\partial t} = \left(\frac{\partial \widetilde{\varphi}}{\partial r} \right)_S. \tag{3.5}$$

En résumant tous les résultats obtenus, on peut présenter l'équation des moments en projection sur l'axe sous la forme:

$$J^* \frac{d\omega}{dt} + \int_S \Phi(\theta) \frac{\partial^2 \zeta}{\partial t^2} \, ds = M, \tag{3.6}$$

où M est le moment des forces extérieures par rapport à l'axe $\vec{\omega}$, et la fonction $\Phi(\theta)$ ainsi que le nombre J^*, étant déterminés par la forme du domaine τ, peuvent être aisément calculés.

On doit ajouter aux équations écrites celle de la constance de pression sur S:

$$\frac{\partial \widetilde{\varphi}}{\partial t} + \frac{d\omega}{dt} \varphi^* - L\zeta = 0, \tag{3.7}$$

$\widetilde{\varphi}$ étant une fonction harmonique dans le domaine dont la dérivée sur Σ s'annule.

Tâchons de trouver la solution du problème sous la forme:

$$\zeta = f(t)\cos\theta. \qquad (3.8)$$

La fonction harmonique $\widetilde{\varphi}$ sera donc définie par des conditions limites telles que:

$$\frac{\partial \varphi}{\partial r} = \begin{cases} 0 & \text{sur} \quad \Sigma \\ f'(t)\cos\theta & \text{sur} \quad S \end{cases}$$

et, par conséquent, peut s'exprimer sous la forme évidente suivante:

$$\widetilde{\varphi} = f'\cos\theta(\widetilde{A}\,r + \widetilde{B}\,r^{-2}) \qquad (3.9)$$

avec

$$\widetilde{A} = -\frac{R_1^3}{R_2^3 - R_1^3},$$

$$\widetilde{B} = -\frac{R_1^3 R_2^3}{R_2^3 - R_1^3}.$$

En appliquant la formule (3.9), on peut exclure la fonction $\widetilde{\varphi}$ de la condition (3.7).

Prenant en considération que $L\cos\theta = 0$, nous ramènerons cette condition à la forme:

$$f''R_1\cos\theta + \omega'\,dR_1\cos\theta = 0 \qquad (3.9)$$

ou bien

$$f'' + \omega'\,d = 0.$$

En utilisant la représentation (3.8), on peut écrire l'équation des moments sous la forme:

$$J^*\omega' + a\,f'' = M, \qquad (3.10)$$

a étant un nombre bien défini.

En excluant f, on trouve

$$(J^* - a\,d)\,\omega' = M. \qquad (3.11)$$

L'équation obtenue résout le problème.

Supposons que le mouvement d'un corps à bulle soit provoqué par des forces extérieures qui ne soient pas des forces de masse, et, par conséquent, ne dépendent pas de la forme de la surface libre. L'équation (3.11) montre alors que le problème du petit mouvement de ce corps se réduit au problème du mouvement d'un autre solide actionné par les mêmes forces.

Il va sans dire que ce résultat n'est juste que pour les mouvements infinitésimaux et pour une enveloppe sphérique.

(b) Problème de la rotation d'un satellite symétrique. 1. Le problème analysé précédemment admettait une étude dans le cadre de la théorie linéaire. La supposition des déplacements peu importants a servi de base à la linéarisation. Il existe encore une série de problèmes où la linéarisation est légitime. Considérons une sphère Σ. Remarquons que tout mouvement rotationel de la sphère Σ autour des axes traversant le centre de celle-ci n'influence pas le liquide. Donc, lorsque le liquide est au repos, le mouvement de la sphère ne provoquera pas le mouvement du liquide et, par suite, ne changera pas la position de la bulle. De ce fait, tout mouvement voisin de celui de rotation autour des axes de symétrie de la sphère admet la linéarisation du problème hydrodynamique.

2. Envisageons le problème de rotation du corps représenté sur le dessin (voir Figure 6). Le corps est supposé posséder un axe de symétrie. Soit ce dernier défini par le vecteur unitaire $\vec{\xi}^0$. On suppose que la projection de la vitesse angulaire instantanée $\vec{\Omega}$ sur son axe de symétrie ω peut avoir n'importe quelle grandeur, tandis que la différence $\vec{\omega}_1 = \vec{\Omega} - \vec{\omega}$ et l'angle de nutation (angle entre $\vec{\xi}^0$ et $\vec{Z}^0$) sont des valeurs petites du premier ordre.

Supposons de même que le centre d'inertie du satellite se déplace d'une manière rectiligne et à vitesse constante. Lions le système inertiel

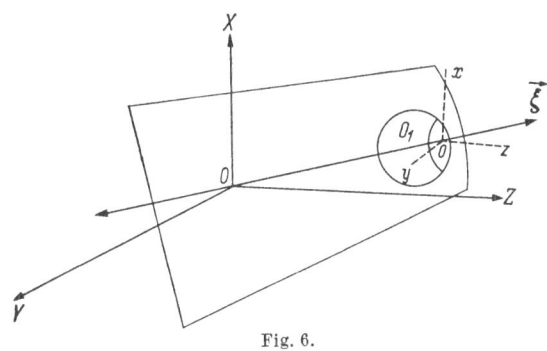

Fig. 6.

de comptage $OXYZ$ au point coïncidant avec le centre d'inertie dans la position d'équilibre. La cavité liquide renferme une bulle que nous allons supposer placée comme il est indiqué sur le dessin (voir Figure 6). Convenons que l'oscillation du satellite est provoquée par le moment:

$$\vec{M} = \eta \vec{Z}^0 \times \vec{\xi}^0. \qquad (3.12)$$

Ce moment peut être de nature gravitationelle ou aèrodynamique. La position d'équilibre du corps étudié correspond à l'angle nul de nutation.

Le mouvement rotationnel autour de l'axe $\vec{\xi^0}$ n'influence pas celui du liquide ni le changement de la position de la bulle. Par suite, dans le cas des petits mouvements de nutation, les conditions limites doivent être réalisées sur la surface sphérique correspondant à la position d'équilibre, par rapport au système des coordonnees $oxyz$.

Ce système est lié au centre de la bulle et avance progressivement, les axes de ce système étant parallèles à ceux du système $OXYZ$.

3. Composons l'expression du moment de la quantité de mouvement par rapport au point o:

$$\vec{G} = \vec{G}_1 + \vec{G}_2, \tag{3.13}$$

où $\vec{G}_1$ est le moment de la quantité de mouvement du solide et $\vec{G}_2$ est relatif au liquide.

Le satellite possédant un axe de symétrie, on a

$$\vec{G}_1 = C\vec{\omega} + A\vec{\omega}_1, \tag{3.14}$$

où C et A sont les moments correspondants d'inertie. Désignons par $\varphi(P, t)$ le potentiel des vitesses du mouvement absolu du liquide, c'est-à-dire du mouvement par rapport au système des coordonnées $OXYZ$. Il en résulte que:

$$\vec{G}_2 = \varrho \int_\tau [(l\vec{\xi^0} + \vec{r}) \times \nabla\varphi]\,d\tau, \tag{3.15}$$

où l est la distance entre le centre de la bulle o et le point O, $\vec{r}$ — le rayon-vecteur du point par rapport au centre de la bulle.

Le potentiel des vitesses φ doit satisfaire sur la surface mouillée Σ à la condition suivante:

$$\frac{\partial\varphi}{\partial n} = v_n \tag{3.16}$$

avec

$$v_n = [(\vec{\omega} + \vec{\omega}_1) \times (l\vec{\xi^0} + \vec{r})]\,\vec{n^0}.$$

Si la surface S était de même rigide, cette condition devrait être réalisée sur S. Le vecteur unitaire de la normale $\vec{n^0}$ étant colinéaire sur S au rayon-vecteur:

$\vec{n^0} = -\vec{r^0}$, on a sur S

$$v_n = (\vec{\omega}_1 \times l\vec{\xi^0})\vec{r^0}.$$

Remarquons que

$$\vec{\omega}_1 = \vec{\xi^0} \times \vec{\xi^0}'. \tag{3.17}$$

L'égalité (3.17) résulte du fait que $\vec{\omega}_1$ est la vitesse angulaire de la rotation du plan de la nutation. On a donc sur S

$$v_n = l \, \vec{\xi^{0\prime}} \, \vec{n^0}. \tag{3.18}$$

En tout point de la surface:

$$v_n = (\vec{\omega} \times \vec{r}) \, \vec{n^0} + (\vec{\omega}_1 \times l \, \vec{\xi^0}) \, \vec{n^0} + (\vec{\omega}_1 \times \vec{r}) \, \vec{n^0}.$$

Comme les centres des sphères Σ et S sont situés sur la droite dont le sens coïncide avec celui du vecteur $\vec{\xi^0}$, on a:

$$(\vec{\omega} \times \vec{r}) \, \vec{n} = 0.$$

Ainsi:

$$(v_n)_\Sigma = l \, \vec{\xi^{0\prime}} \cdot \vec{n^0} + [\vec{\xi^{0\prime}} \, (\vec{r} \, \vec{\xi^0}) - \vec{\xi^0} (\vec{r} \cdot \vec{\xi^{0\prime}})] \, \vec{n^0}. \tag{3.19}$$

Désignons par μ_1, μ_2 et μ_3 les projections du vecteur de la normale extérieure par rapport au domaine occupé par le liquide, et par x, γ, z les composantes du rayon-vecteur $\vec{r} (\vec{\xi^0} = \alpha \, \vec{x^0} + \beta \, \vec{\gamma^0} + \gamma \, \vec{z^0})$.

Transformant le deuxième terme dans la partie droite de l'égalité (3.19), prenons en considération le fait que la grandeur de γ est voisine de l'unité et la dérivée $d\gamma/dt$ une grandeur infinitésimale du deuxième ordre. En gardant dans cette expression seulement les valeurs petites du premier ordre, nous obtenons:

$$v_n = \alpha' \, l \, \mu_1 + \beta' \, l \, \mu_2 + \alpha' \, (z \, \mu_1 - x \, \mu_3) - \beta' \, (z \, \mu_2 - y \, \mu_3). \tag{3.20}$$

Remarquant que sur S les deux derniers termes sont nuls, nous pouvons estimer que, dans le cas où la surface S est impénétrable, le potentiel des vitesses doit satisfaire à la condition (3.20) à la surface $\Sigma + S$.

Introduisons les potentiels de STOKES-YOUKOVSKY φ_i. Les fonctions φ_i sont des fonctions harmoniques satisfaisant sur $\Sigma + S$ aux conditions suivantes:

$$\left. \begin{aligned} &\frac{\partial \varphi_i}{\partial n} = \mu_i \quad i = 1, 2, 3, \\[2mm] &\frac{\partial \varphi_4}{\partial n} = y \, \mu_3 - z \, \mu_2; \qquad \frac{\partial \varphi_5}{\partial n} = z \, \mu_1 - x \, \mu_3, \\[2mm] &\frac{\partial \varphi_6}{\partial n} = x \, \mu_2 - y \, \mu_3. \end{aligned} \right\} \tag{3.21}$$

Ces fonctions peuvent être exprimées explicitement au moyen des fonctions sphériques de LEGENDRE et ne sont définies que par la forme de la cavité Σ et par la forme de la surface libre non-perturbatrice.

Cherchons le potentiel des vitesses sous la forme:

$$\varphi = \varphi^* + \tilde{\varphi} \qquad (3.22)$$

avec

$$\varphi^* = \alpha'\,(l\,\varphi_1 + \varphi_5) + \beta'\,(l\,\varphi_2 - \varphi_4),$$

$\tilde{\varphi}$ étant le potentiel des mouvements ondulatoires.

Procédons au calcul du moment de la quantité de mouvement du liquide $\vec{G}_2$. Envisageons successivement les termes de ce moment:

$$G_{2'} = \varrho\,l\,\vec{\xi^0} \times \int_\tau \alpha'\,l\,\nabla\varphi_1\,d\tau.$$

En appliquant la formule de Green et les conditions (3.21), on trouve que:

$$\int_\tau (\nabla\varphi_1)\,d\tau = \int_\tau (\nabla\varphi_1)^2\,d\tau\,\vec{x^0} + \int_\tau (\nabla\varphi_1\nabla\varphi_2)\,d\tau\,\vec{y^0} + \int_\tau (\nabla\varphi_1\nabla\varphi_3)\,d\tau\vec{z^0}.$$

En raison de la symétrie:

$$\int_\tau \nabla\varphi_i\,\nabla\varphi \cdot d\tau = 0.$$

C'est ainsi que:

$$\int_\tau \nabla\varphi_i\,d\tau = \int_\tau (\nabla\varphi_1)^2\,d\tau\,\vec{x^0}$$

et par suite:

$$\vec{G}_{21} = m_1\,\vec{\xi^0} \times \alpha'\,\vec{x^0}.$$

D'une façon similaire, on peut montrer que:

$$\vec{G}_{22} = \varrho\,l\,\vec{\xi^0} \times \int_\tau \beta\,l\,\nabla\varphi_2\,d\tau = m_1\,\vec{\xi^0} \times \beta'\vec{y^0},$$

$$\vec{G}_{23} = \varrho\,l\,\vec{\xi^0} \times \int_\tau \alpha'\,\nabla\varphi_5\,d\tau = 0,$$

$$\vec{G}_{24} = \varrho\,l\,\vec{\xi^0} \times \int_\tau \beta'\,\nabla\varphi_4\,d\tau = 0.$$

Réunissant ces résultats, on a:

$$\vec{G}_{21} + \vec{G}_{22} + \vec{G}_{23} + \vec{G}_{24} = \varrho\,l\,\vec{\xi^0} \times \int_\tau [\alpha'(l\,\nabla\varphi_1 + \nabla\varphi_5) + $$
$$ + \beta(l\,\nabla\varphi_2 - \nabla\varphi_4)]\,d\tau = m_1\,\vec{\xi^0} \times \vec{\xi^{0\prime}}$$

où m_1 est un nombre quelconque:

$$m_1 = \varrho\,l^2 \int_\tau (\nabla\varphi_1)^2\,d\tau = \varrho\,l^2 \int_\tau (\nabla\varphi_2)^2\,d\tau.$$

Appliquons alors la formule suivante:

$$\int\limits_{\tau} (\vec{r} \times \nabla \varphi)\, d\tau = \int\limits_{\tau} (\nabla \varphi_1 \nabla \varphi_4)\, d\tau\, \vec{x}^0 +$$
$$+ \int\limits_{\tau} (\nabla \varphi \nabla \varphi_5)\, d\tau\, \vec{y}^0 + \int\limits_{\tau} (\nabla \varphi \nabla \varphi_6)\, d\tau\, \vec{z}^0. \quad (3.24)$$

Utilisons cette formule pour calculer le vecteur $\vec{G}_{25}$:

$$\vec{G}_{25} = \varrho \int\limits_{\tau} \left[\vec{r} \times (\alpha'(l\nabla \varphi_1 + \nabla \varphi_5) + \beta(l\nabla \varphi_2 - \nabla \varphi_4)) \right] d\tau.$$

Appliquant cette transformation (3.24) et prenant en considération que, du fait de la symétrie:

$$\int\limits_{\tau} \nabla \varphi_i \nabla \varphi_j = 0, \quad i \neq j,$$

on a:

$$G_{25} = m_2(\alpha' \vec{y}^0 - \beta' \vec{x}^0).$$

Partant, cette expression ne se distingue du produit $m_2 \vec{\xi}^0 \times \vec{\xi}^{0\,\prime}$ que par les petites valeurs du deuxième ordre.

Il s'ensuit que:

$$\vec{G}_{25} = m_2 \vec{\xi}^0 \times \vec{\xi}^{0\,\prime} \quad (3.25)$$

avec

$$m_2 = \varrho \int\limits_{\tau} (\nabla \varphi_4)^2\, d\tau = \varrho \int\limits_{\tau} (\nabla \varphi_5)^2\, d\tau.$$

On a donc

$$\sum\limits_{j=1}^{5} \vec{G}_{2j} = (m_1 + m_2) \vec{\xi}^0 \times \vec{\xi}^{0\,\prime}. \quad (3.26)$$

A présent, il nous reste à calculer les deux derniers termes:

$$\vec{G}_{26} = \varrho\, l \int\limits_{\tau} (\vec{\xi}^0 \times \nabla \tilde{\varphi})\, d\tau,$$

$$\vec{G}_{27} = \varrho \int\limits_{\tau} (\vec{r} \times \nabla \tilde{\varphi})\, d\tau.$$

Répétant les calculs exécutés et utilisant le fait que sur la surface Σ: $\partial \tilde{\varphi}/\partial n = 0$ on va ramener ces expressions à la forme suivante:

$$\left.\begin{aligned}
\vec{G}_{26} &= \varrho\, l\, \vec{\xi}^0 \times \int\limits_{S} \frac{\partial \tilde{\varphi}}{\partial n} [\varphi_1 \vec{x}^0 + \varphi_2 \vec{y}^0 + \varphi_3 \vec{z}^0]\, ds, \\
\vec{G}_{27} &= \varrho \int\limits_{S} \frac{\partial \tilde{\varphi}}{\partial n} [\varphi_4 \vec{x}^0 + \varphi_5 \vec{y}^0 + \varphi_6 \vec{z}^0]\, ds.
\end{aligned}\right\} \quad (3.27)$$

On va décrire la position de la surface libre dans le système de coordonnées $o\,x\,y\,z$ par la fonction $\zeta\,(t,\,r,\,\theta)$. La vitesse du déplacement des points de la surface libre aux petites valeurs du deuxième ordre près est :

$$v_n - \frac{\partial \zeta}{\partial t}.$$

Comme sur S :

$$v_n = \frac{\partial \varphi^*}{\partial n},$$

nous avons :

$$\frac{\partial \zeta}{\partial t} = -\frac{\partial \widetilde{\varphi}}{\partial n}.$$

De ce fait, on peut mettre les expressions (3.27) sous la forme :

$$\vec{G}_{26} + \vec{G}_{27} = \int_S \frac{\partial \zeta}{\partial t}\,\vec{D}\,ds.$$

$\vec{D}$ étant le vecteur-fonction connu et défini d'une manière univoque par la forme du domaine τ dont le sens est perpendiculaire au vecteur $\vec{\xi}^0$ aux petites valeurs du premier ordre près.

D'après l'énoncé, le moment de la quantité de mouvement de l'ensemble du système peut être mis sous la forme :

$$\vec{G} = A^* \, \vec{\xi}^0 \times \vec{\xi}^{0\prime} + c\,\vec{\omega} + \int_S \frac{\partial \zeta}{\partial t}\,\vec{D}\,ds \qquad (3.28)$$

avec

$$A^* = A + m_1 + m_2.$$

4. On va décrire le mouvement du système corps-liquide à l'aide de l'équation des moments par rapport au centre O. Le point O coïncidant avec le centre d'inertie seulement dans la position d'équilibre, il convient d'écrire l'équation des moments sous la forme :

$$\frac{d\vec{G}}{dt} + \mathcal{M}\,\vec{d} \times \vec{v}_0 = M, \qquad (3.29)$$

où $\mathcal{M}$ est la masse de l'ensemble du système, $\vec{d}$ le rayon-vecteur du centre d'inertie du système par rapport au point O, $\vec{v}_0$ la vitesse de ce centre en mouvement relatif. Il est ensuite aisé de démontrer que le deuxième terme de la partie gauche de l'équation (3.29) est une petite valeur du deuxième ordre. C'est pourquoi on écrira cette équation sous la forme :

$$\frac{d\vec{G}}{dt} = \vec{M}, \qquad (3.30)$$

où $\vec{M}$ est le moment des forces extérieures.

Prenant en considération les formules (3.12) et (3.28), on écrira l'équation (3.20) sous la forme:

$$A^* \vec{\xi^0} \times \vec{\xi^0}{''} + c\,\omega' \vec{\xi^0} + c\,\omega \vec{\xi^0}{'} + \int\limits_S \frac{\partial^2 \zeta}{\partial t^2} \vec{D}\,ds = \eta \vec{z^0} \vec{\xi^0}. \quad (3.31)$$

Ecrivons l'équation (3.31) en projection sur le sens de $\vec{\xi^0}$

$$c\,\omega' = 0, \quad (3.32)$$

c'est-à-dire:

$$\omega = \text{const.}$$

Ecrivant l'équation (3.31) en projection sur les sens de $\vec{X^0}$ et $\vec{Y^0}$ et effectuant la linéarisation, nous arrivons au système suivant d'équations:

$$\left. \begin{array}{l} -A^*\,\beta'' + c\,\omega\,\alpha' + \varrho \int\limits_S \frac{\partial^2 \zeta}{\partial t^2}\,\varphi_\beta\,ds + \eta\,\beta = 0, \\[4mm] A^*\,\alpha'' + c\,\omega\,\beta' + \varrho \int\limits_S \frac{\partial^2 \zeta}{\partial t}\,\varphi_\alpha\,ds - \eta\,\alpha = 0. \end{array} \right| \quad (3.33)$$

Il convient d'ajouter à ces équations celle de la surface libre. Si l'on exclue le potentiel des vitesses en introduisant l'opérateur de NEUMANN H, on peut alors présenter l'équation déterminant la surface libre sous la forme suivante:

$$H\,\zeta'' + \beta''\,\varphi_\beta - \alpha''\,\varphi_\alpha + L\,\zeta = 0. \quad (3.34)$$

Le système d'équations (3.33)–(3.34) forme un système fermé d'équations intégro-différentielles.

Afin d'étudier la stabilité, posons:

$$\zeta = e^{\sigma t}\,\zeta_0; \quad \alpha = e^{\sigma t}\,\alpha_0; \quad \beta = e^{\sigma t}\,\beta_0,$$

et introduisons l'opérateur R_σ tel que $R_\sigma f$ est une résolvante de l'équation:

$$(\sigma^2 H + L)\,\zeta = f.$$

Du système (3.33)–(3.34), on peut déduire la relation

$$\sigma^2 \Big[A^* + \varrho \int\limits_S R_\sigma \varphi_\alpha \cdot \varphi_\alpha\,ds\Big]\alpha_0 + \Big[c\,\omega\,\sigma - \sigma^2 \int\limits_S R_\sigma \varphi_\alpha \cdot \varphi_\beta\,ds\Big]\beta_0 - \eta\,\alpha_0 = 0,$$

$$\sigma^2 \Big[A^* \varrho \int\limits_S R_\sigma \varphi_\beta \cdot \varphi_\beta\,ds\Big]\beta_0 + \Big[-c\,\omega\,\sigma - \sigma^2 \int\limits_S R_\sigma \varphi_\beta \cdot \varphi_\alpha\,ds\Big]\alpha_0 - \eta\,\beta = 0.$$

$$(3.35)$$

La condition de perméabilité nous amène à l'équation:

$$\Delta(\sigma) = 0 \quad (3.36)$$

par rapport à σ.

La résolvante étant une fonction méromorphe de la grandeur σ, le problème de stabilité se réduit à la recherche des zéros d'une certaine fonction méromorphe.

La surface Σ étant une sphére, remarquons que tous les coefficients de l'équation (3.36) peuvent être calculés sans recourir aux calculatrices électroniques numériques. Dans des cas plus compliqués le calcul des coefficients de cette équation constitue un problème compliqué de la mathématique de calcul.

En conclusion, notons deux circonstances:

1. Si l'on remplace la surface S par un couvercle solide, il convient de poser dans le système (3.33)–(3.34) $\varphi_\alpha = \varphi_\beta = 0$ et nous aboutirons au problème ordinaire de stabilité d'un gyroscope. φ_α et φ_β se distinguant de zéro, les termes qui comportent ces fonctions fournissent un changement de la quantité de mouvement provoqué par la mouvement ondulatoire du liquide, et sont habituellement assez petits. De ce fait, on peut appliquer la méthode de la théorie de perturbations, afin de calculer les pectre du problème (3.36). Il suffit de n'étudier que les racines des équations (3.36) voisines de celles des équations:

$$\begin{vmatrix} \sigma^2 A^* - \eta & c\,\omega\,\sigma \\ -c\,\omega\,\sigma & A^*\sigma^2 - \eta \end{vmatrix} = 0.$$

2. La bulle se trouvant entièrement à l'intérienr du liquide et étant située concentriquement par rapport à la sphère Σ, on a sur S

$$\varphi_\alpha = h\sin\theta\cos\psi,$$

$$\varphi_\beta = h\sin\theta\sin\psi,$$

où h est une constante.

Dans ce cas:

$$L\,\varphi_\alpha = L\,\varphi_\beta = 0$$

et

$$H\,\varphi_\alpha = \lambda\,\varphi_\alpha,$$

$$H\,\varphi_\beta = \lambda\,\varphi_\beta.$$

Le nombre $\lambda > 0$ ne dépend que de la relation des rayons des sphères Σ et S.

Posant

$$\zeta_0 = a_1\,\varphi_\alpha + a_2\,\varphi_\beta,$$

on trouvera:

$$\zeta_0 = \frac{h}{\lambda}\varphi_\alpha - \frac{h}{\lambda}\varphi_\beta.$$

Utilisant ces égalités et remarquant que:

$$\int_S \varphi_\alpha \cdot \varphi_\beta \, ds = 0,$$

ramenons le système (3.35) à la forme:

$$\sigma^2(A^* + a)\,\alpha_0 + c\,\omega\,\sigma\,\beta_0 - \eta\,\alpha_0 = 0, \left.\right\}$$
$$\sigma^2(A^* + a)\,\beta_0 - c\,\omega\,\sigma\,\alpha_0 - \eta\,\beta_0 = 0. \left.\right\}$$
(3.37)

e étant un nombre qui ne peut être déterminé que par le rapport des rayons des sphères Σ et S et par la densité du liquide.

Alors, dans ce cas, le gyroscope à bulle est équivalent au gyroscope ordinaire, mais possède un moment équatorial d'inertie augmenté. Conformément à cette particularité, la stabilisation du gyroscope en question exige une vitesse angulaire plus grande que ne l'exige le gyroscope dont la surface de la bulle est remplacée par une paroi sphérique:

$$\omega > \frac{4\,\eta\,A + 4\,\eta\,a}{c^2}.$$

La présente communication a traité certains aspects très particuliers de la stabilisation d'un satellite possédant à l'intérieur un liquide. En général, ce problème se heurte à des difficultés mathématiques considérables.

Je ne prétends pas avoir épuisé le problème en cause. Je ne voulais qu'attirer l'attention sur une nouvelle série de problèmes mathématiques particulièrement intéressants qui exigeront des efforts considérables de la part des mathématiciens qui portent attention à la dynamique des satellites.